Preimplantation Genetics

Preimplantation Genetics

Edited by
Yury Verlinsky
and Anver Kuliev
Reproductive Genetics Institute
Chicago, Illinois

Plenum Press • New York and London

Library of Congress Cataloging-in-Publication Data

International Symposium on Preimplantation Genetics (1st : 1990 :
Chicago, Ill.)
Preimplantation genetics / edited by Yury Verlinsky and Anver
Kuliev.
p. cm.
"Proceedings of the First International Symposium on
Preimplantation Genetics, held September 14-19, 1990, in Chicago,
Illinois"--T.p. verso.
Includes bibliographical references and index.

1. Preimplantation genetic diagnosis--Congresses. I. Verlinsky,
Yury. II. Kuliev, Anver. III. Title.
[DNLM: 1. Embryology--congresses. 2. Genetic Techniques-
-congresses. 3. Genetics, Medical--congresses. 4. Preimplantation
Phase--genetics--congresses. QZ 50 I614p]
RG628.3.P74I57 1990
616'.042--dc20
DNLM/DLC
for Library of Congress 91-24088
CIP

Proceedings of the First International Symposium on
Preimplantation Genetics, held September 14-19, 1990,
in Chicago, Illinois

ISBN-13: 978-1-4684-1353-3 e-ISBN-13: 978-1-4684-1351-9
DOI: 10.1007/978-1-4684-1351-9

A Division of Plenum Publishing Corporation
233 Spring Street, New York, N.Y. 10013

PREFACE

Yury Verlinsky and Anver Kuliev

Reproductive Genetics Institute, Illinois Masonic Medical Center, 836 W. Wellington Chicago, IL 60657

Although introduction of a first trimester prenatal diagnosis by chorionic villus sampling (CVS) has considerably improved the possibility for prevention of genetic diseases, it requires a selective abortion in case of an affected fetus. Following the direction of an earlier prenatal diagnosis and to avoid the need for abortion, preimplantation genetic diagnosis has been initiated based on polar body removal and pre-embryo biopsy.

The First International Symposium on Preimplantation Genetics, Chicago, September 17-19, 1990, was organized to explore these important developments, to review the state of knowledge in the field, and to address existing problems to be solved for developing and improving current approaches for preimplantation diagnosis of genetic disorders. A growing interest in the subject was obvious from the wide attendance of the meeting: over 250 scientists from 19 countries participated.

This was the first attempt to put together the advances in different areas of basic and applied research relevant to Preimplantation Genetic Diagnosis, with the multidisciplinary scientific program including the sessions on embryology, micromanipulation and biopsy, genetic analysis of gametes and pre-embryos, IVF, gene expression and gene therapy, and ethical and legal issues.

The deliberations of the Symposium presented in the above mentioned sessions, which comprise the contents of corresponding sections of the Proceedings, open a new area in medical research based on the interaction of IVF and New Genetics. One of the sections of the Proceedings is devoted to the technical aspects of preconception and preimplantation diagnosis and contains the Manuals prepared for the Workshop on Preimplantation Genetic Analysis, held September 14-15, 1990, just prior to the First International Symposium on Preimplantation Genetics.

Being held 12 years after P.S. Steptoe and R.G. Edwards reported the birth of a baby following the transfer of a human egg fertilized _in vitro_, 10 years after the First International Congress on IVF, and 5 years after introduction of polymerase chain reaction (PCR), the Symposium provided intriguing challenges to be resolved and important new data of theoretical and practical value in developing preconception and preimplantation genetic diagnosis. These are further steps in improving and developing more ethically acceptable approaches for prevention and control of genetic disorders, obviating the need for a selective abortion as an obligatory component in preventive services at present.

CONTENTS

INTRODUCTION

Robert G. Edwards

Bourn Hall Clinic
Bourn, Cambridge CB3 7TR, U.K.

The growth of the human embryo *in vitro* has been assessed by several investigators using *in vitro* fertilization (IVF). As expected, many of the processes of growth are similar to those in animals. In some details, however, there are aspects of development peculiar to human embryos which could influence our approaches to the diagnosis of genetic disease before implantation. The first polar body is large, but apparently degenerates within 24 hours and might lose some of its chromosomes during this process. The second polar body is smaller, nucleated and more persistent and it can be mistaken for a blastomere in a cleaving embryo.

It is now widely accepted that 95% of eggs fertilized *in vitro* have two pronuclei, but approximately 5% have three pronuclei, implying that they are dispermic (i.e. fertilized by two spermatozoa) or digynic (a polar body is retained in the egg in addition to the oocyte chromosomes). We do not know if the same situation arises *in vivo*, but the available evidence implies that 5% of these embryos will also be triploid or complex haploids or mosaics, and this is a point to bear in mind for preimplantation diagnosis. It will be essential to check pronuclear histories of any embryo before undertaking diagnostic procedures, or to ensure that the diagnostic tests can cope with triploids. Second, many spermatozoa are attached to the zona pellucida after fertilization *in vitro*, although we are once again unsure of the situation arising with fertilization *in vivo*. The importance of sperm bound to the zona, or partially penetrated into it, concerns the possibility of contaminating biopsies removed from the embryos, thus leading to a false diagnosis. A few cumulus cells also remain attached to the zona pellucida after IVF, and could be another source of contamination in diagnostic procedures if care is not taken in collecting the biopsy.

The zona pellucida is an important membrane, and its constituent glycoproteins have been characterized by

Preimplantation Genetics, Edited by Y. Verlinsky and A. Kuliev, Plenum Press, New York, 1991

Wasserman and his colleagues. It can be dissolved by acids or proteolytic enzymes, or small perforations can be made in it by these enzymes or by physical methods to permit spermatozoa to migrate into the perivitelline space or to extract biopsies from the embryo. It is apparently essential in early growth to hold the individual blastomeres together until compaction establishes an outer epithelial-like trophoblast tissue; it might also be important to protect the embryo against the mother's immune response.

The cleavage of human embryos is regular, with even-sized blastomeres, and compaction takes place after the 8-cell stage. Presumably the cells of cleaving embryos are totipotent or multipotent, as in animal embryos, so the death or loss of one or two of them for diagnosis would be tolerated by the embryo and it would continue to develop normally. Very few embryos seem to develop parthenogenetically, which is a boon for preimplantation diagnosis, although fragmented blastomeres are found in many embryos. This could cause problems if these tissues are mistaken for blastomeres - a possibility with large fragments, because some have micronuclei or no nucleus at all.

The human blastocyst at 5 days post-fertilization has approximately 120 cells, most of them trophoblastic, and an inner cell mass that is usually distinct but can be diffusely spread around the blastocoelic cavity. The blastocoelic cavity is distinct, and there are large secretary-like cells adjacent to this cavity at the base of the inner cell mass. The mitotic rate is high, and the size of individual cells approximates to that of somatic cells in general. There are clearly enough cells to remove some for diagnosis without jeopardizing the continued growth of the embryo.

After day 5, the human blastocyst expands, the zona pellucida becomes thinner and the embryo begins the process of hatching. This is an active process: it makes a small hole in the zona, presumably through the action of an enzyme similar to 'trypsin' described by Wasserman, and then gradually extends through it. These events occur on day 6-7 when approximately one-half of the blastocyst is extruded, while the other half is still expanded and remains within the zona pellucida, giving a "figure 8" appearance. Anomalies in hatching due to an especially hardened zona pellucida might explain the origin of human identical twins. The blastocyst finally hatches completely, leaving the zona pellucida with some cells or debris in it, and then enlarges rapidly as the extraembryonic endoderm differentiates from the inner cell mass. It is possible that these stages of growth might also be valuable for the diagnosis of genetic disease, especially since there is no need to penetrate the zona pellucida to obtain tissue, but the embryos are probably too late in development to be handled easily or give high rates of implantation.

The blastocyst now begins to secrete HCG, attaches to the uterus, the amniotic cavity forms, and the extra-embryonic mesoderm differentiates. These stages signal the onset of pregnancy and of organ differentiation.

It is possible to use simple media for the first 2-3 days of growth, and obtain reasonable rates of implantation. We use Earle's medium with human serus albumin - avoiding bovine serum albumin in view of the risk of inducing hypersensitivity responses in patients. Many other media can be employed including Ham's F10 or F12, BWW medium, oviductal fluid or media based on it, together with serum or feeder layers of cells. Many embryos grow normally to the 8-cell stage in such media.

Sustaining growth to the blastocyst is much more difficult. Only 40% of pronucleate eggs will reach this stage *in vitro*, and media containing serum or bovine albumin seem to be superior to the simpler media. Feeder layers of epithelial cells, often oviductal, have been prepared to help growth to blastocysts, but there seems to be no specificity of action since cells from other tissues are equally effective, and the value of feeder cells might lie in avoiding high pH or in the chelation of heavy metals. It is very important to understand these stages of growth, to gain as much knowledge of them as we can, because the removal of certain tissues might exert major effects on implantation and differentiation.

It is possible to replace embryos at all these stages and obtain pregnancies, although results with blastocysts grown *in vitro* are not as satisfactory as with cleaving embryos. The best rates of implantation are obtained with blastocysts flushed from the uterus. It is possible to establish pregnancies from hatched blastocysts from *in vitro*, but there is insufficient data to judge the effectiveness of this practice.

Embryos in all these stages of growth can be cryopreserved with varying levels of success. Propanediol is used in our clinic for pronucleate and cleaving embryos, and glycerol for blastocysts. Success rates are excellent with pronucleate and cleaving embryos - an overall pregnancy rate of 28% irrespective of age and with the number of embryos replaced varying between 1 and 3. We believe, however, that this high rate of implantation is due to the high fertility of patients who have several frozen embryos, for they have a pregnancy rate of almost 50% with three fresh embryos.

Protein synthesis in the human embryo is encoded by 'maternal' RNA in the oocyte, and the embryonic genome is activated only in the 4-8 cell stage, as described by Tesarik and Braude (in this volume). Transcripts for HCG-ß, for example, are found at this stage of growth. Maternal RNA evidently continues to dominate protein biosynthesis by the embryo, even until the early blastocyst, as shown by the persistently high levels of pyruvate uptake and the continuing high levels of HGPRT in human embryos. This situation contrasts with that in mice, where both of these parameters show sharp reductions in the 4-8 cell embryo, and a sharp rise in morulae and blastocysts as embryonic RNA is transcribed.

In preimplantation diagnosis, it will be necessary to make a decision whether to grow embryos *in vitro*, using IVF procedures, or flush them from the mother's uterus at day 5 approximately. Embryos that have reached blastocysts in their mother's uterus at this time are capable of high rates of implantation, perhaps well over 50%. However, many other embryos are evidently delayed in growth or have degenerated, and no embryos of any sort are collected from many women who have apparently ovulated normally and have had intercourse or artificial insemination. The alternative, to use IVF, has enormous advantage in providing several embryos per cycle after ovarian stimulation, and the methods are getting simpler. Nevertheless, the procedures do make various demands on patients and the implantation rate for each 4-8 cell embryo is perhaps the biggest problem we face, since it is only 15% even in the best clinics, and lower with blastocysts. It will therefore be necessary to diagnose many embryos to get one implantation, even discounting any damage done to the embryos during biopsy, and the need to discard afflicted embryos and perhaps carrier embryos.

Nevertheless, there is a ray of hope on the horizon. Some groups of women seem to be highly fertile when up to 3 embryos are replaced in them, including younger women who produce several oocytes, so permitting several embryos to be frozen. Couples seeking preimplantation diagnosis might fit into this category.

Agonadal women who never had an ovary or those with secondary amenorrhoea, and who are given donated oocytes are also highly fertile; in these women, pregnancy rates of 60% or more can be achieved. The obvious conclusion to draw is that we need much more information to know why implantation is so successful under some circumstances, even though the rate per embryo is still only 20-25% in these highly fertile women.

Biopsies can be taken from eggs or embryos at various stages of growth. The first or second polar body, or both, can be excised from unfertilized or fertilized oocytes, which would offer a means of diagnosis by identifying the distribution of maternal genes in the oocyte and polar bodies. This point will be discussed by Verlinsky in this volume.

Two-cell embryos can be divided into separate blastomeres, to form twin embryos. One of the embryos can then be used for diagnosis to decide if the other should be replaced in the mother. The procedure is straightforward, and the considerable expertise about this procedure in animals has shown a high chance of normal embryonic growth. This was the method used by Epstein and his colleagues in California in the 1970's to produce twin blastocysts, and to show how the embryos could be sexed by calculating the HGPRT/APRT ratio. The major difficulty with 2-cell embryos lies in holding the blastomeres together during cleavage in the absence of a zona pellucida, although the use of semi-solid media substrates, such as agar might help to maintain close contacts between differentiating blastomeres.

4- or 8-cell embryos can be used for preimplantation diagnosis, i.e. before compaction has occurred. The blastomeres are relatively independent and can be loosened by reducing levels of Ca and Mg in the medium or using enzymes, in order to assist the withdrawal of one or two blastomeres. The zona pellucida need not be removed, for small apertures can be made in it using enzymes, a low pH or mechanical means. There is also another advantage with this method, because embryos can achieve a high degree of repair during these stages of growth. This is the time when blastomeres are undergoing their initial differentiation, e.g. in their degree of polarization, but non-polarized initial cells can divide to produce polar and non-polar cells and so reconstitute the embryo. This sort of mechanism might be involved in embryo repair after cryopreservation. Many thawed embryos have fewer blastomeres than they had before freezing, yet they can develop normally; on some occasions, more than half of the cells of a cleaving embryo are destroyed yet the embryo survives as reported from Spain when a human embryo with only three blastomeres left out of eight developed into a normal baby.

Some pliability also remains in blastocysts at the stage when inductive relationship between the inner cell mass and polar trophectoderm immediately above it have been established. There is some evidence to show that these cell types can undergo a limited exchange, and that polar trophoblast cells can migrate and repopulate the polar trophectoderm. The excision of parts of the trophectoderm might therefore be compatible with some capacity for repair.

Tissue can be excised from blastocysts as Gardner and I showed some years ago in the first report on preimplantation diagnosis. Small pieces of mural trophoblast were excised from rabbit blastocysts by withdrawing tissue through a small hole made mechanically in the zona pellucida, or by letting it herniate through the hole. The tissue was then excised, and the tear in the trophoblast was sealed by 'zona-locking', in order to maintain the integrity of the trophoblast in a fold of the zona pellucida and so retain the expansion of the blastocyst. It is also possible to extract cells from the blastocoelic cavity using a very fine needle, a very rapid procedure as introduced by Holland and myself with human embryos. The origin of these blastocoelic cells is uncertain, but some are nucleated, and blastocoelic cells have been removed from blastocysts of other species as shown by Hartshorne. This would be a rapid and easy method, because fine pipettes can be used to penetrate the zona pellucida and mural trophectoderm and aspirate the small cells in the blastocoelic cavity.

In view of all of this knowledge, how shall we approach the diagnosis of embryos today? Which embryological techniques should we use? First of all, it is essential to bear in mind that it may be possible one day to diagnose spermatozoa and perhaps select them for particular characteristics. This would be much simpler than diagnosing embryos, and recent contributions to this field include the application of PCR to amplify DNA in single human spermatozoa, as shown by Arnheim (this volume) and by the analysis of chromosomes in human sperm heads using the hamster test, reported first by Rudak. I would also draw your attention to work done by Clodhill, Johnson and others who can now quantify numbers of X and Y spermatozoa in samples of spermatozoa by fluorescent cell sorting.

Next, a general point about diagnosis. The extraction of polar bodies from human eggs and their analysis by the PCR reaction is a novel idea. The embryos continue to grow normally, and the polar bodies can be analyzed in order to assess the segregation of maternal genes during meiosis. My two reservations about this method are, first, that the diagnosis can only be effective if all the products of meiosis can be located accurately. A second, and more important point because it also applies to the diagnosis of embryos, is that the egg is not diagnosed directly, its gene content being decided by extrapolating from the polar body diagnoses. I believe that, if at all possible, direct information from the embryo should be available before it is replaced. It might be possible, of course, to do a second diagnosis on the embryo after an initial polar body diagnosis has proved unsatisfactory. The point about gaining information directly from the embryo also applies to the work published by Handyside using Y primers to amplify cells excised from 8-cell embryos, for those embryos without any evidence of amplification were considered to be females. This is not the correct way to do preimplantation diagnosis.

It is essential to have the best techniques available to maintain the chances of embryos to recover, implant and continue to grow after biopsy. In our early work, many rabbit blastocysts collapsed inwardly after the excision of trophectoderm, presumably through loss of blastocoelic fluid; a few remained expanded, the proportion rising considerably when 'zona-locking' was used. Initial reports on removing cells from 8-cell human embryos indicate that almost all continue to grow and some will implant (Handyside, this volume).

The organization of the clinic will need attention, and so too will the source of human embryos, i.e. whether they should be flushed from the uterus or grown in vitro using the methods of IVF. How will this work fit into a busy IVF program? Many patients will already be attending for the treatment of infertility, and work on preimplantation diagnosis must fit in with the schedule of the clinic. Biopsies will presumably be taken in the morning, and diagnosed during the day so that the embryo for replacement can be selected in the late afternoon. At present, 8-cell embryos will be the optimal stage for diagnosis.

Embryo cryopreservation is an important component of a program of preimplantation diagnosis since many embryos - ten or more - can be obtained from some patients. Some investigators report that cryopreserving embryos with a fractured zona pellucida entails a high risk of embryonic death; the zona pellucida must nevertheless be fractured to obtain cells for diagnosis. Should cryopreservation come before diagnosis or vice versa? How do we cope with the need to identify one or two embryos free of genetic disease: should all embryos of a particular patient be typed immediately, or should all of them be frozen and diagnosed singly after thawing at some later stage. It is possible to freeze human embryos twice and obtain pregnancies from them, so we have some pliability here.

How many cells should be removed for biopsy? Many human embryos growing _in vitro_ are mosaic, have cell fragments or are triploids or trisomies. Is one cell enough to obtain a genuine diagnosis of an embryo? We know DNA analyses can be achieved on a single cell, but they give comprehensive data only on that single cell, so two or three cells might be needed for a fuller analysis of an embryo. Yet the removal of 2 or 3 cells might impair the growth of cleaving embryos at the 8-cell stage. Questions also arise about diagnosing blastocysts. Will the biopsy of sheets of trophoblast cells be satisfactory for diagnosis? Sheets of cells might not be optimal for diagnosing embryos, since variations between individual cells are obscured. The amplification of several separate cells might give better and more detailed information about the embryo and we would then know if it was uniform or mosaic, or even more complex still. If assays of several single cells prove to be preferable for embryo diagnosis, questions will then be raised about the value of taking a sheet of trophectoderm from blastocysts.

Contamination is another problem, and I refer to cellular rather than bacterial contamination. Cumulus cells or spermatozoa are the most likely causes, although polar bodies or cell fragments raise the same difficulty, and such contaminants can arise even when the greatest care is taken to exclude them. Perhaps we need cell markers in PCR assays which could identify some of these forms of contamination, e.g. by the simultaneous amplification of paternal alleles to ensure that only one is present and so confirm the cell is embryonic. Contamination in the laboratory must also be excluded.

There could be some specific advantages or disadvantages with particular gene systems. One involves loci which display genomic imprinting; it is remarkable that imprinted disease genes transmitted from one parent are not expressed in the embryo whereas those inherited from the other parent are expressed. Would it be necessary to diagnose the embryo if we knew the parental origin of a defective gene would cause it to be inactive?

Finally there is the problem of cell death in embryos. Some cell death is programmed during tissue differentiation and this is a regular feature of embryogenesis. The presence of dead cells might impair the value of some forms of diagnosis e.g. cells removed from the blastocoelic cavity might be dead and of no value for diagnosis. I do not know if DNA in dying cells of this sort would amplify or provide a correct diagnosis of the embryo; some of them might also have undergone fragmentation or other forms of damage causing them to lose DNA.

I trust the above considerations have indicated some of the embryological factors involved in the introduction and interpretation of preimplantation diagnosis. I believe we have a good basis of knowledge to introduce this new form of treatment, based on a reasonable understanding of these embryological events. Time will tell if I am overoptimistic.

PART I

EMBRYOLOGICAL ASPECTS OF PREIMPLANTATION GENETICS

OOCYTE DEVELOPMENT

Joseph Schulman

Genetics and IVF Institute
Fairfax, VA 22031

It is necessary to recall that patterns of development of follicles in the human ovary are under the control of a variety of hormones. Our goal is to obtain oocytes that are appropriate for fertilization, and these are usually obtained from follicles of larger size. In the absence of stimulation, the intraovarian oocyte is in a late meiotic stage known as diplotene. Further oocyte development is regulated by the hypothalamus and the pituitary, and primarily by the hormones FSH and LH which are released from the anterior pituitary gland.

GNRH is a hormone controlling release of gonadotropins from the pituitary. The mechanisms of GNRH effect, of course, are not perfectly understood, but GNRH agonists and antagonists are quite useful in the IVF field, and both FSH and LH release are controlled by this hormone.

Pulsatile rather than continuous GNRH stimulation is required for the pituitary gland to release gonadotropins. This phenomenon is the basis of the apparently paradoxical effect of GNRH agonists, such as leuprolide (Lupron) or Buserilin which by providing a tonic stimulation to the pituitary initially stimulate but then suppress gonadotropin secretion.

In the natural menstrual cycle, the LH peak starts the clock toward ovulation and the production of an oocyte which is in metaphase II, ready for fertilization. Estradiol levels rise during the preovulatory period, and are then maintained before and through the implantation period.

There have been numerous studies measuring steroid concentrations in follicular fluids, with the hope of identifying follicles where oocytes might be particularly successful at forming embryos which will implant. In preovulatory follicles there are high estradiol and progesterone levels and low androgen levels, androstenedione being representative, while in non-ovulatory follicles the reverse is true. There are disorders of ovulation, such as

Preimplantation Genetics, Edited by Y. Verlinsky and
A. Kuliev, Plenum Press, New York, 1991

Stein-Leventhal Syndrome, in which a preponderance of androgen appears to inhibit the production of appropriate oocytes. Unfortunately, no one has been able to identify a follicular fluid marker which can be used to predict exactly which eggs are most healthy, and which are going to have the greatest chance of going on to produce viable embryos and live offspring.

In IVF, one attempts to obtain as many eggs as possible, and form as many embryos as possible.

Natural cycle IVF yielding at most a single embryo has been done, but it would probably be inappropriate in situations where preimplantation genetics is involved.

Unfortunately, many patients (and this will include not only infertility patients but also individuals who are going to consider preimplantation diagnosis) do not respond as well as one would like to the available drugs used for ovarian stimulation. Some patients produce only one dominant follicle. Some have low estradiol values, and this is usually accompanied by a reduced yield of good quality oocytes. In yet others, the estradiol may drop after HCG administration, another sign that the cycle is not optimal. And of course, in some women there will be an endogenous LH surge, usually associated with poorer egg quality. Such surges can be prevented by the use of the GNRH agonists or antagonists.

Even a small rise in LH has an adverse effect on pregnancy outcome, and in our own experience, rises as small as 50-100% over baseline are associated with a significantly decreased pregnancy rate. This observation strengthens the rationale for the use of inhibitors of inappropriate gonadotropin release during IVF stimulations.

Several agents are used for ovarian stimulation. Human menopausal gonadotropins (HMG, or the trade names Pergonal in the United States and Humagon and Pergonal outside the United States) are commonly administered. Typical preparations have an FSH to LH ratio of approximately 1 to 1, and these drugs are administered intramuscularly once or twice a day. Clomiphene citrate, working by a completely different mechanism, can induce multiple ovulation on its own; when combined with gonadotropins, one usually obtains a larger number of eggs than with clomiphene alone. Follicle stimulating hormone (FSH alone known as Metrodin in the United States) is also used in a number of stimulation regimens. Varying sequences and combinations of FSH and the combined gonadotropins are also used.

There is no agreement about the "best" way to do IVF stimulations to obtain embryos. Probably, for different patients, different regimens are optimal. The following are some examples of popular sequences. In one, Clomiphene citrate is typically given early in the cycle and then there may be concurrent or later administration of gonadotropins. In other regimens, Pergonal is used alone or in combination with FSH (sometimes to achieve FSH enrichment during the early phases of stimulation). GNRH analogs, such as leuprolide, are often used now in many IVF programs, along with stimulation with gonadotropins to optimize the quality. There is some negatives in using GNRH; their suppression of the ovary may reduce the response to gonadotropins and increase the need for administering large doses of gonadotropins. In France, for example, it is very common to use two weeks or more of suppression of the ovary prior to stimulation, and massive doses of gonadotropins are often required thereafter; although the results are good, the cost of IVF is greatly increased.

When drugs which stimulate the ovaries are administered, estradiol, progesterone and LH levels are monitored often on a daily basis, with periodic ultrasound examinations. Each day the dose of gonadotropin used is individually determined for the patients based on these results. Preprogrammed stimulation sequences are an alternative.

Then, at an appropriate time, HCG must be administered as it is established clearly that ovulation in humans occurs within approximately a 36-38 hour period from the administration of HCG. The criteria for administration of HCG in terms of follicle size and hormone levels vary with different stimulating regimens. In general, at HCG one hopes to have one or more preovulatory follicles with a size of 15-17 mm. and an estradiol level suggesting that there are a number of healthy follicles. IVF can be accomplished and eggs obtained in simple settings, outside of the operating room, using ultrasound guided techniques. An intravaginal ultrasound probe makes egg recovery extremely simple and safe for patients, and that simplicity and safety will make preimplantation diagnosis far more reasonable for patients to consider than if these techniques still required surgery. It is remarkably simple to puncture through the vaginal wall, under local anesthesia with sedation, and aspirate follicles one by one, and get the eggs required.

In conclusion, a few words about the impact of age are essential. Age is the great enemy in the field of human reproduction, as older women reproduce with greatly reduced efficiency compared to younger ones. Preimplantation genetic diagnosis may turn out to be, in its application, almost a mirror image of conventional prenatal diagnosis. More of the latter is generally done as people get older. I think preimplantation diagnosis is going to be difficult to achieve with reasonable efficiency in women in their 40s, whereas with young women it will probably be very valuable. There is approximately a five fold decline in the efficiency of human reproduction between the mid-20s and early 40s. This is shown by data from a number of different studies in non contracepting populations. In IVF we see exactly the same phenomenon. In our own experiences with IVF patients, under the age of 39, approximately 15% of patients got pregnant in a single initiated cycle; this includes many male factor patients and others where improved IVF technology would still not alter the outcome. With patients who are in their 40s, there is approximately a threefold average reduction in the rate of attaining an IVF pregnancy.

In addition, miscarriages, important factors in reducing the efficiency of preimplantation genetics methods, come into play as a function of age as well. The miscarriage rate of pregnancies that have been established in IVF settings will be under 10% below age 29, but with women who are in their 30s the rate is about 25% and in the early 40s reaches nearly 40%.

EXPERIMENTAL CYTOGENETICS OF PREIMPLANTATION DEVELOPMENT

A.P. Dyban

Department of Embryology, Institute for Experimental Medicine Academy of Medical Sciences, Leningrad, USSR

INTRODUCTION

The progress in preimplantation diagnosis depends mostly on the knowledge in basic science, especially on the achievements of developmental biology, developmental cytogenetics and molecular biology, i.e. on the studies of the controlling mechanisms of early mammalian development. Although preimplantation development is quite different in different mammalian species, there are many common features which characterize these stages of embryogenesis.

Cleavage, morula and blastocyst formation are the main processes which take place before implantation. Cleavage is a rather special type of mitotic division because after every cleavage (cell division) blastomeres become smaller and smaller until they are very close in size to the typical somatic cells, i.e. the cells of peri- and postimplantation embryo. Number of blastomeres, rate of cell division and mitotic index are the main criteria of normal cleavage. Morula is formed as a result of cells' shape changes and their interaction, i.e. compaction and polarization of blastomeres. The formation of blastocyst strongly depends on the cell-cell interaction and the development of a tight junction between the population of blastomeres located outside (trophectoderm). There are two cell lineages at this stage, the inner cell mass and the trophectoderm, which are committed already at the late morula stage. A cavity (blastocell) filled with fluid is formed at this stage and this process is called cavitation.

Thus, there are very reliable criteria for evaluation of normality of morula and blastocyst development in any mammals including humans. These criteria must be used for examining the influence of chromosomal abnormalities on preimplantation development not only in laboratory animals but human embryos as well. This approach was used in our study of numerical and structural chromosomal aberrations during early mammalian development starting from fertilization.

Preimplantation Genetics, Edited by Y. Verlinsky and A. Kuliev, Plenum Press, New York, 1991

It must be stressed that in any mammalian species the genes are not expressed at the very beginning of development. In mice, the gene transcription starts at the two cells stage, but in other mammals it occurs later, i.e. in pig - at the four cells stage, in cow and sheep - between the eight and sixteen cells stage, and in humans - at the eight cells stage (for review see: Magnuson, Epstein, 1981, First, Barnes, 1989, Telford, Watson, Schultz, 1990). Thus, it seems that early development in mammals is fully dependent on maternal genetic information stored in ovum in the form of mRNA and/or proteins whereas at later stages genetic activity of chromosomes must play an important role in controlling morphogenetic processes. Therefore, the study of the influence of chromosomal abnormalities on preimplantation development can provide substantial data for detecting the participation of chromosomes' genetic activity at these stages of embryogenesis (for review see: Epstein, 1985, 1987, Gropp, Winking, 1981, Dyban, Baranov, 1987, Searle, 1989).

In this paper only some of these data are summarized, which support the hypothesis (Dyban 1987, 1988) that early stages of mammalian embryogenesis are dependent on maternal genetic information as well as on transcription of genes which are not scattered randomly via genome, but are clustered in X-chromosome and some autosomes.

GENOMIC ABERRATIONS

The influence of genomic abnormalities on mice embryonic development, especially on preimplantation stages, is well studied.

Haploidy

Embryos with a haploid set of chromosomes may arise in the course of parthenogenesis, gynogenesis, or androgenesis (for review see: Kaufman, 1983, Dyban, Noniashvili, 1986a). No data are available on chromosomal complements of gynogenetic and androgenetic embryos, cytogenetic studies being performed only on parthenogenetic mice embryos.

Two types of haploid embryos (genetically uniform and mosaic) and two types of diploid parthenogenones (with two haploid pronuclei and with one diploid pronucleus) can be obtained in mice. Parthenogenesis can arise spontaneously, causing some types of ovarian teratomas not only in mice, but probably also in humans. The reasons why oocytes enter into parthenogenetic development inside the follicle (in ovary) are not known. Probably, there are some gene mutations which increase the sensitivity of oocytes to activation stimuli and remove the meiosis block at the metaphase I or II. In favor of such assumption are the observations on mutant strain of mice (LT/Sv) with spontaneous parthenogenesis of itrafollicular and ovulated oocytes.

The technique of experimental production of parthenogenetic mammalian embryos provides reproducible results (Kaufman, 1983). Many agents (physical, chemical, mechanical, etc.) can induce parthenogenesis if applied to ovulated ova _in vivo_ and _in vitro_. The best results can be obtained by ethyl alcohol or hyperthermia (heat shock).

The action of ethyl alcohol and the heat shock, i.e. short action of elevated temperature (+39-41°C) on mice ova was studied in our laboratory in detail (Dyban, Noniashvili, 1986b, Dyban, Noniashvili, 1986c). Parthenogenetic development was observed at least in 30 percent of the oocytes when alcohol was given to female mice shortly before or soon after ovulation (Dyban, Khozhai, 1980). Although it is difficult to extrapolate the data obtained in mice on humans, it is not wise to ignore this finding. There are reasons to suggest that heavy drinking can be the cause of parthenogenetic activation of human ova. These ova might not be fully activated and retain their fertilizing ability. If such ova are used in IVF program, the development will be abnormal because of abnormal paternal or maternal chromosome behavior.

The heat shock is as effective as ethyl alcohol and can induce in vitro all types of parthenogenetic development of mice ova. It leads to the assumption that short hyperthermia in humans might also induce parthenogenetic development. Therefore, some fever conditions may fully or partially activate human ovulated ova thus presenting a serious obstacle in IVF if such ova are used. Whether human ova _in vivo_ or _in vitro_ are sensitive to elevated temperature or ethyl alcohol and whether these agents are able to induce parthenogenetic development of human ova is not known. The answer to this question may be important for improving the efficiency of IVF programs.

The cytogenetic analysis of parthenogenetic mice embryos performed in our laboratory (Dyban, Noniashvili, 1986d) has shown chromosomal abnormalities in some of them. These abnormalities arrived during the activation of the oocytes as a result of non-disjunction of chromosomes at the first or second meiotic metaphase. Some structural chromosomal aberrations were also observed (chromosome and chromatid breaks, isochromosomes, dicentrics, etc.).

There is a strong evidence that haploid karyotype is unstable and at the late preimplantation stages or at early postimplantation stages the haploid set of chromosomes changes to diploid or to polyploid.

Although karyotype peculiarities in gynogenetic and androgenetic embryos have not been studied yet, it seems that the haploid karyotype in all these cases will be unstable and chromosomal aberrations (structural and numerical) may be observed.

Preimplantation and postimplantation development of parthenogenetic mice embryos was studied in different laboratories. These data are in good accordance and can be summarized as follows.

The duration of one cell stage in haploid embryos appears to be sufficiently shorter than in fertilized ova. The rate of cleavage, normal at the very beginning, becomes slower from the eight cells stage on. There are some cases of asynchronous cleavage, which seem typical for haploid embryos. As a rule, in haploid embryos a substantial delay in compaction of blastomeres is observed, morula being formed later than in normal embryos. Moreover, some haploid embryos start to be decompacted again after compaction has taken place. This process (compaction-decompaction) can be observed once or twice before the blastocyst is formed as a cause of a delay in cavitation and formation of a typical blastocyst. In most cases haploid blastocysts contain a reduced number of blastomeres in inner cell mass. These data lead to the conclusion that the development of haploid embryos is far behind the diploid fertilized ova.

One unexpected finding was observed in our study on ribosomal genes activity in haploid mice embryos (Dyban, Severova, unpublished). Using Ag-NOR staining, we showed that all five nucleolus organizer regions are stained with silver in haploid cleaving embryos, indicating high transcription activity of all ribosomal cystrons. Intensification of ribosomal genes transcription in haploid blastomeres may be due to a compensation of gene doses. However, better understanding of gene activity in haploidy compared to diploidy in mammalian embryogenesis is still needed.

Triploidy

In mammals triploidy may result from fertilization of ova with two sperms (diandry) as well as from suppression of polar body formation (digyny). In the latter, both sets of chromosomes remain in the oocyte fertilized with one sperm. Both types of triploidy have similar influence on embryonic development, though substantial differences in survival of triploid embryos in different mammalian species are observed. In humans, some triploid fetuses can survive until birth or even later (Kuliev, 1972, 1975), whereas in mice, this genomic abnormality is lethal during early postimplantation stages or during early organogenesis.

Triploid mice embryos have some peculiarities found even during preimplantation stages: there is a substantial reduction in cleavage rate, which is especially evident after the eight cells stage when the length of cell cycles is substantially prolonged.

Triploid mice embryos enter compaction earlier than diploid ones, which is especially evident during the culture *in vitro*. In our experiments with cultured embryos obtained from LT/Sv female, all embryos, entering compaction earlier appeared to be triploid (digynic triploids). They started cavitation earlier than diploids, blastocysts having fewer cells than normal embryos, and the number of blastomeres in inner cell mass being normal or a bit lower than in diploids. Our preliminary data indicate a lower ribosomal gene activity in triploid preimplantation embryos.

Further development of triploid embryos strongly depends on genetic background, ranging from death shortly after implantati embryonic arrest at the later stages. However, a typical phenotype for triploidy was also described, though only in some genetic strains of mice.

Tetraploidy

Developmental peculiarities of tetraploid embryos produced by different means in experiments or arrived at spontaneously are strongly manifested. A substantial reduction in cleavage rate, leading to a decreased number of tetraploid blastomeres compared to diploids of the same age is probably due to a pronounced increase in the length of cell cycles in tetraploids. In addition, tetraploid blastomeres have a higher ability to compact and most of these embryos start compaction earlier than diploid embryos. The cavitation is also accelerated, blastocyst being formed from fewer number of blastomeres. As only a few blastomeres are progenitors of inner cell mass lineage, the formation of embryonic body is substantially delayed and fewer cells can participate in organogenesis. Similar to triploids, the preimplantation development of tetraploids might be close to normal, depending on genetic background. Although most of these embryos die during the early organogenesis, some tetraploids can develop during the second half of gestation and die soon after birth.

Summarizing the data on genomic aberrations it may be concluded that despite some abnormalities in development, mice embryos with haploidy, triploidy and tetraploidy survive preimplantation stages reaching blastocyst and may induce decidual reaction in uterus and start implantation. Implantation progresses normally in some of them, while in others (especially in haploidy) embryos die during the implantation.

NUMERICAL ABERRATIONS OF INDIVIDUAL CHROMOSOMES IN MICE EMBRYOS

The influence of trisomy of autosomes on mice development is comparatively better studied than that of other types of numerical aberrations (for review see: Dyban, Baranov, 1987, Searle, 1989). These types of chromosomal aberrations in mice embryos can be obtained by different approaches, especially by the use of the carriers of Robertsonian translocations (Gropp, Winking, 1981). There is a strong evidence that an extra copy of any autosome or gonosome in mice has no harmful effect on preimplantation development. At the same time the influence of some trisomies on the rate of cleavage was observed (Severova, Dyban, 1984), which was not lethal but could interfere with cell interaction during implantation and formation of egg cylinder (gastrulation), leading to the arrest during early mice organogenesis.

Some malformations described are typical for trisomy of a certain autosome, but the expression strongly depends on the genetic background of the embryos: cyclopy in trisomy 1, exencephaly in trisomy 12 and/or 14, cleft palate in trisomy 19. All these observations were made on some genetic strains of mice but not reproduced on the others.

Developmental effects of autosomal monosomy in mice were not sufficiently studied. We are still almost completely ignorant of the embryonic manifestation of monosomies 4,7,8,10,11,13 and 18, and only few cases of monosomies 1,2,3,5,6,and 9 have so far been examined in preimplantation mice development. Among those better studied are autosomal monosomies 15, 17 and 19, the latter being investigated by different experimental methods.

In our study, we performed karyotyping of microsurgically removed isolated blastomeres of monosomy and trisomy. The biopsy of a single blastomere was performed at the four cells stage so that the remaining three blastomeres from each embryo could be cultured in vitro for 44 or 68 hours before they reached blastocyst stage. This is the first successful attempt to determine the karyotype in living embryo for evaluation of the influence of trisomy or monosomy on preimplantation development (Severova, Dyban, 1984, for review see: Dyban, Baranov, 1987).

In table 1, the mean cell count in euploid mouse embryos is compared with trisomy or monosomy 15 and in table 2, with trisomy or monosomy 17.

These experiments strongly suggest that trisomy 15 can delay cleavage while monosomy 15 has quite different effect on cleavage than monosomy 17.

Summarizing the data available, it is possible to conclude that monosomies 1,3,6,9,12,14,15,16,19 do not interfere with cleavage, compaction and blastocyst formation, all monosomic embryos dying during or shortly after implantation.

Table 1. The cleavage of euploid mice embryos, compared with trisomy or monosomy 15 (from Dyban, Baranov, 1987)

Culture conditions	Karyotype	Duration of culture interval			
		42-44 h		66-68 h	
		No. of embryos	Mean cell count	No. of embryos	Mean cell count
Four-cell embryos, cultured intact (non-biopsied)	Euploid	9	26.4+1.9	11	78.0+5.0
	Trisomy	-	-	3	80.0+11.1
	Monosomy	_	_	1	33
Three blasto meres of four-cell embryos, cultured after biopsy (removing of one blasto-mere)	Euploid	48	24.4+0.7	21	49.8+2.9
	Trisomy	15	20.7+0.8	5	41.8+11.4
	Monosomy	11	19.2+1.2	5	31.4+7.7

Table 2. The cleavage of euploid mice embryos, compared with trisomy or monosomy 17 (from Dyban, Baranov, 1987)

Culture conditions	Karyotype	Duration of culture interval			
		42-44 h		66-68 h	
		No. of embryos	Mean cell count	No. of embryos	Mean cell count
Four-cell embryos, cultured intact (non-biopsied)	Euploid	25	54.2+2.37	-	-
	Trisomy	4	41.3+3.4	-	-
	Monosomy	2	28.3	-	-
Three blasto meres of four-cell embryos, cultured after biopsy (removing of one blasto-mere)	Euploid	60	33.6+1.4	12	66.7+2.5
	Trisomy	10	28.1+4.1	4	63.2+5
	Monosomy	9	14.9+1.6	1	29

Quite different is the expression of monosomies 2, 5, 17, which stop the development at the eight - sixteen cells stage (early morula). The expression of monosomics appeared to be strongly dependent on genetic background, i.e. on the parental genotype as well as on the origin of monosomies - whether the chromosome was lost in oogenesis or spermatogenesis.

Nullisomy provides a unique opportunity to study the actual impact of any chromosome or its segment on development, as both homologous chromosomes (or alleles) will be missing. However, there are serious obstacles in producing such mice embryos.

According to limited data available only for nullisomies 15 and 16, this type of chromosomal abnormality is lethal at the very early cleavage (at the two-four cells stage). X-nullisomy (OY karyotype) seems to be lethal at the four-eight cells stage, while no data exist regarding nullisomies of other chromosomes.

There are data on the effects of some chromosomal loci (nullisomy of different chromosomal segments) which block cleavage at the homozygous state. The effect strongly depends on the genetic background of the embryo, the size of the deletion and the chromosome it is located on. For example, the deletion in chromosome 7 stops cleavage at the two-six cells stage, while the deletion in chromosome 17 (t-locus) - at the morula, blastocyst or egg cylinder stage.

DEVELOPMENT OF DIPLOID <-> ANEUPLOID CHIMAERAS

This approach has not been widely used, but the data obtained are rather promising (Epstein, 1985). Using different combinations of blastomeres with normal and abnormal karyotypes put together in the same embryo (chimaera) it was demonstrated that an unknown mechanism, working during embryogenesis, selects cells with normal karyotype against the abnormal karyotype. The selective pressure starts during preimplantation development, but is expressed stronger after implantation, during gastrulation and organogenesis. This may explain the fact why so few cells with abnormal karyotypes were detected in fetal tissue of diploid-aneuploid chimaeras.

CONCLUSION

Genomic aberrations in mice (haploidy, triploidy, tetraploidy) are compatible with preimplantation development. The same is true for autosomal trisomies and monosomies of some chromosomes (1,3,6,9,12,14,15,16,19), however monosomies 2, 5, 17,autosomal nullisomies and X-nullisomy (OY karyotype) are lethal in early cleavage stage. The developmental expression of genomic and chromosomal aberrations in mice strongly depends on genetic background and origin, and this probably applies also to humans.

Cytogenetic as well as some molecular biological data support the assumption that preimplantation development depends on maternal genetic information and genetic activity of loci ("early genes") clustered in some autosomes as well as in X-chromosome. While one dose of these genes is sufficient for cleavage, morula and blastocyst formation, implantation and postimplantation mice development require normal function of a whole diploid set of chromosomes.

Although the data obtained in experiments with mammalian embryos could not be directly extrapolated to the clinical practice and to human cytogenetics, they are of direct relevance to IVF and preimplantation genetic diagnosis because preimplantation development of human embryos has much in common with that of other mammals.

REFERENCES

Dyban, A.P., 1987, Studies of cellular, chromosomal and molecular mechanisms of very early mammalian embryogenesis, in: "Sov. Sci. Rev. F. Physiol. Gen. Biol., " vol. 1., T. Turpayev, ed., Harwood Academic Publishers, UK.

Dyban, A.P., 1988, "Early Development of Mammals," Nauka, Leningrad, (in Russian).

Dyban, A.P. and Baranov, V.S., 1987, "Cytogenetics of Mammalian Embryonic Development," Oxford University Press, UK.

Dyban, A.P. and Khozhai, L.E., 1980, Parthenogenetic development of ovulated mouse ova induced by ethyl alcohol, Byull. eksp. Biol. Med., 89:487-489 (in Russian).

Dyban, A.P. and Noniashvili, E.M., 1986a, Parthenogenesis in mammals, Ontogenesis (Sov. J. Dev.Biol.), 17:4, 280-296.
Dyban, A.P. and Noniashvili, E.M., 1986b, The study of factors determining rate and type of parthenogenetic development of mice ova activated by ethyl alcohol, Ontogenesis (Sov.J. Dev. Biol.), 17:2, 165-175.
Dyban, A.P. and Noniashvili, E.M., 1986c, Parthenogenetic development of mice ova activated by heat shock, Ontogenesis (Sov.J.Dev.Biol.), 17:6, 587-591.
Dyban, A.P. and Noniashvili, E.M., 1986d, Cytogenetic aspects of mammalian parthenogenesis, in: "Developmental Biology and Regulation of Heredity, " K. Belyaev, V.A. Strunnikov, eds., Nauka, Moscow.
Epstein, C.J., 1985, Mouse monosomies and trisomies as experimental system for studying mammalian aneuploidy, Trends in Genetics, 1:129-134.
Epstein, C.J., 1987, "The Consequences of Chromosome Imbalance: Principles, Mechanisms and Models," Cambridge University Press, New York.
First, N. L. and Barnes, F.L., 1989, Development of preimplantation mammalian embryos, in: "Development of Preimplantation Embryos and Their Environment," ed., Alan R. Liss, New York.
Gropp, A. and Winking, H., 1981, Robertsonian translocation: cytology, meiosis, segregation patterns and biological consequences of heterozygosity, Symp. Zool. Soc., London, 47:141-181.
Kaufman, M.H., 1983, "Early Mammalian Development: Parthenogenetic Studies," Cambridge University Press, Cambridge, UK.
Kuliev, A.M., 1972, Phenotype-karyotype interrelationship in embryogenesis of man, Genetica, 8:140-153.
Kuliev, A.M., 1975, Phenotypical peculiarities of chromosomal embryolethals in man, Doctoral Thesis, Moscow.
Magnuson, T. and Epstein, C.J., 1981, Genetic control of very early mammalian development, Biol.Review, 56:369-408.
Searle, A.G., 1989, Chromosomal variants, in: "Genetic Variants and Strains of the Laboratory Mouse," M.Grayson, A.Searle, ed., Oxford University Press, UK.
Severova, E.L. and Dyban, A.P., 1984, Selection of mice embryos in living condition according to sex and peculiarities of karyotype, Ontogenesis (Sov.J.Devel.Biol.), 15:6, 585-591.
Telford, N.A., Watson, A.J. and Schultz, G.A., 1990, Transition from maternal to embryonic control in early mammalian development: a comparison of several species, Molec.Reprod. Devel. 26:90-100.

CHANGING PHYSIOLOGICAL SYSTEMS IN THE PREIMPLANTATION EMBRYO

John D. Biggers

Harvard Medical School
Department of Cellular and Molecular Physiology
and Laboratory of Human Reproduction and
Reproductive Biology
Boston, MA 02115

INTRODUCTION

Two well-known changes in preimplantation development are the transfer of the control of development from the maternal to the embryonic genome, a genetic change, and compaction, a morphological change (see First and Barnes [1989], and Johnson and Pratt [1983], for reviews]. Both changes have important physiological consequences. Less well known changes during preimplantation development involve the entry and exit of substances from the embryo - the so-called transport functions. To emphasize the developmentally dynamic state of this period of the life cycle, three phenomena involving transport processes will be discussed: the physiological correlates of compaction, the regulation of intracellular pH (pH_i) of blastomeres, and the transtrophectoderm potential. These processes are of considerable interest in that they may provide the opportunity to study the genetic control of development with a relatively simple physiological model whose components and functions are known. Knowledge of the physiological changes which occur in preimplantation development is also of practical importance if we are logically to design artificial media for supporting in vitro development of preimplantation embryos.

PHYSIOLOGICAL CONSEQUENCES OF COMPACTION

In 1935 Lewis and Wright wrote of the mouse preimplantation embryo:

> *"The 8-cell stages exhibit varying degrees of compactness, or adhesion of the cells to one another."*

Preimplantation Genetics, Edited by Y. Verlinsky and
A. Kuliev, Plenum Press, New York, 1991

Forty years later, Ducibella and Anderson (1975) coined the word *compaction* to denote this phenomenon. However, the key observation about compaction was made twelve years earlier by Enders and Schlafke (1965). They observed that junctional complexes develop between the outer cells of several species of mammalian embryos at the time compactness appears. The physiological consequences of this morphological change were recognized by Enders, who wrote in 1971:

> "*Formation of continuous junctional complexes around the free borders of the trophoblast cells can be considered the necessary prerequisite to blastocyst formation, since this process converts what is essentially an intercellular space into a space surrounded by an epithelium. It is at this point that the blastocyst becomes an organism with an external and internal environment rather than just a collection of cells.*"

Fig. 1 illustrates this fundamental physiological change by a theoretical model which shows the alteration of the transport systems that occurs at the time of compaction. Compaction results in a switch from simple cellular exchanges across the cell membrane of single cells to a complex system of transport that includes the vectorial transport of substances across an epithelium. As a result, the preimplantation embryo acquires the capability of controlling the composition of its extracellular fluid. This capability has been demonstrated experimentally by culturing 2-cell mouse embryos to the blastocyst stage in a chemically defined medium and comparing the composition of samples of the blastocoel fluid obtained by micropuncture with that of the external culture medium (Borland et al., 1977). The results are shown in Table 1.

Major functional changes occur at the time of compaction because of a spatial change in the distribution of a single enzyme - Na^+/K^+-ATPase (the sodium pump) [see Biggers et al., (1991), for a review]. This enzyme has a ubiquitous role in organisms by maintaining large Na^+ and K^+ gradients across cell membranes. The potential energy provided by these electrochemical gradients is then used to transport a variety of substances in and out of the cell [see Lechene (1988), for a review]. All attempts to demonstrate Na^+/K^+-ATPase prior to compaction by cytochemical and immunochemical methods in the mouse and rabbit have failed. However, evidence for its presence in the unfertilized ovum and 2-cell embryo of the mouse has been obtained by isotopic experiments (Powers and Tupper, 1977). It is possible, however, that Na^+-gradients play only a minor role in the precompacted embryo, since it has been shown recently in the 2-cell mouse embryo that two major transport systems involving Na^+ are absent (Baltz et al., 1991). These are the Na^+/H^+ antiport and the Na^+/Cl^- exchanger.

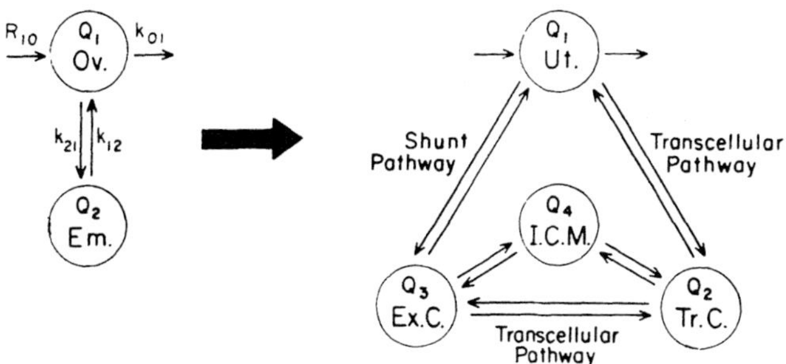

Fig. 1. Theoretical models showing changes in exchange pathways in mammalian preimplantation embryo as a result of compaction. Two stages of development are a precompacted 8-cell stage, represented by a 2-compartment model (left) and a compacted morula or blastocyst, represented by a 4-compartment model (right). Q_1, amount of substance in oviduct compartment; Q_2, amount of substance in embryonic compartment; k_{12}, rate constant for transport into oviduct compartment from embryonic compartment ($time^{-1}$); k_{21}, rate constant for transport into embryonic compartment from oviduct compartment ($time^{-1}$); R_{10}, rate of transport of unlabeled substance into oviduct compartment from mother; k_{01}, rate constant for transport from oviduct compartment to mother and proximal and distal regions of reproductive tract ($time^{-1}$); Ov, oviduct; Em, embryo; Ut, uterus; TrC, trophoblast cell; ExC, extracellular compartment; ICM, inner cell mass. (From Biggers et al, 1988, reprinted by permission.)

Table 1. Concentrations of Na, Cl, K, Ca, Mg, S, and P in blastocoel fluid of mouse embryos cultured form 2-cell stage in a chemically defined medium.

Element	Blastocoel fluid	Concentration,mM — Culture medium: Initial (n=7)	Final (n=3) Mean	Final (n=3) Range
Na	172±2	151±3	150	(148-152)
Cl	147±3	121±4	147	(135-156)
K	13±1	6.2±0.11	6.02	(5.76-6.51)
Ca	8.94±0.43	1.21±0.21	2.86	(1.9-4.4)
Mg	2.61±0.15	0.98±0.13	1.60	(1.48-1.76)
S	6.04±0.28	3.22±0.36	3.37	(1.90-4.82)
P	1.96±0.22	1.08±0.16	2.06	(1.43-3.09)

Values are means ± SE; n, no. of experiments. Because embryos were cultured under high-density conditions, changes in concentrations of Cl, Ca, Mg, and P may have been caused by developing embryos themselves or by a few degenerating embryos. (From Borland et al., 1977.)

When cells are aggregated into epithelial sheets, Na^+/K^+-ATPase is restricted to the basolateral side of the tight junctions. The local Na^+ gradient produced by this spatial arrangement of the enzyme controls in part the vectorial transport of water across the epithelium. Since the predominant molecule that has to be transported across the trophectoderm to form blastocoel fluid is water, it is not surprising that Na^+/K^+-ATPase is concentrated in the basolateral membranes of the trophoblast cells. The first suggestion that Na^+/K^+-ATPase is involved in blastocoel formation came from the studies of Smith (1970) and Gamow and Daniel (1970) on the rabbit, and DiZio and Tasca (1977) on the mouse, who showed that ouabain, a specific inhibitor of Na^+/K^+-ATPase, reduced the accumulation of blastocoel fluid. Later Biggers et al. (1978) and Benos and Biggers (1981), studying Na^+ transport in the rabbit blastocyst, provided evidence that the main site of action of ouabain was on the juxtacoelic side of the trophectoderm, thus locating Na^+/K^+-ATPase to the basolateral side of the trophectoderm junctional complexes. This distribution of the enzyme was subsequently confirmed in the mouse by immunocytochemical methods (Watson and Kidder, 1988).

One important question to ask about the differentiation of the preimplantation embryo is how the spatial distribution of Na^+/K^+-ATPase is controlled before and after compaction [see Biggers et al., 1988, 1991; Wiley et al.,(1990) for reviews]. Na^+/K^+-ATPase is synthesized in the endoplasmic reticulum and then transported to its site of action (McDonough et al., 1990). Since the half life of Na^+/K^+-ATPase is only about 18 h (Karin and Cook, 1986), compaction must result in a redirection of the supply of enzyme from its site of synthesis in the endoplasmic reticulum to the basolateral membrane. This change is temporally correlated in the mouse with the synthesis of at least two proteins which are involved in the formation of the junctional complexes - ZO1 (Fleming et al., 1989) and uvomorulin (Watson et al., 1990). Compaction in the mouse is also correlated with the appearance of several glycoconjugates in the cell membranes of the blastomeres [see Fenderson et al. (1990), for a review]. It has been shown recently in the nervous system that at least one glycoprotein (AMOG) is associated with the α_2 subunit of Na^+/K^+-ATPase (Wilkin and Curtis, 1990). Thus, a series of compounds that appear at the time of compaction may influence the redistribution of Na^+/K^+-ATPase. It will be of interest to determine whether a cluster of genes controls the production of the proteins involved in this complex process. There is some evidence that different isoforms of Na^+/K^+-ATPase operate before and after compaction (Watson et al., 1990).

PRECOMPACTED EMBRYO

One of the important homeostatic mechanisms in cells is the maintenance of a constant intracellular pH (pH_i). The pH_i is maintained at a set point by a balance between transport systems in the cell membrane that compensates for increased acidity or alkalinity inside the cell. The common systems that combat acidity are the Na^+/H^+ antiport and the Na^+-dependent HCO_3/Cl^-exchanger. The common system that protects against alkalinity is the Na^+-independent HCO_3/Cl^-exchanger. Recently, the combined application of two techniques has permitted the study of these three transport systems in the 2-cell mouse embryo (Baltz et al., 1990, 1991a). One technique is the fluorometric measurement of the pH_i of single blastomeres, using 2',7'-Bis(2-carboxyethyl)-5(and 6)-carboxy fluorescein acetoxymethul ester, a pH-sensitive dye; the other is the ammonium chloride pulse technique introduced by Boron and DeWeer (1976) to study pH_i transients in giant squid axons. The effect of exposing a 2-cell mouse embryo to ammonium chloride is shown in Fig. 2. This salt produces an immediate intracellular rise in pH (alkalinity), after which the pH slowly falls. On removal of the external ammonium chloride the pH rapidly falls beyond the initial resting level (acidity) and then it gradually returns to the initial level. In most cells the final return to the initial pH is mediated by the Na^+/H^+ antiport and the Na^+-dependent HCO_3/Cl^- exchanger. In 2-cell mouse embryos, however, the final recovery occurs in Na^+-free media, indicating the non-functionality of these two transport systems. The lack of the Na^+/H^+ antiport and the Na^+-dependent HCO_3/Cl^- exchanger is confirmed by the ineffectiveness of the inhibitors ethylisopropylamiloride (EIPA) and 4,4'-diisothiocyanostilbene (DIDS) respectively. In contrast the Na^+-independent HCO_3/Cl^-exchanger, which protects against alkalinity, is functional (Baltz et al., 1991b). The difference in the transport systems that regulate pH_i between a typical adult cell and the blastomere of a 2-cell mouse embryo is shown in Fig. 3.

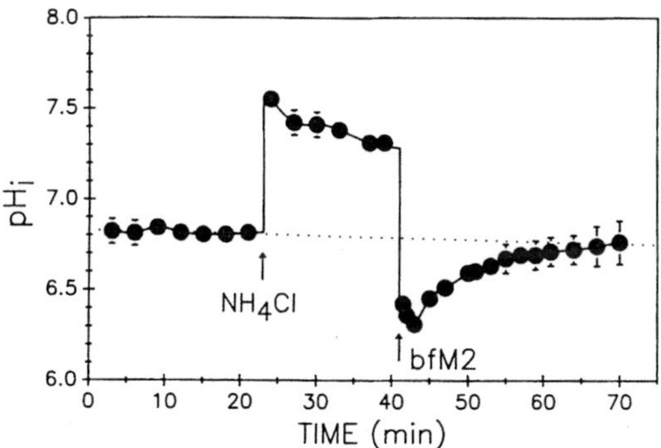

Figure 2. Acid-loading and recovery of 2-cell embryos in medium bfM2. Ten 2-cell embryos were acid-loaded by the NH_4Cl pulse technique (pulse between arrows). The initial medium (before first arrow) was bfM2. Recovery occurred in bfM2 (after second arrow). Each point is the mean of pH_i determinations of the 20 blastomeres of the 10 embryos. Error bars indicate one standard deviation. (From Baltz et al, 1990, reprinted by permission).

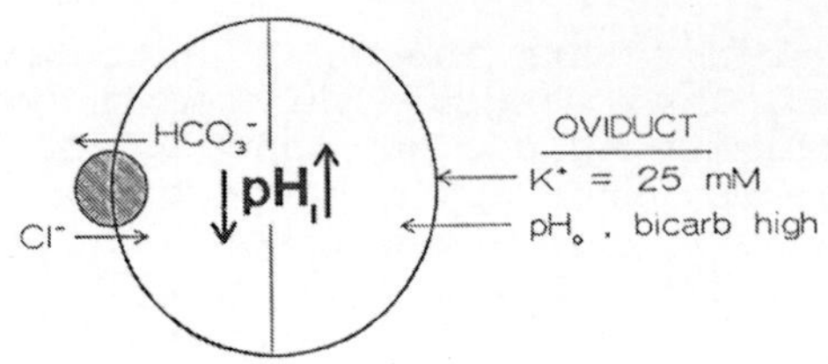

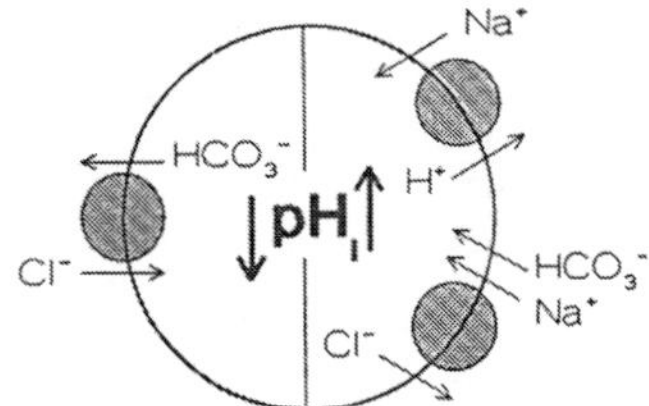

Figure 3. The transport systems involved in regulating intracellular pH in the 2-cell mouse embryo and a generalized cell.

As yet, it is unknown when the transport systems that protect against acidity become functional. Indirect evidence suggests that the Na^+/H^+ antiport is functional in the mouse blastocyst, since Na^+-flux into the blastocoel is inhibited by EIPA (Manejwala et al., 1989). As a working hypothesis it may be assumed that the protein transport systems are degraded or turned off during oogenesis and resynthesized or reactivated before the blastocyst forms. Whether the restoration of the transport systems is due to gene activation or to some other indirect mechanism before compaction is still unknown.

TRANSPORT SYSTEMS AFTER COMPACTION

After compaction a potential difference is established across the trophectoderm - the so-called transtrophectoderm potential (Δ_t). Fig.4 shows the values of Δ_t observed by Powers et al. (1977) across the rabbit trophectoderm from day 4-7 post coitum (pc). Initially Δ_t has a small negative value of approximately -5 mV. At about 6.5 days the potential reverses and reaches a positive potential of about +20 mV. When a 6.5 day rabbit blastocyst is exposed to increasing concentrations of amiloride, the Δ_t is depressed to a value comparable to that observed before the potential reverses (Powers et al., 1977; Biggers and Powers, 1979). The inhibitory effect of amiloride implicates Na^+ as having a role in the reversal of Δ_t, which occurs between day 6 and 7 in the rabbit blastocyst. This phenomenon provides a clear demonstration that transport systems in the embryo may shift between compaction and implantation. Other examples of transport systems that change in the rabbit trophectoderm prior to implantation are the transport of methionine (Benos, 1981) and the furosamide-sensitive Na^+/Cl^- symport (Benos and Biggers, 1983). It will be of interest to determine whether these changes are brought about by gene action or by local environmental controls.

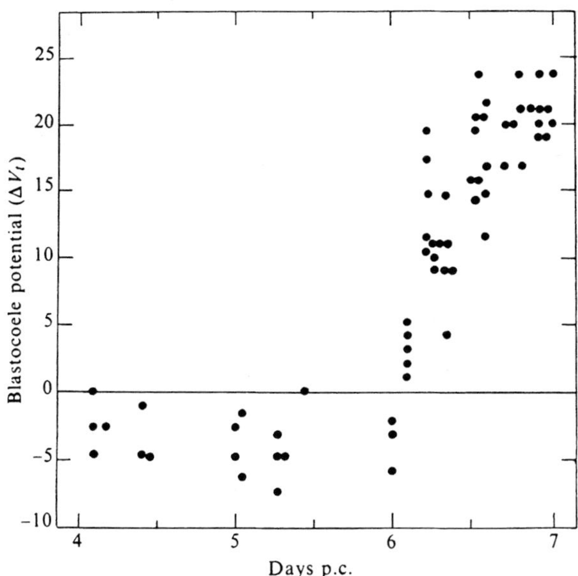

Figure 4. Developmental changes in the transtrophectodermal potential difference (ΔV_t) measured in freshly collected rabbit blastocysts on days 4, 5, 6 and 7 p.c. The ΔV_t were measured at 37°C in modified F10 medium supplemented with 20% foetal calf serum. The pH of the medium was 7.4. (From Powers et al., 1977, reprinted by permission.).

IMPLICATIONS FOR THE DESIGN OF CULTURE MEDIA

The changes in the transport systems that occur in the preimplantation embryo are but one aspect of a more complex system involved in early pregnancy. During the preimplantation phase of pregnancy the embryo is unattached to the genital tract of the mother and travels from the site of fertilization to the site of implantation. To understand this period functionally we have to analyze a complex system that consists of several parts (Fig. 5). Among these parts are the irreversible developmental changes, both morphological and functional, such as the transport changes described in this paper, and the composition and formation of the local microenvironments around the embryo as it travels through the female genital tract. In view of the changes that occur in the embryo as well as its location we may well ask: Can a single culture medium be developed which supports development maximally throughout the preimplantation period? Should the composition of the medium be changed to correspond with the physiological changes which occur as development proceeds? Recent work on the mouse has shown that several media of quite different composition may support a high degree of development throughout the preimplantation stages. For example, medium CZB (Chatot et al., 1989) and medium aKH_2PO_4 (Lawitts and Biggers, 1991), which differ widely in their content of glucose and phosphate, give comparable results (Table 2: Fig. 6). These results suggest that several media may exist which support maximal development, thus raising the possibility that preimplantation embryos may readily adapt to foreign environments.

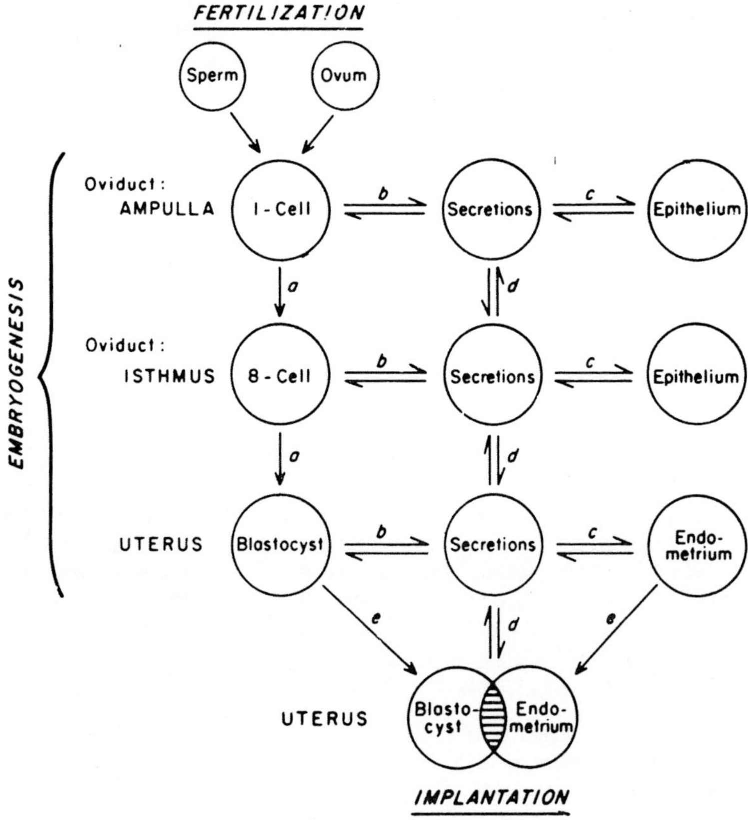

Figure 5. A theoretical model showing the changes between various compartments involved in the establishment of pregnancy. (From Biggers, 1981, reprinted by permission).

Table 2. Percentage of zygotes developing to ≥4 cells and blastocysts in media aKH_2PO_4 and CZB (from Lawitts and Biggers, 1991).

Medium	≥4 cells	Blastocysts
aKH_2PO_4	91	71
CZB	98	80

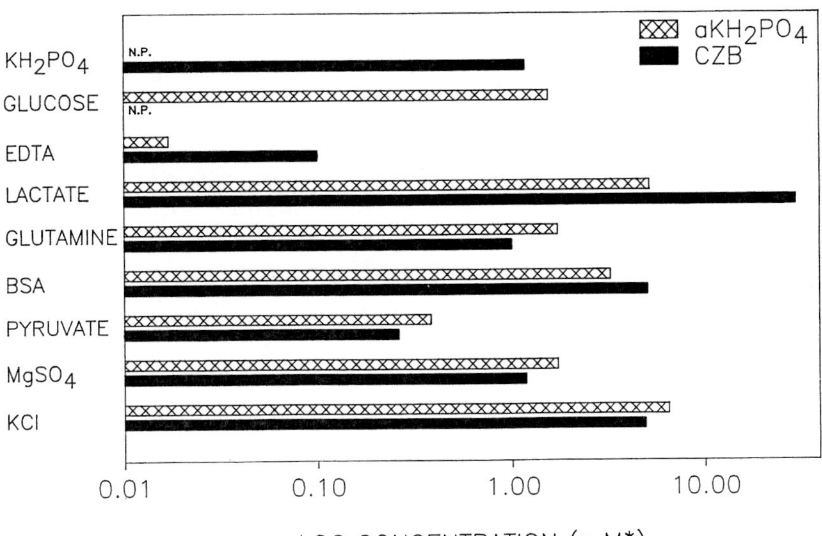

Figure 6. The composition of medium aKH_2PO_4 (Lawitts and Biggers, 1991) and medium CZB (Chatot et al, 1989).

Acknowledgements: The preparation of this manuscript was supported in part by the National Cooperative Program on Non-Human In Vitro Fertilization and Preimplantation Development with funding from the National Institute of Child Health and Human Development, NIH, through cooperative agreement HD21988, and by NIH research grant HD21581. I am indebted to Dr. Betsey S. Williams for valuable criticism of the manuscript and Carol Kountz for its preparation.

REFERENCES

Baltz, J.M., Biggers, J.D. and Lechene, C., 1990, Apparent absence of Na^+/H^+ antiport activity in the two-cell mouse embryo, Devel. Biol., 138: 421-429.

Baltz, J.M., Biggers, J.D. and Lechene, C., 1991, Two-cell-stage mouse embryos appear to lack mechanisms for alleviating intracellular acid loads. J. Biol. Chem. 266:6052-6057.

Baltz, J.M., Biggers, J.D. and Lechne, C., 1991b, Relief from alkaline load in 2-cell-stage mouse embryos by bicarbonate/chloride exchange. J. Biol. Chem. In Press.

Benos, D.J., 1981, Developmental changes in epithelial transport characteristics of preimplantation rabbit blastocysts. J. Physiol. Lond. 316: 191-202.

Benos, D.J. and Biggers, J.D., 1981, Blastocyst fluid formation. In: "Fertilization and Embryonic Development In Vitro," Ed. Mastroianni, L. and Biggers, J.D. pp.283-297. Plenum, New York.

Benos, D.J. and Biggers, J.D., 1983, Sodium and chloride cotransport by preimplantation rabbit blastocysts. J. Physiol. Lond. 342: 23-33.

Biggers, J.D., 1981, Cell biology of the developing egg. In: "Cellular and Molecular Aspects of Implantation, " S.R. Glasser and D.W. Bullock, Eds., Plenum Publishing, New York, 39 - 42.

Biggers, J.D. and Powers, R.D., 1979, Na^+ transport and swelling of the mammalian blastocyst: effect of amiloride. In: "Amiloride and Epithelial Sodium Transport," A.W. Cuthbert, G.M. Fanelli, and A. Scriabine, Eds., Urban and Schwarzenberg, Baltimore, pp. 167-179.

Biggers, J.D., Baltz, J.M. and Lechene, C., 1991, Ions and preimplantation development. In: "Animal Applications of Basic Research in Mammalian Development," Current Communications in Cell and Molecular Biology Series, Cold Spring Harbor Laboratory, Cold Spring Harbor (NY) pp. 121-146.

Biggers, J.D., Bell, J.E. and Benos, D.J., 1988, Mammalian blastocyst: transport functions in a developing epithelium. Am. J. Physiol. 255 (Cell Physiol. 24): C419-C432.

Biggers, J.D., Borland, R.M. and Lechene, C.P., 1978, Ouabain-sensitive fluid accumulation and ion transport by rabbit blastocysts. J. Physiol., Lond. 280: 319-330.

Borland, R.M., Biggers, J.D. and Lechene, C.P., 1977, Studies on the composition and formation of mouse blastocoele fluid using electron probe microanalysis. Dev. Biol. 55: 1-8.

Boron, W.F. and DeWeer, P., 1976, Intracellular pH transients in squid giant axons caused by CO_2, NH_3 and metabolic inhibitors. J. Gen. Physiol. 67: 91-112.

Chatot, C.L., Ziomek, C.A., Bavister, B.D., Lewis, J.L. and Torres, I., 1989, An improved culture medium supports development of random-bred 1-cell mouse embryos in vitro. J. Reprod. Fert. 86: 679-688.

DiZio, S.M. and Tasca, R.J., 1977, Sodium-dependent amino acid transport in preimplantation mouse embryos. III. Na^+/K^+-ATPase-linked mechanisms in blastocysts. Devel. Biol. 59:198-205.

Ducibella, T. and Anderson, E., 1975, Cell shape and membrane changes in the eight-cell mouse embryo: prerequisites for morphogenesis of the blastocyst. Devel. Biol. 47: 45-58.

Enders, A.C., 1971, The fine structure of the blastocyst. In: "The Biology of the Blastocyst," Ed. R.J. Blandau. pp. 71-94. University of Chicago Press.

Enders, A.C. and Schlafke, S.J., 1965, The fine structure of the blastocyst: some comparative studies. In: "Preimplantation Stages of Pregnancy," Ed. G.E.W. Wolstenholme and M. O'Connor. pp 29-59. Churchill, London.

Fenderson, B.A., Eddy, E.M. and Hakomori, S., 1990, Glycoconjugate expression during embryogenesis and its biological significance. BioEssays 12: 173-179.

First, N.L. and Barnes, F.L., 1989, Development of preimplantation mammalian embryos. In: "Development of Preimplantation Embryos and their Environment," Ed. Yoshinaga, K. and Mori, R. pp. 151-170. Liss, New York.

Fleming, T.P., McConnell, J., Johnson, M.H. and Stevenson, B.R., 1989, Development of tight junctions de novo in the mouse early embryo: control of assembly of the tight junction-specific protein, ZO-1. J. Cell Biol. 108: 1407-1418.

Gamow, G. and Daniel, J.C., 1970, Fluid transport in the rabbit blastocyst. Wilhelm Roux Arch. Entwicklungsmech. Org. 164: 261-278.

Johnson, M.H. and Pratt, H.P.M., 1983, Cytoplasmic localizations and cell interactions in the formation of the mouse blastocyst. In: "Time, Space, and Pattern in Embryonic Development," Ed. Jeffery, W.R. and Raff, R.A. pp.287-312. Liss, New York.

Karin, N.J. and Cook, J.S., 1986, Turnover of the catalytic subunit of Na^+/K^+-ATPase in HTC cells. J. Biol. Chem. 261: 10422-10428.

Lawitts, J.A. and Biggers, J.D., 1991, Optimization of mouse embryo culture media using simplex methods. J. Reprod. Fert. 91:543-556.

Lechene, C., 1988, Physiological role of the Na-K pump. Prog. Clin. Biol. Res. 2688: 171-194.

Lewis, W.H. and Wright, E.S., 1935, On the early development of the mouse. Contrib. Embryol. Carneg. Inst. 25: 114-143.

Manejwala, F.M., Cragoe, E.J. and Schultz, R.M., 1989, Blastocoel expansion in the preimplantation mouse embryo: Role of extracellular sodium and chloride and possible apical routes of their entry. Devel. Biol. 133: 210-220.

McDonough, A.A., Geering, K. and Farley, R.A., 1990, The sodium pump needs its beta-subunit. FASEB J. 4: 1598-1605.

Powers, R.D. and Tupper, J.T., 1977, Developmental changes in membrane transport and permeability in the early mouse embryo. Devel. Biol. 56: 306-315.

Powers, R.D., Borland, R.M. and Biggers, J.D., 1977, Acquisition of amiloride-sensitive rheogenic Na^+-transport in the rabbit blastocyst. Nature 270:603-604.

Smith, M.W., 1970, Active transport in the rabbit blastocyst. Experientia Basel 26: 736-738.

Watson, A.J. and Kidder, G.M., 1988, Immunofluorescence assessment of the timing and appearance and cellular distribution of Na^+/K^+-ATPase during mouse embryogenesis. Devel. Biol. 126: 80-90.

Watson, A.J., Pape, C., Emanuel, J.R. Levenson, R. and Kidder, G.M., 1990, Expression of Na^+/K^+-ATPase alpha and beta subunit genes during preimplantation development of the mouse. Devel. Genetics 11: 41-48.

Wiley, L.M., Kidder, G.M. and Watson, A.J., 1990, Cell polarity and development of the first epithelium. BioEssays 12: 67-73.

Wilkin, G.P. and Curtis, R., 1990, Cell adhesion molecules and ion pumps - do ion fluxes regulate neuronal migration? BioEssays 12: 287-288.

PART II

MICROMANIPULATION AND BIOPSY OF GAMETES AND PREEMBRYOS

BIOPSY OF HUMAN GAMETES

Yury Verlinsky

Reproductive Genetics Institute, Illinois
Masonic Medical Center, 836 W. Wellington
Chicago, IL 60657

INTRODUCTION

Biopsy of gametes opens an intriguing possibility for preconception diagnosis of inherited diseases, as genetic analysis of biopsied gamete material will make it realistic to select gametes containing the unaffected allele for fertilization and subsequent transfer. In this way, not only selective abortion of affected fetus but also fertilization involving affected gametes is avoided as an option for couples at risk for conceiving genetically abnormal fetus.

Although preconception genetic diagnosis could be achieved by genotyping either oocytes or sperm, the latter approach is not realistic at the present time. Despite extensive studies, claims to have successfully separated X and Y spermatozoa have not been confirmed to use this method in prevention of X-linked disorders.

No method also exists to retain viability of sperm after analyzing its genotype, although polymerase chain reaction (PCR) now permits genotyping individual human sperm (Li et al, 1988; Coutelle et al, 1989; Arnheim et al, 1990). Development of methods of culturing the primary spermatocytes and spermatogonia with following genetic analysis of maturated spermatides may theoretically be possible, however this still remains an elusive goal.

The only approach for preconception diagnosis at present is genotyping oocytes which will be addressed in this paper based on our experience on micromanipulation and biopsy of human oocytes with following genetic analysis of biopsied material.

BIOPSY OF OOCYTES

It is possible to obtain the genotype of an oocyte by direct analysis. However, this analysis destroys the

Preimplantation Genetics, Edited by Y. Verlinsky and
A. Kuliev, Plenum Press, New York, 1991

oocyte. We have developed a way of genotyping the oocyte by removing the first polar body and performing genetic analysis. In the absence of crossing-over, the first polar body will be homozygous for the allele not contained in the oocyte and second polar body (Fig. 1A and 1B). If crossing-over occurs, the oocyte will be heterozygous and the eventual genotype of the oocyte cannot be predicted (Fig. 1C). The frequency of crossing-over will vary with the distance between the locus and the centromere. For telomeric genes the crossover frequency approaches 50%, while for genes close to centromere the frequency may be close to zero.

Even with telomeric genes, 50% of polar bodies will be homozygous, and half of these (25%) will be homozygous for the mutant allele indicating that the oocytes can be used for fertilization and transfer. Since preconception genetic diagnosis involves induction of super-ovulation, several follicles are retrieved making polar body removal (PBR) feasible. Following PBR and analysis, 1 out of four oocytes can be fertilized and transferred.

Theoretically, for the 50% of oocytes with heterozygous polar bodies, the genotype of the oocyte could be determined by aspiration of the second polar body.

The method for PBR is described in detail in technical section of this volume (Verlinsky et al, this volume) and presented in figure 2.

To apply PBR for preconception genetic diagnosis the following problems have to be solved: (1) effect of PBR on fertilization and preimplantation development and (2) reliability of genetic diagnosis.

Effect of PBR on Fertilization and Preimplantation Development

Examination of any possible detrimental effect of the micromanipulation and biopsy on fertilization and embryo development is of vital importance since pregnancy rates in IVF are already low, and if micromanipulation further reduced the likelihood of pregnancies, preimplantation diagnosis could be impractical.

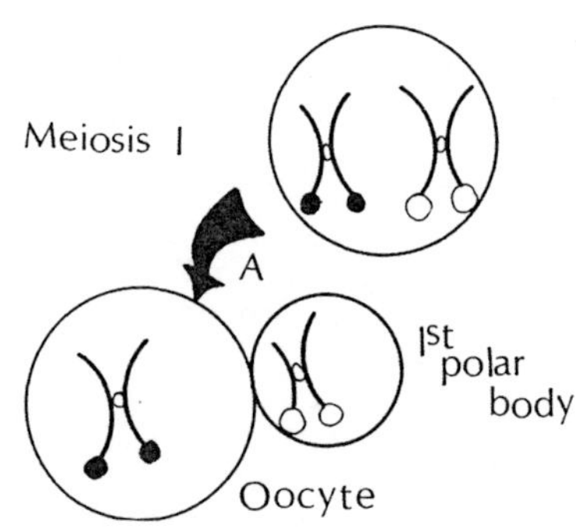

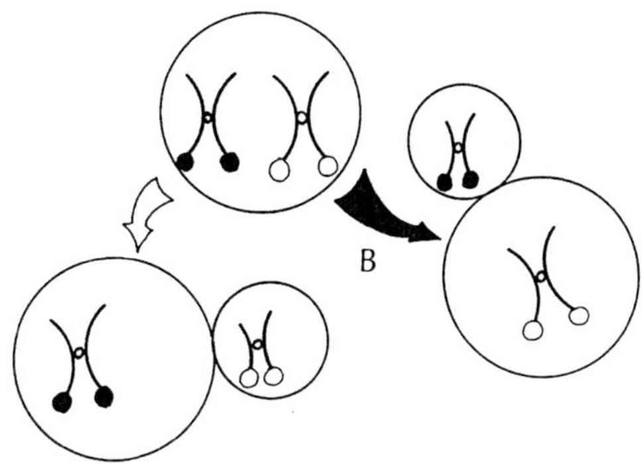

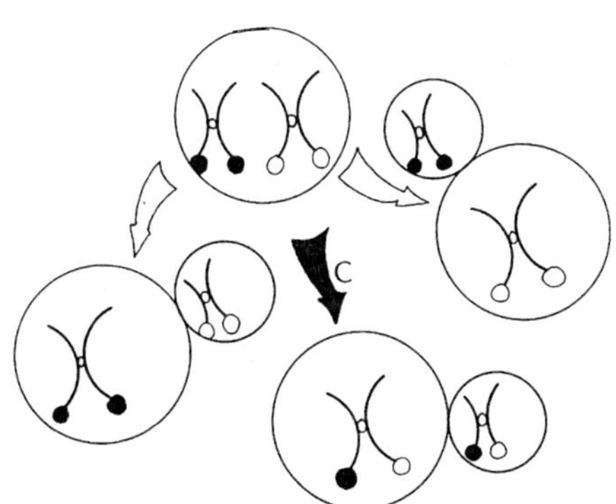

Figure 1. Genotype of oocyte and polar body following Meiosis I. (Affected alleles are shown as dark dots): A - In the absence of crossing-over if first polar body is normal, the oocyte is affected; B - In the absence of crossing-over if first polar body is affected, the oocyte will be normal. C - Crossing-over occurs leading to both polar body and oocyte being heterozygous.

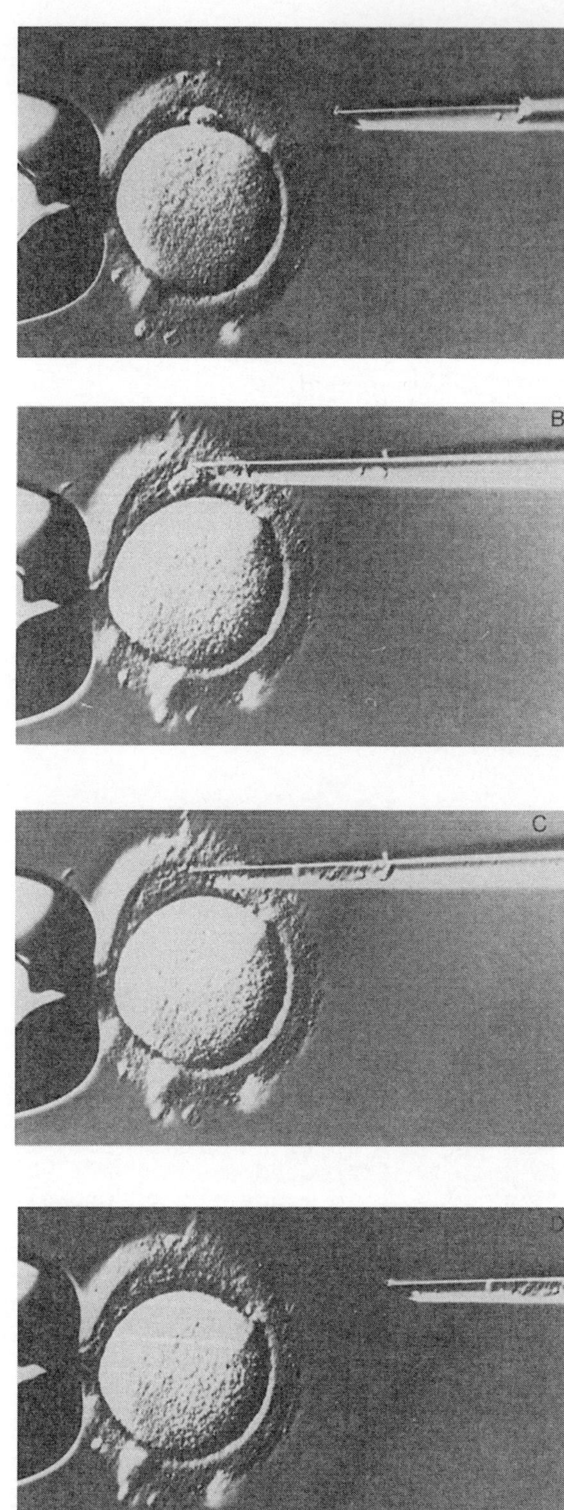

Figure 2. Polar Body removal (Description in the text)

Table 1 shows our data on the effect of PBR on <u>in vitro</u> fertilization and development in comparison with intact control oocytes, retrieved from 40 IVF patients randomly chosen of which 20 became pregnant. PBR was performed on a total of 92 non-atretic oocytes, including those retrieved from women who were carriers for CF, carriers for alpha-1-antitrypsin deficiency, or hemophilia A. Eighteen of these oocytes were found to be immature on the day of aspiration so that PBR and insemination were performed 24 hr later. In an attempt to remove the polar body from some of the oocytes it was found that the polar body was still attached to the ooplasm. In these cases upon an additional 3 hr of incubation the polar body was detached and successfully removed.

As seen from Table 1, there was no decrease in the fertilization rate for oocytes following PBR as compared to control oocytes (66.3 and 69.3% respectively), the percentage of embryos entering cleavage being the same in oocytes subjected to PBR compared to control oocytes (100% vs 97.4%). There was no increase in the percentage of polyspermic embryos in the PBR oocytes with respect to control oocytes (7.6 vs 5.1%).

Therefore, following PBR oocytes demonstrated no significant differences in fertilization rate, polyspermia, or cleavage indicating that PBR had no deleterious affect on fertilization or pre-embryonic development.

This is in agreement with the assumption that the 1st polar body does not play a role in embryo development. In fact, it begins to degenerate shortly following its extrusion (Rodman, 1971) and it was observed that normal babies have been born following the destruction of the 1st polar body during zona pellucida dissection (PZD) as a treatment for male factor infertility (J. Cohen, personal communication). Although we do not share the suggestion that the first polar body can be fertilized and develop along with its corresponding oocyte as a triploid "twin" (Bieber et al, 1981), removal of the polar body eliminates such a possibility.

Table 1

FERTILIZATION AND DEVELOPMENT
OF HUMAN OOCYTES
WITHOUT AND FOLLOWING POLAR BODY REMOVAL

CYCLES	OOCYTES	FRACTURED/ ATRETIC	FERTILIZATION MONOSPERMIC	DISPERMIC	TOTAL	CLEAVAGE
FOLLOWING PBR						
10	121	29	54/92 58.7%	7/92 7.6%	61/92 66.3%	54/54* 100%
WITHOUT PBR						
40	266	30	151/236 64.2%	12/236 5.1%	163/236 69.3%	147/151* 97.4%

* Triploid embryos were not included

It has been found that partial zona dissection (PZD) before insemination increases the risk of polyspermia (Gordon et al., 1988; Malter and Cohen, 1989). Although the zona pellucida had been punctured during PBR, there was no significant increase in the rate of polyspermy. In fact, the slit in the zona remaining after the withdrawal of the microneedle can hardly be considered as a "dissection" made according to the PZD protocol. However, we cannot dismiss the possibility, that the puncture in the zona pellucida during PBR may cause an increase in polyspermy, though our data failed to demonstrate such a possibility.

To analyze a possible detrimental effect of PBR beyond cleavage all the embryos predicted as affected have been cultured up to blastocyst stage. In these cases in addition to PBR blastomere biopsy was performed either to confirm the diagnosis in case polar body appeared to be normal or to select unaffected embryos for transfer if polar body appeared to be heterozygous. Blastomere biopsy was performed in 32 cases following PBR leading to the identification of 7 embryos for transfer. From the remaining 25 embryos, 16 (64%) developed to blastocyte stage - highly acceptable ratio for human embryos, suggesting no detrimental effect of micromanipulations and biopsy mentioned on pre-embryo development.

Reliability of Preconception Genetic Analysis

Our clinical trial on the application of preconception genetic diagnosis includes 8 couples at risk for conceiving children with autosomal recessive and X-linked disorders:CF, 4 couples, alpha-1-antitrypsin deficiency - 3 couples and hemophilia - 1 couple.

Table 2 shows the results of genetic analysis of a total of 83 polar bodies: 22 for delta-F508 mutation of CF, 33 for alpha-1-antitrypsin deficiency (AAT-ZZ) and 28 for hemophilia (HbaI polymorphism in the factor VIII gene). Successful analysis has been achieved in 18 (81.8%) polar bodies for delta-F508 mutation, in 30 (90.9%) for α-1-antitrypsin deficiency and 21 (75%) for hemophilia.

Of the successful analysis of polar bodies for hemophilia, 6 (28.6%) appeared to be normal, i.e. showing the presence of HbaI polymorphic site corresponding to the normal gene in this family (hybridized to 7,3 probe), 13 (61.9%) were lacking HbaI site (hybridized to 7,4 probe), indicating that corresponding oocyte was normal and 2 - were heterozygous (hybridized to both 7.3 and 7.4 probes).

Of 30 successful analysis of polar bodies for AAT-Z mutation, 3 (17%) - were identified to be AAT-M, 14 (46%) heterozygous, while 11 (37%) appeared to be affected (AAT-Z) indicating that corresponding oocytes are free from the abnormal gene.

Of the successful analysis of polar bodies for delta-F508 mutation, 5 (28%) were homozygous normal, 5 (28%) - heterozygous, indicating that an odd number of crossover events had occurred, and 8 (44%) - homozygous for delta-F508 (Fig. 3).

Table 2

PRECONCEPTION GENETIC ANALYSIS*

	NORMAL	HETEROZYGOUS	AFFECTED	FAILED PCR	TOTAL
DELTA F-508	5	5	8	4	22
**HAEMOPHILIA	6	2	13	7	28
***α-1-ANTITRYPSIN DEFICIENCY	5	14	11	3	33
TOTAL	16	21	32	14	83

* The numbers show the number of polar bodies studied.

** Analysis of the HbaI polymorphism in amplified DNA (The absence of HbaI site in this family, i.e. hybridization to probe 7.4, correspond to Haemophilia mutation, indicating the presence of mutant allele)

*** Scanning the AAT-Z mutation

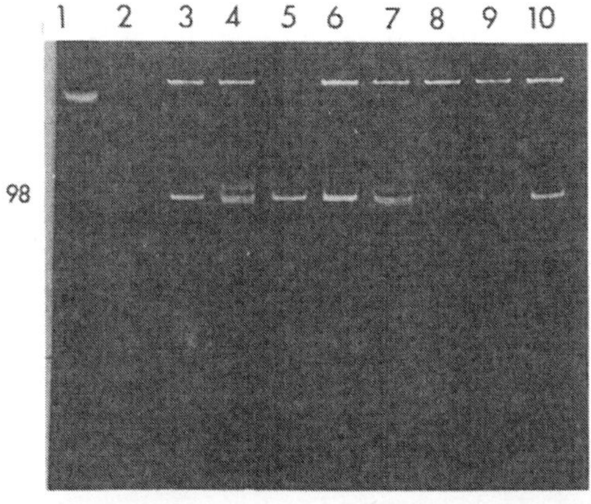

Figure 3. Analysis of Polar Bodies for Delta-F508 mutation. Polar bodies were removed from two oocytes and subjected to 40 cycles of PCR. Lane 1: Size standards (pBR 322/HaeIII digestion); Lane 2: Buffer only, control; Lane 3: 10 ng Homozygous normal control DNA; Lane 4: 10 ng Heterozygous control DNA; Lane 5: Polar body #1; Lane 6: Polar body #1 mixed with homozygous normal DNA; Lane 7: Polar body #1 mixed with homozygous delta-F508 DNA; Lane 8: Polar body #2; Lane 9: Polar body #2 mixed with homozygous normal DNA; Lane 10: Polar body #2 mixed with homozygous delta-F508 DNA.

One oocyte, predicted to be homozygous normal by polar body analysis did not fertilize and was analyzed and found to have the normal allele, thus confirming the polar body analysis for delta-F508 mutation. An embryo predicted to be homozygous delta-F508 by polar body analysis resulted in a pre-embryo whose genotype was determined to be homozygous delta-F508 by blastomere biopsy (Strom et al, 1990). In 5 cases heterozygote status of the embryo also was confirmed by blastomere biopsy. Reliability of diagnosis in cases of hemophilia was monitored by blastomere biopsy and gender determination using X and Y specific primers. In all cases when the embryos were predicted to be affected the diagnosis was confirmed (both X and Y sites were present).

ADVANTAGES AND DISADVANTAGES OF PRECONCEPTION GENETIC DIAGNOSIS

Preconception diagnosis by PBR has clear advantages compared to preimplantation genetic analysis by blastomere biopsy. First of all, the first polar body has no known function in embryonic development. Second, PBR does not involve any damage for developing embryo as no embryonic material is removed. Finally, the polar body is removed within 24 hours following oocyte retrieval and genetic analysis is completed by the second day, with no risk of missing the implantation window as embryo development _in vitro_ occurs at a slower rate than _in vivo_ (Bavister, 1988).

However, PBR will not make it possible to establish genetic diagnosis of the embryo in a number of situations: (1) paternal alleles cannot be analyzed (2) the test might not be efficient for telomeric loci as the crossing-over will occur with high frequency leading to the heterozygote state of the polar body; (3) gender determination is not possible. In these situations blastomere biopsy becomes an important adjunct to PBR.

The fact that by PBR the genotype of the oocyte is inferred rather than directly determined has also been suggested as an additional weakness. However, our experience so far demonstrates an accurate prediction of the genotype of the corresponding oocyte (Verlinsky et al, 1990b, Strom et al, 1990).

Therefore, there seems to be no need for further information about embryos whose 1st polar bodies were found to be homozygous for the abnormal recessive gene. As mentioned these embryos represent between 25% and 50% of all embryos (depending on the distance between the gene being analyzed and the centromere). If the mother is a carrier for a recessive or a dominant gene, fertilization of such oocytes will result in an unaffected fetus. The analysis of the polar bodies could theoretically identify all oocytes that contain the unaffected gene. However, in recessive diseases, 50% of oocytes containing the abnormal gene will be fertilized with sperm carrying the normal gene and will be unaffected. Therefore, polar body analysis will theoretically only be able to identify about 2/3 of all normal embryos in recessive diseases, the remaining unaffected embryos to be identified by embryo biopsy and genotype analysis during preimplantation development.

In conclusion, it should be noted that for application into clinical practice both preconception and preimplantation genetic analysis should be available to enable genotyping and confirmation of preimplantation diagnosis for couples at risk for autosomal recessive, dominant and X-linked disorders.

REFERENCES

Arnheim, N., Li, H. and Cui, X. (1990) PCR analysis of DNA sequences in single cells; single sperm gene mapping and genetic disease diagnosis. Genomics 8:415-419.

Bavister, B.D. (1988): Role of oviductal secretions in embryonic growth in vivo and in vitro Theriogenology 29:143-154.

Bieber, F.R., Nance, W.E., Morton, C.C., Brown, J.A., Redwine, R.O. (1981): Genetic studies in acardiac monster: evidence of polar body twinning in man. Science 213:775-777.

Coutelle, C., Williams, C., Handyside, A., Hardy, K., Winston, R., and Williamson, R. (1989): Genetic analysis from DNA from single human oocytes: A model for preimplantation diagnosis of cystic fibrosis. Br. Med. J. 299:22-24.

Gordon, J.W., Grunfeld, L., Garrisi, G.J., Talansky, B.E., Richards, C., Laufer, N. (1988): Fertilization of human oocytes by sperm from infertile males after zona pellucida drilling. Fert. Steril. 50:68-73.

Li, H., Gyllensten, U., Cui, X., Saiki, R., Erlich, H. and Arnheim, N. (1988): Amplification and analysis of DNA sequences in single human sperm. Nature 335:414-417.

Malter, H.W., Cohen, J. (1989): Partial zona dissection of the human oocyte: a nontraumatic method using micromanipulation to assist zona pellucida penetration. Fert. Steril. 51:139-148.

Rodman, T.C. (1971): Chromosomes of the first polar body in mammalian meiosis. Exptl. Cell Res. 68:205-210.

Strom, C.M., Verlinsky, Y., Milayeva, S., Evsikov, S., Cieslak, J., Lifchez, A., Valle, J., Moise, J., Ginsberg, N., Applebaum, M. (1990): Preconception genetic diagnosis for cystic fibrosis by polar body removal and DNA analysis. Lancet 336:306-308.

Verlinsky, Y., Ginsberg, N., Lifchez, A., Valle, J., Moise, J., Strom, C.M. (1990): Analysis of the first polar body: Preconception genetic diagnosis. Human Reproduction 5(7):826-829.

Verlinsky, Y., Milayeva, S., Evsikov, S., White, M., Cieslak, J., Lifchez, A., Valle, J., Moise, J., Strom, C.M. (1991): Preconception and preimplantation diagnosis for cystic fibrosis. Prenatal Diagnosis In Press.

MANIPULATION TECHNIQUES WITH POTENTIAL USE IN ANIMAL AGRICULTURE

Neal L. First

University of Wisconsin
Madison, WI

INTRODUCTION

Animal breeding is entering an exciting new era in which it will have the tools to rapidly multiply animals and to make new strains, precisely tailored to meet product and environmental demands. Many of these new tools depend on micromanipulation of embryos. The new tools include the ability to transfer new genes into the genome of domestic animals, the genetic screening of embryos by DNA marker assisted selection to directly select for genes encoding desirable traits and against genetic defects, as well as the use of embryo bisection or cloning to produce identical copies of embryos from a genetically tested high performance embryo line.

The major effort to develop these technologies has been with cattle where a viable commercial embryo transfer industry exists and where the economic payback is expected to be profitable.

These technologies are also expected to enhance traditional animal breeding programs. For example, it is estimated that twin offspring from embryo bisection would increase the power of selection in a MOET (multiple ovulation embryo transfer) selection scheme by 1.3 to 1.8 times (Gearheart et al., 1989) and use of cloned offspring would result in a 5-fold increase (Smith, 1989).

This review will focus on four micromanipulation techniques: 1) the bisection of embryos to produce twins, 2) nuclear transfer of cells from a valuable donor embryo into enucleated oocytes to produce clonal lines, 3) biopsy of embryos for sexing and genetic screening and 4) pronuclear microinjection of DNA to introduce new genes into the genome of the embryo. These technologies are not viewed as separate entities. For example, a sample of 2 to 4 cells' residual from embryo bisection or nuclear transfer is sufficient for sexing or genetic screening. Embryonic cells may have new genes introduced before they are used to make cloned embryos and cloning of embryos from cells may be used to rapidly multiply rare transgenic animals.

MULTIPLICATION OF EMBRYOS

The ability to produce multiple copies of an embryo provides a powerful selection and propagation tool, especially useful for traits of low heritability. A large number of genetically identical embryos provides a means for embryo selection wherein clonal lines descendent from one embryo could be selected by progeny test for clonal multiplication to large numbers. This system approaches phenotypic rather than genetic selection by taking advantage of both additive and nonadditive inheritance. It could permit rapid change in selected characteristics such as meat or milk production and when combined with in vitro production of embryos it could provide a way to produce large numbers of high quality embryos for frozen storage and commercial embryo transfer.

Multiplication of Embryos by Bisection

Embryos of cattle (Ozil et al., 1982; Williams et al., 1982; Baker and Shea, 1985; Williams and Moore, 1988; Leibo, 1988), sheep (Willadsen, 1982; Herr et al., 1990a) and swine (Willadsen, 1982; Rorie et al., 1985) have been bisected at the morula or blastocyst stage to produce twin offspring by a technique developed originally by Willadsen (1979). This technique is now used by numerous cattle embryo transfer companies to nearly double the number of embryos available for transfer. In one large study the pregnancy rate was 56.5% from transfer of 515 intact blastocyst stage bovine embryos and 52.4% from transfer of 842 half embryos (Leibo, 1988).

The embryo is bisected with a microknife (Williams et al., 1982) or pulled apart with two glass needles (Ozil et al., 1982; Willadsen and Godke, 1984). Highly efficient modifications of these techniques have now been developed (Rorie et al., 1985; Williams and Moore, 1988; Herr et al., 1990a). The most effective is a technique using only one micromanipulator in which the embryo is held in place by attachment to an electrostatic culture dish and the sectioning knife, a fragment of razor blade, is forced through the embryo as a guillotine. The cells left on the splitting knife have been used to sex the embryo (Herr et al., 1990a,c). This method is shown in Figure 1 and is sold commercially by the ABT Company of Australia and Seattle USA. When the blastocyst is split, care must be taken to bisect the inner cell mass into two equal halves. The bisection reduces the number of cells by half. There are still sufficient cells for normal embryo development unless another half or more are killed later by freezing, or quartering or any other procedure further reducing cell number. Pregnancies have resulted from bisected embryos frozen by techniques causing little damage (Lehn-Jensen and Willadsen, 1983) and from quartered embryos (Voelkel et al., 1985, 1986) but the pregnancy rate is less than normal when embryos are frozen immediately after bisection.

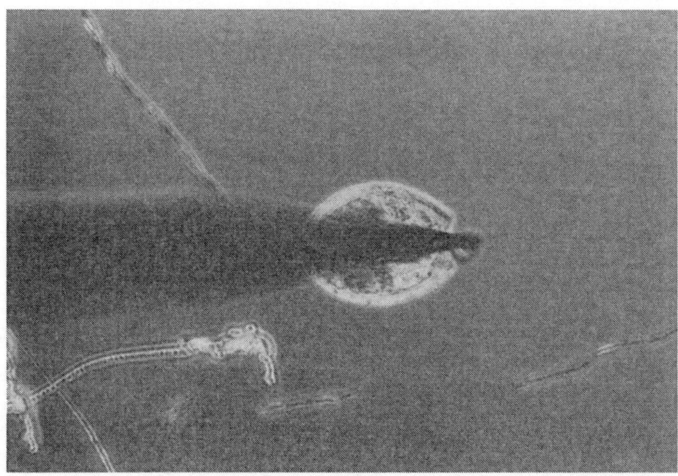

Figure 1. Bisection of a blastocyst stage bovine embryo held at the bottom of an electrostatic culture dish. Photo courtesy of Dr. Charles Herr, A.B. Technology, PT4 Limited, Australia.

Embryos have been bisected and transferred both with and without zona pellucida with little difference in pregnancy rate (Warfield et al., 1987). In litter bearing species the bisected embryos must be placed in a zona pellucida before transfer because the zona-free embryos will aggregate and fuse forming large chimeric embryos. Thus far there has been no evidence for increased incidence of birth defects or abnormal offspring from this procedure. The frequency of an additional splitting and spontaneous twinning has, however, been increased by bisection (Seidel, 1985). In sheep and cattle this technique is being combined with sexing of the embryo wherein a few cells (2-8) are removed from the trophoblast at the time of bisection and sexed before transfer 3-7 hr later (Herr et al., 1990a,c).

Nuclear Transfer

The second method of producing multiple copies of an embryo is by nuclear transplantation. Nuclear transfer is a tool which allows nearly clonal propagation of animals. Besides potentially providing large numbers of near identical offspring, it allows selection of clonal lines whose productive value can be well characterized before nuclear transfer.

A nuclear transplantation procedure has recently been shown successful in producing viable embryos and offspring in cattle (Prather et al., 1987; Willadsen, 1989; Barnes et al., 1990; Bondioli et al., 1990), sheep (Willadsen, 1986; Smith and Wilmut, 1989), swine (Prather et al., 1989) and rabbits (Stice and Robl, 1988).

The nuclear transplantation procedure is a modification of a procedure originally developed and shown successful for the amphibian xenopus (Briggs and King, 1952). The procedure (Fig. 2) involves transfer of a blastomere from a valuable embryo of a multicellular stage into an enucleated metaphase II oocyte with subsequent development to a multiple cell stage and use as a donor in a serial recloning.

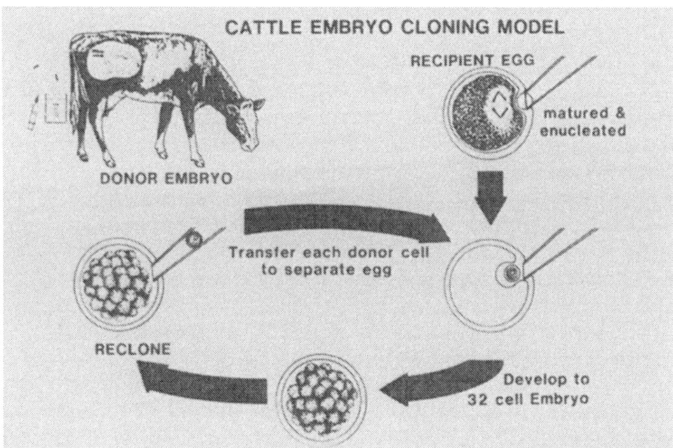

Figure 2. Procedures used to multiply bovine embryos by nuclear transfer. Late stage multicellular embryos of high value are recovered by nonsurgical flush from the uterus of an inseminated cow. The individual cells or blastomeres are removed from the valuable donor embryo and transferred into enucleated oocytes and the resulting new embryos developed to the morula stage when the process can be repeated (presented originally as a report to W.R. Grace and Company by N.L. First).

For cattle at least four commercial companies are developing nuclear transfer programs and two companies are already marketing embryos resulting from nuclear transfer. Although a few thousand offspring have been produced thus far most of the clones produced are sets of two to four with the greatest number being eight (Bondioli et al., 1990). The frequency of developed blastocysts and maintained pregnancies after nuclear transfer have been less than normal with approximately one quarter of the nuclear transfers resulting in late morula or blastocysts (Prather et al., 1987; Barnes et al., 1990; Bondioli et al., 1990) and approximately 30% of the morula or blastocysts transferred into recipient cows resulting in established and maintained pregnancies (Bondioli et al., 1990). The frequency of morula and blastocysts resulting from nuclear transfer is higher, approximately 35 to 50% in sheep (Willadsen, 1986; Smith and Wilmut, 1989) and rabbits (Stice and Robl, 1989) and lower for swine (Prather et al., 1989).

Major variables contributing to success of nuclear transfer relate to: 1) the donor nucleus, 2) the recipient oocyte and its enucleation and 3) the cell fusion and activation process.

The Donor Nuclei

A major variable concerning the nucleus of the donor blastomere is its totipotency. This seems to differ with species and may relate to the stage of embryo development at which each species transition from maternal to embryonic control of transcription. Nuclei from xenopus embryos when transferred into enucleated oocytes or eggs are totipotent until the gastrula stage, 4,000 cells (Gurdon, 1986), whereas blastomere transfer is generally unsuccessful in mice when donor stages later than 2 cells are used (McGrath and Solter, 1984; Robl et al., 1986). Cell division is dependent on embryonic transcription after the 2-cell stage in mice and 4,000 cell stage in xenopus. The totipotency limits are as yet unknown for blastomeres of domestic animals. Thus far successful nuclear transfers have been performed with stages as late as 4 cells in swine (Prather et al., 1989) and the 48 to 64 cell stage in cattle (Bondioli et al., 1990) and the inner cell mass of the sheep blastocyst (Smith and Wilmut, 1989). In rabbits cells of the blastocyst inner cell mass were totipotent but at a lower frequency than cells of morulae whereas trophoblast cells were not totipotent (Collas and Robl, 1991).

The 64-cell stage and inner cell mass of the blastocyst are not far from the blastocyst stage at which totipotent embryonic stem cells of mice and hamsters can be obtained and multiplied to large numbers under differentiation inhibiting conditions (Evans and Kaufman, 1981; Doetschman et al., 1988; Rossant and Joyner, 1989). Thus far cultured cell lines from blastocysts of domestic animals have been identified as pluripotent but not totipotent (Evans et al., 1990). If cell lines with totipotency can be isolated and cultured their nuclei would allow an infinite number of identical offspring to be produced by nuclear transfer. These cells might also be sexed, subjected to genetic screening or receive new genes by transfection, infection or microinjection.

Evidence in amphibians suggests that the stage of the donor cell cycle especially for slowly dividing cells affects development after nuclear transfer (McAvoy et al., 1975; Ellinger, 1978; Von Beroldington, 1981). Development is at a higher rate if the nuclear donor is at G_2 of the cell cycle (Von Beroldington, 1981). Nuclei from G_1 cells when transferred into metaphase II oocytes often undergo incomplete DNA replication and propagate missing information until development fails (King, 1966). Effects of stage of the cell cycle are unknown for domestic animals.

The Recipient Oocyte

Based on amphibian studies the stage of oocyte commonly enucleated as the recipient cell in mammalian nuclear transfer is the metaphase II oocyte (Willadesn, 1986, 1989; Stice and Robl, 1988; Prather et al., 1987, 1989; Smith and Wilmut, 1989; Bondioli et al., 1990). Most of the above studies have used aged metaphase II oocytes probably because a higher frequency of oocyte activation occurs with the aged oocyte (Ware et al., 1989). Unfortunately, fertilization of aged oocytes is known to result accelerated first cell cycle and in lower pregnancy and embryo survival (Barrett, 1948; Casida, 1950; reviewed by Boerjan and deBoer, 1990).

Another oocyte variable is the efficiency of oocyte enucleation. Enucleation has commonly been done by one of two methods either bisecting the oocyte and discard of the half adjacent to a polar body or by aspiration of the polar body and a variable amount of adjacent cytoplasm. Both methods rely on the expectation that the metaphase plate (nuclear material) is adjacent to the recent divisional product, a visible polar body.

Willadsen (1986) showed that very few embryos developed after nuclear transfer if nuclear material remained in the cytoplasm. By these methods enucleation has commonly been complete for 85 to 95% of the attempts. Recently developed methods using DAPI or Hoechst 33258 fluorescence staining to identify the nuclear material to be removed or after removal (Westhusin et al., 1990) have resulted in essentially 100% enucleation.

Cell Fusion

Fusion of a donor blastomere into an enucleated oocyte is commonly accomplished by electrofusion.

Electro-cell-fusion is accomplished by providing a pulse of electricity that is sufficient to cause a transient breakdown of the plasma membrane. This breakdown of the plasma membrane is very short and the membrane reforms very rapidly. If two adjacent membranes are induced to breakdown and upon reformation the lipid bilayers intermingle, then small channels will open between the two cells. Due to the thermodynamic instability of such a small opening, it enlarges until the two cells become one. Efficient fusion depends on healthy cell membranes, physical contact of both oocyte and blastomere membranes to be fused, and direction of the fusion current primarily through the point of membrane contact (Zimmerman and Vienken, 1982).

Earlier experiments indicated that the efficiency of cell fusion decreased rapidly as smaller and later stage donor blastomeres were used (cattle, Prather et al., 1987; sheep, Smith and Wilmut, 1989). However, more recent reports show only a small reduction in fusion efficiency with donor cells as late as the 40- to 64-cell stage (Bondioli et al., 1990). Present methods allow fusion efficiencies of approximately 70% (Bondioli et al., 1990).

In summary, cloned offspring are being produced from nuclear transfer and primarily in cattle. In cattle the recipient oocyte can be derived from ovaries of commercially slaughtered cattle and matured in vitro, the donor embryos can be stored frozen and successful recloning has been performed (Barnes et al., 1990). The latest stage embryo providing donor cells has been inner cell mass of the blastocyst and there are hopes for use of cultured stem cells as the nuclear donors.

The overall efficiencies of producing morulae and blastocysts and maintained pregnancies are less than desired, being approximately 25 and 30%, respectively.

BIOPSY OF EMBRYOS FOR SEXING OR GENETIC SCREENING

There has been little research in domestic animals with the biopsy of embryos for genetic screening because only a few gene candidates for screening have been identified. The principal genetic screening practiced has been on the sexing of embryos. A large number of genetic markers are being developed for use in cattle and other domestic animals with the hope some will be associated with desirable or undesirable genotypes (Georges et al., 1990). Some use is being made of: a marker associated with milk production in a specific family of cattle (Cowan et al., 1990), markers for a milk casein variant associated with cheese yield and milk production called kappa casein (Lien et al., 1990), a marker for a milk protein, β-lactoglobulin (Lien et al., 1990) and a genetic defect, the weaver syndrome, also associated with milk production in Brown Swiss cattle (Hoeschale and Meinert, 1990). Except for sexing embryos these genetic markers have not been applied thus far to genetic screening of embryos. In sexing embryos the genetic screening was done on cells left on the splitting knife (Herr et al., 1990a,b,c) or cells from a biopsy of the trophoblast of the blastocyst (Bondioli et al., 1989). Little difficulty is expected for cells from the morula or blastocyst stage since as previously discussed as much as half the embryo could be used for genetic screening with half embryos resulting in pregnancy rates nearly equivalent to whole embryos (Leibo, 1988).

There is evidence that development is of reduced frequency but not totally impaired when blastomeres are removed from cleavage stage embryos of rodents and that mouse embryos biopsied at the 4 or 8 cell stage show greatly impaired development if more than one blastomere is removed (Kim et al., 1990); reviewed by Papaioannou, 1989). In domestic animals the tolerance for bisection at cleavage stages appears to be much higher probably because their commitment to differentiation is much later than rodents. For example, quartering the sheep embryo at the 8-cell stage has produced quadruplets and a high rate of development through the blastocyst stage has resulted when blastomeres were divided at the 2-cell stage (Willadsen, 1979, 1980; Gatica et al., 1984).

Isolation and culture of single cells have resulted in approximately two thirds the normal frequency of development to the blastocyst stage in rabbits (Yang et al., 1990). Porcine blastomeres cultured individually from the 8-cell or 16-cell stage on fibronectin, krebs ringer bicarbonate and 10% lambs' serum developed to the blastocyst stage at frequencies of 38.6 and 25%, respectively (Saito and Niemann, 1990).

Since the cattle embryos transferred commercially are morula and blastocyst it is expected that cattle genetic screening and sexing will be at the morula or blastocyst stage.

Methods for sexing bovine embryos by identification of Y chromosome DNA sequences have been developed. The cloning and sequencing of repetitive DNA elements from the Y chromosome have been reported for bovine by Khandekar et al. (1986), Leonard et al. (1987), Ellis et al. (1988), Popescu et al. (1988), Vaiman et al. (1988), Reed et al. (1988), Bondioli et al. (1989), Herr et al. (1989a,b) and Aasen and Medrano (1990). Sheep and goats (Reed et al., 1988; Aasen and Medrano, 1990; Herr et al., 1990a,b) and Barbary sheep, sitatunga and oryx (Lord, 1989).

These sequences have been assayed by ^{32}P dot blot hybridization to cells from bovine embryo biopsies (Ellis et al., 1988; Bondioli et al., 1989) or in situ hybridization (Leonard et al., 1987, Vaiman et al., 1988). PCR has been used to amplify the signal sufficiently that only a few cells are needed for assay of cattle or sheep embryos (Herr et al., 1989a,b, 1990a,b,c; Kirzenbaum et al., 1989). This sexing procedure is improved if the assay identifies separately DNA from the X as well as the Y chromosome as published for domestic animals by Aasen and Medrano (1990). There are other yet unpublished reports for use of similar systems with cattle embryos and swine embryos. When genes are identified for which genetic screening is desirable, it is expected that the cattle embryo transfer industry will utilize genetic screening more extensively. For review of domestic animal embryo sexing see Bondioli et al. (1989) and Herr et al. (1990c).

GENE TRANSFER

Methods for transfer of genes into the genome of cattle (Bondioli et al., 1991), sheep (Rexroad, 1990,1991; Wilmut et al., 1990a,b), swine (Pursel, 1990; Pursel et al., 1990; Pinkert et al., 1991), rabbits (Hammer et al., 1985) and rats (Heideman, 1991) are similar to those discussed for mice in this book. The principal differences are that the eggs and recipient animals are much more expensive, the zona pellucida is often more difficult to penetrate and the eggs of cattle and swine must be centrifuged at 10,000 to 15,000 x g as shown in Figure 3 to clear cytoplasm over the pronuclei. The experiments reporting successful gene transfer in domestic animals and their efficiency have been summarized recently and are shown in Tables 1-3. Thus far the only method used for gene transfer in these species has been microinjection of highly concentrated DNA into an egg pronucleus. As in mice the frequency of integration depends on filling pronuclei to the point of rupture with highly concentrated DNA of a construct containing DNA sequences known to result in high integration and usually with introns intact.

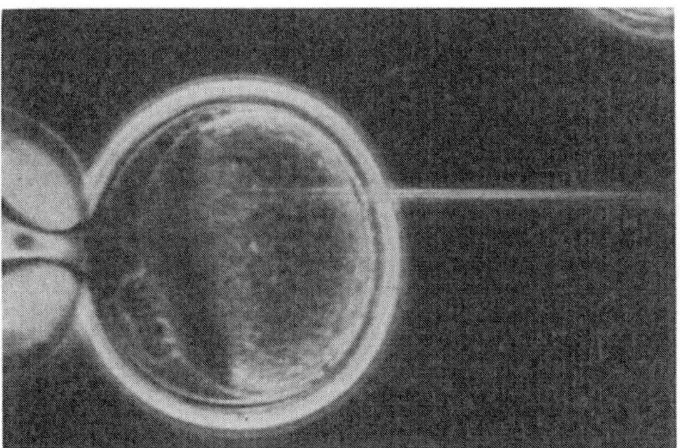

Figure 3. Microinjection of DNA into a pronucleus of a bovine egg after centrifugation of the egg to remove cytoplasm obscuring the pronuclei (x 400). The holding pipette inside diameter is approximately 40 μm and the microinjection pipette is approximately 1-2 μm. Presented originally as part of a report to W.R. Grace and Co.-Conn. by N.L. First.

Table 1. Genes transferred into swine

Growth	Mammary	Disease
mMT-bGH & -hGH mTF-bGH & -hGH mALB-hGRF, mMT-hGRF hMT-pGH, CMV-pGH MLV-pGH & -rGH PEPCK-bGH, PRL-bGH mMT-hIGF-1, pβ^c GLOB MSV-cSKI	mWAP	mlgA

From V. Pursel (1990).

Abbreviations: m - mouse; MT - metallothionein; b - bovine; GH - growth hormone; h - human; TF - transferrin; ALB - mouse albumin; GRF - growth hormone releasing factor; CMV - cytomegolo virus; p - porcine; MLV - murine leukemia virus; r - rabbit; PEPCK - phosphoenol pyruvate carboxy kinase; PRL - prolactin; IGF - insulin-like growth factor; β^c GLOB - immunoglobulin; MSV - murine sarcoma virus; cSKI - an oncogene.

Table 2. Genes transferred into ruminants

Species	Growth	Mammary	Disease
Cow	MTV-bGH & -hIGF-1 cSKA-hER & -hIGF-1 mTF-bGH		BPV
Sheep	mMT-bGH & hGH oMT-oGH, mTF-bGH mALB-hGRF, mMT-hGRF	oβLG-hFIX & -hα1AT mWAP	mlgA
Goat		mWAP-hTPA	

From C. Rexroad (1990)

Abbreviations: MTV - mammary tumor virus; b - bovine; GH - growth hormone; h - human; IGF - insulin like growth factor; cSKA-hER - human estrogen receptor; mTF - mouse transferrin; MT - metallothionine; m - mouse; oβLG - ovine beta lactoglobulin; h-FIX - human clotting factor IX; lgA - immunoglobulin; o - ovine; TF -transferrin; hα1AT - human α1 antitrypsin; mALB-hGRF - mouse albumin - human growth hormone releasing factor; mWAP-hTPA - a mouse whey acetic protein promoter driving expression of a human tissue plasminogen activator gene.

Genes and promoter sequences used have been summarized by Pursel (1990) and by Rexroad (1990) and are shown in Tables 1 and 2. Thus far transgenic embryonic stem cells have not been used to transfer genes into embryos of domestic animals because totipotent stem cells have not yet been identified in species other than mouse and hamster (Doetschman, 1991) and viral vectors developed thus far have not been approved for environmental release.

Table 3. Efficiency of Production of Transgenic Livestock by Microinjection

Species	Expts. #	Embryo #	Born %	TG %	Expressing #	%
Cows	3	1330	10.8	4.2	1/3	33
Pigs	15	13046	9.9	8.6	46/104	44
Sheep	10	4736	9.6	8.8	13/30	43

From C. Rexroad (1990).

Gene transfer is expected to be used in domestic animals for at least three purposes: 1) the introduction of desired genes that do not exist in the species or are of very low frequency, 2) through use of tissue specific promoters the targeting of genes to specific tissues such as muscle (Shani, 1991) or the mammary gland (Wilmut et al., 1990a,b; Hennighausen et al., 1991) and 3) where stem cells are available the use of homologous recombination to delete genes or add genes to specific sites.

Several genes have already been targeted to the mammary gland by mammary specific promoters such as mouse mammary tumor virus or promoter sequences from genes coding for milk proteins such as whey acidic protein, β-lactoglobulin, casein and α-lactalbumin (reviewed by Hennighausen et al., 1991; Wilmut et al., 1990a,b).

The targeting of genes to the mammary gland has been for purposes of: 1) changing the composition of milk especially its caseins as they relate to the yield and composition of cheese and 2) the production of valuable pharmaceutical proteins by the mammary gland.

Gene transfer is considered a powerful tool for improving growth, lactation, reproduction, wool production, meat quality and disease resistance of domestic animals. Unfortunately the genes controlling these traits are not sufficiently understood to allow commercial applications.

When useful transgenics are made their initial number will be small. The technologies of artificial insemination, embryo cloning and transfer are expected to be used extensively to rapidly multiply the transgenic.

These are the tools of the future for livestock breeding. Their application will be primarily through commercial embryo transfer and through artificial insemination. The extent of their use will be determined by the financial benefit each returns to the livestock breeder.

References

Aasen, E. and Medrano, J.F., 1990, Amplification of the ZFY and ZFX genes for sex identification in humans, cattle, sheep and goats, Biotechnology 8:1279.

Baker, R.D. and Shea, B.F., 1985, Commercial splitting of bovine embryos, Theriogenology, 23:3.

Barnes, F.L., Westhusin, M.E. and Looney, C.R., 1990, Embryo cloning: principals and progress, in: "Proceedings 4th World Congress on Genetics Applied to Livestock Production," Vol. XVI, pp. 323-333.

Barrett, G.R., 1948, Time of insemination and conception rates in dairy cows. Ph.D. Thesis, University of Wisconsin-Madison.

Boerjan, M.L. and deBoer, P., 1990, First cell cycle of zygotes of the mouse derived from oocytes aged postovulation in vivo and fertilized in vitro, Mol. Reprod. Dev., 25:155.

Bondioli, K.R., Ellis, S.B., Pryor, J.H., Williams, M.W. and Harpold, M.M., 1989, The use of male-specific chromosomal DNA fragments to determine the sex of bovine preimplantation embryos, Theriogenology, 31:95.

Bondioli, K.R., Westhusin, M.E. and Looney, C.R., 1990, Production of identical bovine offspring by nuclear transfer, Theriogenology, 33:165.

Bondioli, K.R., Biery, K.A., Hill, K.G., Jones K.B. and De Mayo, F.J., 1991, Production of transgenic cattle by pronuclear injection, in: "Transgenic Animals," N.L. First and F.P. Haseltine, ed., Butterworth-Heinemann, Boston.

Briggs, R. and King, T.J., 1952, Transplantation of living nuclei from blastula cells into enucleated frogs' eggs, Zoology, 38:455.

Casida, L.E., 1950, The repeat breeder cow, Vlaams Diergen Tijdschr, 19:273.

Collas, P. and Robl, J., 1991, Development of rabbit nuclear transplant embryos from morula and blastocyst stage donor nuclei, Theriogenology, 35:190.

Cowan, C.M., Dentine, M.R., Ax, R.L. and Schuler, L.A., 1990, Structural variation around prolactin genes linked to quantitative traits in an elite Holstein sire family, Theor. Appl. Genet., 79:577.

Doetschman, T., Williams, P. and Maeda, N., 1988, Establishment of hamster blastocyst-derived embryonic stem (ES) cells, Dev. Biol., 127:224.

Doetschman, T.C., 1991, Gene targeting in embryonic stem cells, in: "Transgenic Animals," N.L. First and F.P. Haseltine, ed., Butterworth-Heinemann, Boston.

Ellinger, M.S., 1978, The cell cycle and transplantation of blastula nuclei in Bambiana orientalis, Dev. Biol., 65:81.

Ellis, S.B., Bondioli, K.R., Williams, M.E., Pryor, J.H. and Harpold, M.M. 1988, Theriogenology, 29:242.

Evans, M.J. and Kaufman, M.H., 1981, Establishment in culture of pluripotential cells from mouse embryos, Nature, London, 294:154.

Evans, M.J., Notarianni, E., Laurie, S. and Moor, R.M., 1990, Derivation and preliminary characterization of pluripotent cell lines from porcine and bovine blastocysts, Theriogenology, 33:125.

Gatica, R., Boland, M.P., Crosby, T.F. and Gordon, I., 1984, Micromanipulation of sheep morulae to produce monozygotic twins, Theriogenology, 21:555.

Gearheart, W.W., Smith, C. and Teepher, G., 1989, Multiple ovulation and embryo manipulation in the improvement of beef cattle: relative theoretical rates of genetic change, J. Anim. Sci. 67:2863.

Georges, M., Mishra, A., Sargeant, L., Steele, M., Zhao, X. and Fries, R., 1990, Highly polymorphic DNA markers for use in cattle breeding, in: "Mapping the Genomes of Agriculturally Important Animals," J. Womack, ed., Cold Spring Harbor Laboratory Press.

Gurdon, J.B., 1986, Nuclear transplantation in eggs and oocytes, J. Cell Sci., 4(Suppl):287.

Hammer, R.E., Pursel, V.G., Rexroad, C.E., Jr., Wall, R.J., Bolt, D.J., Ebert, K.M., Palmiter, R.D. and Brinster, R.L., 1985, Production of transgenic rabbits, sheep and pigs by microinjection, Nature, 315:680.

Heideman, J., 1991, Transgenic rats: A discussion, in: "Transgenic Animals," N.L. First and F.P. Haseltine, ed., Butterworth-Heinemann, Boston.

Hennighausen, L., Westphal, C., Sankaran, L. and Pittius, C.W., 1991, Regulation of expression of genes for milk proteins, in: "Transgenic Animals," N.L. First and F.P. Haseltine, ed., Butterworth-Heinemann, Boston.

Herr, C., Matthaei, K.I. and Reed, K.C., 1989a, Accuracy of a rapid Y-chromosome-detecting bovine embryo sexing assay, Australia and New Zealand Society for Cell Biology, 8th Annual Meeting, University of Melbourne, Feb. 8-10, 1989.

Herr, C., Matthaei, K.I., Holt, N. and Reed, K.C., 1989b, Field implementation of a rapid Y-chromosome-detecting bovine embryo sexing assay, Australia and New Zealand Society for Cell Biology, 8th Annual Meeting, University of Melbourne, Feb. 8-10, 1989.

Herr, C., Holt, N., Petrzak, U., Old, K. and Reed, K.C., 1990a, Increased number of pregnancies per collected embryo by bisection of blastocyst stage ovine embryos, Theriogenology, 33:1.

Herr, C., Matthaei, K.I., Petrzak, U. and Reed, K.C., 1990b, A rapid Y-chromosome-detecting ovine embryo sexing assay, Theriogenology, 33:1.

Herr, C.M., Matthaei, K.I. and Reed, K.C., 1990c, Rapid, accurate sexing of livestock embryos, in: Proceedings 4th World Cong. on Genetics Applied to Livestock Production, Vol. XVI, pp. 323-333.

Hoeschale, I. and Meinert, T.R., 1990, Association of genetic defects with yield and type traits: a major production gene is linked to weaver, J. Dairy Sci., (in press).

Khandekar, P., Talwar, G.P. and Chaudhury, R., 1986, Cloning of Y derived DNA sequences of bovine origin in Escherichia coli, J. Biosci., 10:481.

Kim, J.G., Roudebush, W.E., Dobson, M.G. and Minhas, B.S., 1990, Murine embryo biopsy and full term development following transfer of biopsied embryos, Theriogenology, 33:266.

King, T.J., 1966, Nuclear transplantation in amphibia, Meth. Cell Biol., 2:1.

Kirzenbaum, M., Vaiman, M., Cotinot, C. and Fellous, M., 1989, PCR sexing bovine embryos using Y chromosome specific sequences, J. Cell Biochem., (Suppl)13E:293.

Lehn-Jensen, H. and Willadsen, S.M., 1983, Deep-freezing of cow 'half' and 'quarter' embryos, Theriogenology, 19:49.

Leibo, S.P., 1988, Bisection of mammalian embryos by micromanipulation, in: The American Fertility Society Regional Postgraduate Course. Hands-on IVF, Cryopreservation and Micromanipulation, April 25-29, Madison, WI.

Leonard, M., Kirszenbaum, M., Cotinot, C., Chesne, P., Heyman, Y., Stinnakre, M.G., Bishop, C., DeLouis, C., Vaiman, M. and Fellous, M., 1987, Sexing bovine embryos using chromosome specific DNA probes, Theriogenology, 27:248.

Lien, S., Ragne, S., Steine, T., Longsrud, T., Vegarudl, G. and Alestrom, P., 1990, Methods for K-casein and β-lactoglobulin genotyping of bulls, in: Genetic Engineering of Animals, W. Hansel and B.J. Weir, eds., J. Reprod. Fert., 41(Suppl):211.

Lord, E.A., 1989, Ph.D. Thesis, Australian National University (Canberra ACT).

McAvoy, J.W., Dixon, K.E. and Marshall, J.A., 1975, Effects of differences in mitotic activity, stage of the cell cycle and degree of specialization of donor cells on nuclear transplantation in xenopus laevis, Dev. Biol., 45:330.

McGrath, J. and Solter, D., 1984, Inability of mouse blastomere nuclei transferred to enucleated zygotes to support development in vitro, Science, 226:1317.

Ozil, J.P., Heyman, Y. and Renard, J.P., 1982, Production of monozygotic twins by micromanipulation and cervical transfer in the cow, Vet. Rec., 110:126.

Papaioannou, V.E., 1989, Regulative potential of micromanipulated embryos, in: Medically Assisted Conception - An Agenda for Research. Report of a Study by a Committee of the Institute of Medicine and National Research Board on Agriculture, National Academy Press, Washington, D.C., pp. 304-318.

Pinkert, C.A., Kooyman, D.L. and Dyer, T.J., 1991, Enhanced growth performance in transgenic swine "Transgenic Animals," N.L. First and F.P. Haseltine, ed., Butterworth-Heinemann, Boston.

Popescu, C.P., Cotinot, C., Boscher, J. and Kirszenbaum, M., 1988, Chromosomal localization of a bovine male specific probe, Ann. Genet., 31:39.

Prather, R.S., Barnes, F.L., Sims, M.L., Robl, J.M., Eyestone, W.H. and First, N.L., 1987, Nuclear transfer in the bovine embryo: assessment of donor nuclei and recipient oocyte, Biol. Reprod., 37:859.

Prather, R.S., Sims, M.M. and First, N.L., 1989, Nuclear transplantation in early pig embryos, Biol. Reprod., 41:414.

Pursel, V.G., 1990, Transgenic swine. in: "Proceedings OTA Workshop on Biotechnology," May 1990.

Pursel, V.G., Hammer, R.E., Bolt, D.J., Palmiter, R.D. and Brinster, R.L., 1990, Integration, expression, and germline transmission of growth-related genes in pigs. in: "Genetic Engineering of Animals," W. Hansel and B.J. Weir, ed., J. Reprod. Fert., 41(Suppl):77.

Reed, K.C., Matthews, M.E. and Jones, M.A., 1988, Sex determination in ruminants using Y-chromosome specific polynucleotides. Published Patent Application. Patent Cooperative Treaty No. WO88/01300.

Rexroad, C., 1990, Transgenic ruminants. in: "Proceedings OTA Workshop on Biotechnology," May 1990.

Rexroad, C., 1991, Production of sheep transgenic for growth hormone genes. in: "Transgenic Animals," N.L. First and F.P. Haseltine, ed., Butterworths-Heinemann, Boston.

Robl, J.M., Gilligan, B., Crister, E.S. and First, N.L., 1986, Nuclear transplantation in mouse embryos: assessment of recipient cell stage, Biol. Reprod., 34:733.

Rorie, R.W., Voelkel, S.A., McFarland, C.W., Southern, L.L. and Godke, R.A., 1985, Micromanipulation of day-6 porcine embryos to produce split-embryo piglets, Theriogenology, 23:225.

Rossant, J. and Joyner, A.L., 1989, Towards a molecular-genetic analysis of mammalian development, Trends in Genetics, 5:227.

Saito, S. and Niemann, H., 1990, In vitro development of blastomeres isolated from 8 and 16 cell porcine embryos on extracellular matrices, Theriogenology, 33:317 (abst.).

Seidel, G.E., Jr., 1985, Are identical twins produced from micromanipulation always identical? in: "Proc. Ann. Conf. Artif. Insem. and Embryo Transfer in Beef Cattle (Denver)," Nat. Assoc. Anim. Breeders, Columbia, MO, pp. 50-53.

Shani, M., 1991, Application of germline transformation to the study of myogenesis, in: "Transgenic Animals," N. First and F. Haseltine, ed., Butterworth-Heinemann, Boston.

Smith, C., 1989, Cloning and genetic improvement of beef cattle, Anim. Prod., 49:49.
Smith, L.C. and Wilmut, T., 1989, Influence of nuclear and cytoplasmic activity on the development in vivo of sheep embryos after nuclear transplantation, Biol. Reprod., 40:1027.
Stice, S.L. and Robl, J.M., 1988, Nuclear reprogramming in nuclear transplant rabbit embryos, Biol. Reprod., 39:657.
Stice, S.L. and Robl, J.M., 1989, Nuclear reprogramming in nuclear transplant rabbit embryos, Biol. Reprod., 39:657.
Vaiman, M., Cotinot, C., Kirszenbaum, M., Leonard, M., Chesne, P., Heyman, Y., Stinnakre, M-G, Bishop, C. and Fellous, M., 1988, Sexing of bovine embryo using male-specific nucleic acid probes, in: "Proc. World Sheep and Cattle Breeding Congress", pp. 93-105.
Voelkel, S.A., Viker, S.D., Johnson, C.A., Hill, K.G., Humes, P.E. and Godke, R.A., 1985, Multiple embryo-transplant offspring produced from quartering a bovine embryo at the morula stage, Vet. Rec., 117:528.
Voelkel, S.A., Rorie, R.A., McFarland, C.W. and Godke, R.A., 1986, An attempt to produce quarter embryos from non-surgically recovered bovine blastocysts, Theriogenology, 25:207.
Von Beroldington, C.H., 1981, The developmental potential of synchronized amphibian cell nuclei, Dev. Biol., 81:115.
Ware, C.B., Barnes, F.L., Maiki-Laurila, M. and First, N.L., 1989, Age dependence of bovine oocyte activation, Gamete Res., 22:265.
Warfield, S.J., Seidel, Jr., G.E. and Elsden, R.P., 1987 Transfer of bovine demi-embryos with and without the zona pellucida, J. Anim. Sci., 65:756.
Westhusin, M.E., Levanduski, M.J., Scarborough, R., Looney, C.R. and Bondioli, K.R., 1990, Utilization of fluorescent staining to identify enucleated demi-oocytes for utilization in bovine nuclear transfer, Biol. Reprod., 42:Suppl 1, abstract 407.
Willadsen, S.M., 1979, A method for culture of micromanipulated sheep embryos and its use to produce monozygotic twins, Nature, 227:298.
Willadsen, S.M., 1980, The viability of early cleavage stages containing half the normal number of blastomeres in the sheep, J. Reprod. Fert., 59:357.
Willadsen, S.M., 1982, Micromanipulation of embryos of the large domestic species, in: Mammalian Egg Transfer, C.E. Adams, ed., CRC Press, Boca Raton, Florida, pp. 185-210.
Willadsen, S.M. and Godke, R.A., 1984, A simplified procedure for the production of identical sheep twins, Vet. Rec., 114:240.
Willadsen, S.M., 1986, Nuclear transplantation in sheep embryos, Nature, London 320:63.
Willadsen, S.M., 1989, Cloning of sheep and cow embryos, Genome, 31:956.
Williams, T.J. and Moore, L., 1988, Quick-splitting of bovine embryos, Theriogenology, 29:477.
Williams, T.J., Elsden, R.P. and Seidel, G.E., 1982, Identical twin bovine pregnancies derived from bisected embryos, Theriogenology, 17:114.
Wilmut, I., Archibald, A.L., Harris, S., McClenaghan, M., Simons, J.P., Whitlaw, C.B.A. and Clark, A.J., 1990a, Methods of gene transfer and their potential use to modify milk composition, Theriogenology, 33:113.
Wilmut, I., Archibald, A.L., Harris, S., McClenaghan, M., Simons, J.P., Whitelaw, C.B.A. and Clark, A.J., 1990b, Modification of milk composition, in: Genetic Engineering of Animals, W. Hansel and B.J. Weir, eds., J. Reprod. Fert. LTD, pp. 135-146.
Yang, X., Zhang, L., Kovacs, A., Tobback, C. and Foote, R.H., 1990, Isolation and culture of single 8-cell blastomeres in rabbits with a novel approach, Theriogenology, 33:353 (abstract).
Zimmerman, U. and Vienken, J., 1982, Electric field-induced cell-to-cell fusion, J. Membrane Biol., 67:165.

ENHANCEMENT OF FERTILIZATION AND HATCHING USING MICROMANIPULATION

Jacques Cohen, Mina Alikani, Henry Malter, Beth Talansky, Michael Tucker, Sharon Wiker and Graham Wright

The Center for Reproductive Medicine and Infertility
The New York Hospital-Cornell Medical Center
New York, N.Y.
and Reproductive Biology Associates, Atlanta, GA

INTRODUCTION

Over 50 babies have now been born following the application of micromanipulation to oocytes and embryos. The current proceedings deal almost uniquely with the genetic diagnosis of biopsied polar bodies, blastomeres and nuclei from blastomeres or trophoblastic cells, however, micromanipulation procedures in assisted reproductive technology are performed for a variety of other reasons. Although the purpose of preimplantation diagnosis may differ from that of other microsurgical technologies, these methods have in common that the zona pellucida is breached. When normal fertilization is absent due to abnormalities in the sperm or oocytes, micromanipulation has been used to promote sperm-egg fusion (Malter and Cohen, 1989a; Ng et al, 1990). When polyspermic fertilization occurs, micromanipulation may be used to return the zygote to a genetically normal, viable state (Gordon et al, 1989; Malter and Cohen, 1989b). Also, micromanipulation has been used to promote implantation by assisting the hatching process (Cohen et al, 1990a). However, micromanipulation imposes artificial conditions on the gametes and embryos. Gaps are created in the zona pellucida. Sperm cells are compressed and possibly damaged in the lumen of sperm injection needles. Enucleation needles enter and disrupt the ooplasmic cytoskeleton. These and other potentially negative factors must be considered in the development and management of clinical micromanipulation. This paper will discuss some current micromanipulation methods and attempt to address the potential problems, which may have consequences for the safe application of biopsy procedures.

Preimplantation Genetics, Edited by Y. Verlinsky and A. Kuliev, Plenum Press, New York, 1991

MICROMANIPULATION TO ASSIST SPERM-EGG FUSION

Human in vitro fertilization (IVF) can be considered a treatment for reduced gamete interaction since sperm and eggs are placed in close proximity under carefully optimized conditions (Cohen et al, 1985). However, the presence of sperm and possibly oocyte abnormalities as well, often results in a complete failure of fertilization when using the most carefully optimized application of the standard technique. Various micromanipulation strategies have been suggested for the promotion of sperm-egg fusion. These fall into three basic categories: the direct injection of a single spermatozoon into the ooplasm (microinjection); the placement of a single or multiple spermatozoa into the perivitelline space (subzonal insertion or SZI); and the breaching of the zona pellucida to provide an opening through which fertilizing sperm can more easily gain access to the egg (Partial zona pellucida dissection or PZD). Of these methods, only the last two have been applied successfully in the human (Cohen et al, 1988; Ng et al, 1988; Fishel et al, 1990). Direct sperm injection into the ooplasm has produced offspring in rabbits (Hosoi et al, 1988). However, this method is highly traumatic and reports from human application studies would seem to indicate that it is inappropriate for immediate clinical use. PZD and SZI will now be discussed.

Zona pellucida dissection

Gordon and Talansky (1986) demonstrated that when the zona pellucidae of mouse eggs were partially opened by localized acidic digestion, sperm-egg fusion and fertilization were promoted. When sperm concentration was greatly reduced, "zona drilling" resulted in a marked increase in fertilization as compared to control eggs with intact zona pellucidae. Furthermore, offspring were routinely produced following the transfer of embryos resulting from zona drilling. Zona drilling was then attempted in the human clinical laboratory. Fertilization was obtained in cases where sperm quality was suboptimal (Gordon et al, 1988). However, to this date, no pregnancies have been established following the transfer of embryos resulting from zona drilling. When zona drilling was attempted on human eggs which failed fertilization during routine IVF, the acidic medium appeared to cause damage to the oocyte and compromise preimplantation development. Zona drilling with acidic medium was, therefore, abandoned in the human in favor of a mechanical method termed partial zona dissection (PZD) (Malter and Cohen, 1989a). The PZD protocol involves forcing a sharp microneedle through the zona pellucida to create a slit-like opening. The increased perivitelline space needed for this procedure is obtained by shrinking the oocyte in a sucrose solution. PZD is a very simple and apparently non-traumatic procedure. Less than 1% of the oocytes are damaged during this micromanipulation procedure (Malter and Cohen, 1989a).

In previous studies (Cohen et al, 1990c; Tucker et al, 1990) performed in Atlanta, we demonstrated that PZD in male factor patients increased the incidence of fertilization by approximately 15-20% compared to that of zona-intact control oocytes. However, 20% (75/383; Table 1) of the zona-intact oocytes fertilized, indicating that some of the selected patients were capable, to some extent, of unassisted zona penetration.

The patient selection criteria for inclusion in this study are described elsewhere (Cohen et al, 1989; Tucker et al, 1990). The effect of a previous failure of fertilization in a conventional IVF cycle (without micromanipulation) on the prognosis of PZD has been described recently (Cohen et al, 1990d). More than 40% of the patients who do not fertilize their partner's oocytes during a conventional IVF cycle were successful when PZD was performed and 27% of these couples achieved pregnancy.

Using Kruger's strict criteria (Kruger et al, 1988) for sperm morphology analysis, we have performed PZD in another 58 patients at Cornell. These teratozoospermic patients had less than 10% normally shaped spermatozoa. Half of those had previously failed fertilization using conventional IVF techniques (Table 2). Elsewhere, we demonstrated that PZD embryos derived from oocytes inseminated with sperm from men with severe teratozoospermia (less than 5% normal forms) rarely implanted, whereas the implanting capacity of embryos from couples in which the male had moderate teratozoospermia was higher than 18% (Cohen et al, 1990d).

Subzonal insertion

SZI refers to the direct placement of spermatozoa into the perivitelline space surrounding the oocyte. This technique involves both a physical and a physiological breaching of the zona pellucida, and is therefore considered a more invasive form of micromanipulation than PZD. The need for SZI in clinical IVF is threefold. First, it may be applied

Table 1. Results of PZD trials with ZP intact controls (117 cycles in male factor couples) performed at Reproductive Biology Associates (RBA, Atlanta, Georgia) between March 1988 and September 1990.

Parameter	Cohen et al, 1990c	Tucker et al, 1990	Total
Fertilization of ZP intact (ZI) eggs	41/129 (33%)	33/254 (13%)	75/383(20%)
Fertilization (PZD) eggs	75/281 (54%)	73/281 (26%)	148/419 (35%)
Replacements/cycles	37/47 (79%)	30/70 (43%)	67/117 (57%)
#ZI embryos replaced	24	20	44
#PZD embryos replaced	55	40	95
# fetuses/ # embryos replaced	15/79 (19%)	17/60 (28%)	32/139 (23%)
Clinical Pregnancy	10	16	26
Ongoing/delivered	8	13	21
Preg. from PZD exclusively	2	4	6
Preg.PZD/ZI mixed	8	12	20
Clin. preg./cycle	10/47 (21%)	16/70 (23%)	26/117 (22%)
Clin. preg./ transfer	10/37 (27%)	16/30 (53%)	26/67 (39%)

to cases in which sperm counts are severely reduced and/or motility is impaired. In such instances, sperm samples are of such poor quality that PZD would be an inefficient method for assisting fertilization. For instance, the success of PZD is largely dependent on the ability of the sperm to "find its way" through the artificially produced gap in the zona pellucida. Thus, severely compromised sperm samples (low motility and/or reduced counts) might not be able to fertilize oocytes whose zonae were subjected to PZD. It might, therefore, be necessary to implement SZI in such cases of severe male factor infertility in which direct placement of sperm in the vicinity of the vitellus is needed to assist fertilization.

SZI may also be suitable for those cases in which standard IVF consistently results in polyspermic fertilization. This may be due to an ineffective block of extranumerary sperm penetration at the level of the zona pellucida or to additional unknown factors. Here, PZD would be inadvisable since it requires large numbers of sperm in the insemination suspension. Thus, SZI of very few spermatozoa, or ideally, a single sperm cell might be used to reduce or completely eliminate the chance of polyspermic fertilization. This leads to a third indication for the application of SZI. Cases of severe male factor infertility for which PZD yields only polyspermic fertilization may require SZI of spermatozoa. Since it is virtually impossible to control the number of sperm contacting the oocyte which has been subjected to PZD, it may be necessary to employ a micromanipulation technique in which limited populations of sperm are brought into direct contact with the egg.

PZD and SZI were compared in 25 couples with severe male factor infertility using sibling oocytes. SZI was exclusively performed on the oocytes of 12 other couples, since too few spermatozoa were available for PZD. The results of both groups of patients are presented in Table 2. Previously, we demonstrated that fertilization following SZI increased when more spermatozoa were inserted (Cohen et al, 1990c). However, the incidence of polyspermy also increased, especially when more than 2 live sperm cells were introduced into the perivitelline space. Insertion of more than 8 spermatozoa resulted in 100% polyspermy indicating that sperm cells derived from infertile men are capable of multiple fusion with the oolemma. The number of spermatozoa inserted was therefore limited to a maximum of 3 or 4 per oocyte in half of these couples. No differences in monospermic fertilization were found between PZD and SZI (Table 2). Another 10 to 15% of the eggs were polyspermically fertilized. None of the zona intact or PZD embryos implanted in this particular trial. Three of the 33 (10%) SZI embryos implanted. Two pregnancies were achieved in patients who had only SZI embryos available for transfer. The third patient had three SZI and one PZD embryo replaced. Two of these pregnancies are still too early to consider definitely viable.

General considerations of microsurgical fertilization

At least 12 centers worldwide have obtained ongoing pregnancies from the replacement of PZD embryos. Four

Table 2. Results of Microsurgical Fertilization (MF) trials (95 cycles in male factor couples with teratozoospermia) performed at The Center for Reproductive Medicine and Infertility, The New YOrk Hospital - Cornell Medical Center (NYH-CUMC, New York, New York), between October 1989 and September 1990.

	PZD and occasional internal ZP intact control	SZI and occasional PZD	Combined
Fertilization (PZD)	83/276 (31%)	21/116 (18%)	104/392 (27%)
Fertilization (SZI)		38/204 (19%)	38/204 (19%)
Replacements/Cycles	37/58 (64%)	25/37 (68%)	62/95 (65%)
#ZI embryos replaced	36	8	44
# PZD [SZI] embryos replaced	51 [0]	14 [33]	65 [33]
# fetuses / # embryos	16/87 (18%)	3/55 (5%)	21/142 (15%)
Biochemical	1	1	2
Clinical pregnancy	13	3	16
Ongoing or delivered	11	≥1	≥12
Preg. from MF only	5	3	8
Pregnancy MF/ZI mixed	5	-	25
Clinical pregnancy/cycle	13/58 (22%)	3/37 (8%)	16/95 (17%)
Clinical pregnancy/ transfer	13/37 (35%)	3/25 (12%)	16/62 (26%)

[a] from SZI embryos only

centers have reported viable pregnancies from the transfer of SZI. The rate of implantation of the latter embryos appears very low. A number of explanations for this phenomenon can be given, all of which are hypothetical. It is possible that sperm cells are damaged during the insertion into the microneedle or their expulsion into the perivitelline space. Nevertheless as many as 16% of individual spermatozoa from infertile men and 59% of spermatozoa from fertile men can fuse with the oolemma and decondense within the cytoplasm (Cohen et al, 1990e). Another possibility is that the cytoskeleton of the oocyte is damaged as the volume of the perivitelline space is increased during SZI. Free radicals attached to the sperm membrane or freely present in the sperm suspension among other cytotoxic elements may interfere with normal embryonic development. Finally, some abnormally shaped sperm cells may be forced to fuse with the oocyte, possibly affecting the normal organization of the embryo. The incidence of abnormal karyotypes following SZI was, however, not increased in a small group of investigated embryos (Kola, 1990). We have noticed delayed fertilization in a number of SZI oocytes. Spermatozoa released into the perivitelline space often remain vigorously motile as their tails are able to beat freely, possibly limiting the duration of gamete interactions. Perivitelline spermatozoa are rarely seen after PZD. Only five oocytes had perivitelline spermatozoa when 77 unfertilized PZD oocytes were flattened and stained with acetocarmine. This indicates that most spermatozoa are caught with their tail in the PZD incisions, limiting head movement. This phenomenon resembles that of natural fertilization and PZD may therefore have an advantage over SZI. Also in rodents it was demonstrated recently, that spermatozoa preferentially fertilize at the site of the artificial hole in the zona pellucida (Talansky et al, 1990).

The initial results from microsurgical fertilization of the first 212 cycles in our programs have been satisfactory, despite the considerations above and the severity of male factors studied (Table 3). Sixteen babies were born and many pregnancies are still ongoing.

POSSIBLE DISADVANTAGES OF ZONA MICROMANIPULATION

Despite the fact that gamete micromanipulation has great potential for increasing the chances for successful gamete interaction, it is not without potential disadvantages and risks. These hazards are inherent to any method which disturbs the structural integrity of the oocyte, and may affect the unfertilized oocyte, the zygote, and the cleaving embryo (Table 4).

Polyspermy

Any procedure which involves the physical disruption or circumvention of the oocyte's natural protective barrier, or zona pellucida, will leave the egg's plasma membrane susceptible to multiple sperm. This is undesirable since it has been suggested that in the human oocyte, it is the zona pellucida and not the plasma membrane which is largely responsible for limiting supernumerary sperm penetration (Gordon et al, 1988b; Malter and Cohen, 1989; Cohen et al, 1990d). Opening the zona pellucida, such as in zona drilling or PZD, may jeopardize the egg's principal selective barrier and the oocyte may become fertilized by multiple spermatozoa. This polyspermic state may result in an excessive genetic complement and a non-viable embryo. Potential techniques by which such polyploid embryos may be restored to a "normal" diploid state are currently under investigation (Gordon et al, 1989; Malter and Cohen, 1989b; Cohen et al, 1990c).

Table 3. Results of microsurgical fertilization trials from RBA and Cornell combined (March 1988 - September 1990).

Patient cycles	212	(a)
Replacements	129 (61% of a)	(b)
Positive pregnancy tests	44	
Clinical pregnancies	42 (20% of a; 33% of b)	
Ongoing pregnancies or delivered	35 (17% of a; 27% of b)	
Pregnancies from SZI embryos only	2	
Pregnancies from a mixture of PZD/SZI	1	
Pregnancies from PZD embryos only	11	
Pregnancies from zona intact and PZD	26	
Pregnancies from zona intact embryos	2	

It is difficult to prevent polyspermy when performing PZD since it is necessary to inseminate the oocyte with as high a sperm concentration as possible. Although this will vary according to the requirements of individual cases, a general risk of PZD is that multiple sperm will enter the gap. If SZI is used as an alternative means by which to achieve fertilization, the risks of polyspermy could probably be more easily controlled, since exact sperm numbers are determined by the individual performing micromanipulation. However, this is a complex issue. That is, within a population of sperm, whether from a patient with normal or compromised fertility, apparently "normal" cells may be incapable of completing capacitation and the acrosome reaction. Consequently, such spermatozoa will be unable to fuse with an oocyte. It is necessary to artificially induce capacitation and the acrosome reaction in entire populations of sperm to be used for SZI in order to ensure selection of a fertilizable cell. However, even if very low numbers of sperm are inserted into the perivitelline space, there is still no guarantee that polyspermy will be avoided. On the other hand, since SZI of a single sperm can be an inefficient method of achieving gamete fusion, it will generally be necessary to introduce multiple sperm.

Table 4. Possible disadvantages of zona micromanipulation [Based on Malter and Cohen (1989b) and Cohen et al. (1990a and c)].

	before cleavage	after cleavage	during transfer	after transfer
polyspermy	+			
non-selective fertilization	+	+		+
seminal toxins	+	+		
environmental toxins	+	+	+	+
manipulation damage	+	+	+	+
abnormalities of hatching			?	+

Non-selective fertilization

The selection process which precedes fertilization is equally if not more important than capacitation and acrosome reaction. Although millions of sperm are present at the distal region of the female reproductive tract, only a few of these eventually contact the vitelline membrane. The outer vestments of the egg such as the follicular cells, cumulus complex and zona pellucida all contribute to the selection process by various biological mechanisms, some of which are poorly understood. By disrupting the integrity of the zona, we are bypassing the oocyte's natural ability to exert selection on the overwhelming numbers of approaching sperm and are therefore allowing a heterogenous population of spermatozoa direct access to the oocyte.

The ability to penetrate an oocyte and form a male pronucleus does not prove that a given spermatozoon is genetically fit. The role of the sperm cell is certainly not complete at the moment of fertilization. In fact, it has been shown that in the human embryo, transcription is not initiated until a point between the 4-8 cell stage of development (Tesarik, 1987). Therefore, since the contribution of the paternal genome is not immediately evident, a successful fertilization may not necessarily be predictive of a genetically normal embryo. Some form of selection for genetically healthy sperm is probably functional at the level of the various oocyte barriers. Since it creates a direct channel to the oolemma, zona micromanipulation by PZD somewhat reduces the oocyte's natural control over sperm selection. However, it has been observed in our laboratory that most spermatozoa in the perivitelline space of PZD oocytes had normal morphology. This possibly indicates that the hole can only be traversed by relatively normal spermatozoa. SZI may be more problematic, since the role of the zona pellucida in the selection process is completely bypassed. There is a greater chance for an abnormal sperm to fuse with the egg after SZI, since it would be selected from a relatively heterogeneous population. Therefore, not only does natural selection play a more influential role in PZD, but the increased numbers of sperm favor fertilization by normal sperm.

Environmental toxins

Invasion of toxins through the artificial gap in the zona may occur at any point after early embryonic cleavage. For instance, even after replacement, the manipulated embryo is still susceptible to invasion by any foreign cells present in the uterine environment. The possibility of immune cell invasion through the micromanipulated zona pellucida has been investigated (Cohen et al, 1990b). In order to reduce the possibility of immune cell interaction with the embryo, patients whose oocytes were subjected to micromanipulation received low doses of corticosteroids for 4 days following egg collection. Results demonstrated a significantly higher rate of implantation among those patients receiving cortiscosteroids and this protocol has become incorporated into several IVF programs in which micromanipulation is performed. The mechanisms by which immune cells, toxins, bacteria or other foreign bodies may disrupt the oocyte and early embryo remain undefined. What we can hypothesize at this time, however, is that the presence of a gap in the zona, whether created by PZD or SZI, may interfere with normal embryonic development.

The importance of the embryo replacement procedure

The last step of the IVF laboratory protocol in which embryos are handled is the replacement. Although often considered a trivial, simple and standardized procedure, it is actually a critical factor in the success of IVF. During the replacement, mucus, blood clots, and endometrial epithelium may obstruct the end of catheter; possibly trapping the embryos. Examination of embryos recovered from the external os following transfers in which the catheter was distorted revealed normal morphology. However, this is only true for zona-intact embryos, for it has become increasingly evident that embryos with gaps in their zonae are quite susceptible to damage during replacement. It appears as though passage through a narrow catheter, often occluded with mucus or tissue, exerts excessive pressure on the embryo. In the case of a non-manipulated embryo, the intact zona is malleable and can easily withstand environmental perturbations. However, manipulated embryos are less suited to withstand external pressure during the replacement procedure. Transfers of PZD embryos have occasionally resulted in the loss of blastomeres, most likely through the gap in the zona. In one particular study, completely empty zonae were recovered from the replacement catheter in two patients in which the zona pellucidae were micromanipulated shortly before embryo replacement to assist the hatching process. The tips of both catheters were filled with mucus, tissue and blood cells (Fehilly, personal communication).

Several IVF programs which have incorporated PZD into their procedures have reported poor results. Although the embryos resulting from PZD have usually normal morphology, it is distressing that transfer of these embryos often does not result in pregnancy. It is feasible to suggest that the replacement procedure, in part, may determine the fate of micromanipulated embryos. Specifically it is the type of catheter employed and the vigor with which the replacement is performed, which seem to be critical factors in the survival and subsequent implantation of micromanipulated embryos.

IMPROVEMENT OF IMPLANTATION BY ASSISTING HATCHING

An additional element to be considered when evaluating developmental potential of oocytes subjected to micromanipulation is the hatching process, a prerequisite to implantation. Both the timing and morphology of the hatching process in vitro mouse and human PZD embryos are altered (Malter and Cohen, 1989b; Cohen et al, 1990a). The zona does not thin as it does normally, and hatching occurs earlier and at a higher frequency. Though the sample size is still small, it appears that human PZD embryos implant to a higher extent than zona-intact embryos (Table 1 and 2). PZD may therefore be a useful technique with which to enhance IVF results. A number of findings support the hypothesis that human IVF embryos sometimes lack the ability to produce zona lysins. Cleaved embryos with thinned area in their zonae appear to implant more frequently than embryos with uniform zonae (Cohen et al, 1989). The possible advantages of small gaps in zonae of all IVF embryos was therefore investigated in a prospective study using consenting IVF patients (Cohen et al, 1990c). The use of micromanipulation to promote hatching following IVF by introducing a PZD incision in the zona pellucida of early cleaved embryos just prior to replacement was outlined elsewhere (Cohen et al, 1990a). This procedure has been called "Assisted Hatching". The zonae of a total of 144 fresh two to eight-cell embryos were micromanipulated. All embryos appeared intact following micromanipulation and none of the blastomeres were damaged. Positive ßhCG was confirmed in 17/51 (34%) of the control patients who received zona intact embryos, while 24/48 (50%) of the "Assisted Hatching" patients became biochemically pregnant. The incidence of clinical pregnancy increased from 26% to 46% per embryo replacement, a significant improvement ($p<0.05$; chi-square test). Moreover, embryonic implantation increased from 13 to 22% ($p<0.05$). Half the pregnancies in the experimental group were either twin or triplet pregnancies. The same procedure was performed in patients who received freeze-thawed embryos. Again, embryonic implantation increased markedly, from 14% to 23%. It was therefore postulated that a substantial number of IVF embryos are unable to breach the zona at the time of hatching and that many can be rescued by opening their zona several days earlier. Similar high rates of implantation were obtained by Handyside et al (1990) following biopsy of eight-cell embryos for gender selection in patients who were at risk for genetic disease.

The improved capacity of human embryos to hatch and implant has only been apparent in studies involving the creation of large openings, like PZD or embryo biopsy. Techniques in which the zonae are only pierced by a microneedle, such as SZI or polar body removal, may not benefit from this phenomenon. The creation of a long narrow slit following PZD of four-cell human embryos shortly before transfer may be advantageous, but could lead to trophoblast constriction or cell loss prior to the formation of the gap and tight junctions as described above. In addition, narrow gaps in the zona pellucida may cause constriction of the hatching blastocyst, trophoblast incompetence, monozygotic and possibly even siamese twinning. The creation of a relatively large round opening in eight-cell embryos with acidic medium and consequent transfer may be an advantageous alternative, allowing for a safer application of assisted hatching techniques (Handyside et al, 1990). This approach may be beneficial as it avoids the partial loss of blastomeres when cellular junctions are being formed. In addition, it may also avoid trophoblast constriction during hatching of the blastocyst as the artificial gap is enlarged. The disadvantages of acidic Tyrode's for zona drilling of oocytes should not necessarily be extrapolated to the fertilized human egg. Zona hardening improves the efficiency of acidic dissolution markedly and the total volume of the perivitelline space in zygotes and embryos is higher than that of oocytes. These differences may be sufficient to reconsider the application of acidic dissolution of the embryonic zona pellucida.

REFERENCES

Cohen, J., Edwards, R.G., Fehilly, C.B., Fishel, S.B., Hewitt, J., Purdy, J.M., Rowland, R.F., Steptoe, P.C., and Webster, J.B., 1985, In vitro fertilization: a treatment for male infertility, Fertil. Steril., 43:422.

Cohen, J., Malter, H., Fehilly, C., Wright, G., Elsner, C., Kort, H., and Massey, J., 1988, Implantation of embryos after partial opening of oocyte zona pellucida to facilitate sperm penetration, Lancet, 2:162.
Cohen, J., Malter, H., Wright, G., Kort, H., Massey, J., and Mitchell, D., 1989, Partial zona dissection of human oocytes when failure of zona pellucida penetration is anticipated, Hum. Reprod., 4:435.
Cohen, J., Elsner, C., Kort, H.I., Malter, H., Massey, J., Mayer, M.P., and Wiemer, K., 1990a, Impairment of the hatching process following IVF in the human and improvement of implantation by assisting hatching using micromanipulation, Hum. Reprod., 5:7.
Cohen, J., Malter, H., Elsner, C., Kort, H.I., Massey, J., Mayer, M.P., 1990b, Immunosuppression supports implantation of zona pellucida dissected human embryos, Fertil. Steril., 53:662.
Cohen, J., Malter, H., Talansky, B.E., Tucker, M., Wright, G., 1990c, Gamete and embryo micromanipulation, in: "Seminars in Reproductive Endocrinology: The impact of IVF on the infertile couple," Marrs, R.P., Speroff, L., eds, Thieme Inc., New York, in press.
Cohen, J., Talansky, B.E., Malter, H., Alikani, M., Adler, A., Reing, A., Berkely, A., Graf, M., Davis, O., Liu, H., Bedford, J.M., and Rosenwaks, Z., 1990d, Microsurgical fertilization and teratozoospermia, Hum. Reprod., in press.
Cohen, J., Alikani, M., Malter, H.E., and Talansky, B.E., 1990e, Most motile human spermatozoa can fuse with oolemma. Submitted for publication.
Fishel, S., Antinori, S., Jackson, P., Johnson, L., Lisi, F., Chiariello, F., and Versaci, C., 1990, Twin birth after subzonal insemination, Lancet, i:772.
Gordon, J.W., and Talansky, B.E., 1986, Assisted fertilization by zona drilling: a mouse model for correction of oligospermia, J. Exp. Zool., 239:347.
Gordon, J.W., Grunfeld, L., Garrisi, G.J., Talansky, B.E., Richards, C. and Laufer, N., 1988, Fertilization of human oocytes by sperm from infertile males after zona pellucida drilling, Fertil. Steril., 50:68.
Gordon, J.W., Grunfeld, L., Garrisi, G.J., Navot, D., and Laufer, N., 1989, Successful microsurgical removal of pronucleus from tripronuclear human zygotes, Fertil. Steril., 52:367.
Handyside, A.H., Kontogianni, E.H., Hardy, K., and Winston, R.M.L., 1990, Pregnancies from biopsied human preimplantation embryos sexed by Y-specific DNA amplification, Nature, 344:768.
Kola, I., Lachan, O., Jansen, R.P.S., Turner, M., and Trounson, A., 1990, Chromosomal analysis of human oocytes fertilized by microinjection of spermatozoa into the perivitelline space, Human Reprod., 5:575.
Malter, H.E., and Cohen, J., 1989a, Partial zona dissection of the human oocyte: a nontraumatic method using micromanipulation to assist zona pellucida penetration, Fertil. Steril., 51:139.
Malter, H.E., and Cohen, J., 1989b, Embryonic development after microsurgical repair of polyspermic human zygotes, Fertil. Steril., 52:373.

Ng, S.C., Bongo, A., Ratnam, S.S., Sathananthan, H., Chan, C.L.K., Wong, P.C., Hagglund, L., Anandakumar, C., Wong, Y.C., and Goh, V.H.H., 1988, Pregnancy after transfer of multiple sperm under the zona, Lancet., 2:790.

Ng, S.C., Bongo, A., Sathananthan, H., and Ratnam, S.S.,1990, Micromanipulation: Its relevance to human in vitro fertilization, Fertil. Steril., 53:203.

Hosoi, Y., Miyake, M., Utsumi, K., and Iritani, A., 1988, Development of rabbit oocytes after microinjection of spermatozoa, in : "Proc.11th Int. Congress Anim. Reprod. Artif."

Kruger, T.F., Acosta, A.A., Simmons, K.F., Swanson, R.J., Matta, J.F., and Oehinger, S., 1988, Predictive value of abnormal sperm morphology in in vitro fertilization, Fertil. Steril., 49:112.

Talansky, B.E., Malter, H.E., and Cohen, J., 1990, A preferential site for sperm-oolemma fusion in rodents, Molec. Reprod. Dev., in press.

Tesarik, J., 1987, Gene activation in the human embryo developing in vitro, in: "Future aspects in human in vitro fertilization," W. Fechtinger, P. Kemeter, Springer-Verlag, Berlin and Heidelberg.

Tucker, M.J., Bishop, F.M., Cohen, J., Wiker, S.R., and Wright, G., 1990, Routine application of partial zona dissection for male factor infertility, submitted for publication.

BIOPSY OF HUMAN CLEAVAGE STAGE EMBRYOS AND SEXING BY DNA AMPLIFICATION

Alan H. Handyside

Institute of Obstetrics and Gynaecology
Royal Postgraduate Medical School
Hammersmith Hospital
Du Cane Road
London W12 0NN, UK

INTRODUCTION

For the preimplantation diagnosis of inherited disease by DNA analysis, a cell or cells must be biopsied from each embryo. With established techniques for *in vitro* fertilization (IVF), embryo culture and transfer, this could be accomplished, theoretically, at any stage between the 2-cell and blastocyst stages. Biopsy of some of the outer trophectoderm (TE) cells of the blastocyst has several advantages. Most importantly, at this stage the embryo reaches a maximum number of cells before implantation. The number of cells in human blastocysts *in vitro* increases from an average of 58 to 126 cells between days 5 to 7 post insemination and the majority of these are TE cells (Hardy et al., 1989a). Thus, it has been possible to biopsy between 10 and 30 TE cells from individual blastocysts on day 6 (Dokras et al., 1990 and this volume). Clearly, the more cells available for analysis the more reliable any diagnosis is likely to be. TE biopsy also has the advantage that these cells are thought, by analogy with detailed lineage studies of rodent embryos, to be extraembryonic and contribute mainly to the placenta after implantation; the inner cell mass (ICM) of the blastocyst, from which the fetus is later derived, is not affected. In effect, therefore, this would be the equivalent of an early chorion villus biopsy and may be ethically more acceptable.

The major problem with this approach, however, is the disappointing pregnancy success rate after transfer of blastocysts during routine IVF (Dawson et al., 1988; Bolton et al., 1991) which coupled with the low proportion of embryos reaching this stage (Hardy et al., 1989a) is likely to reduce the number of unaffected embryos available for transfer and the prospects of establishing a pregnancy.

CLEAVAGE STAGE BIOPSY

Although there are fewer cells at earlier stages, the only alternative, at the present time, is to biopsy embryos at cleavage stages when pregnancy success rates after transfer are optimal. At these early stages, each cell of the mammalian embryo remains totipotent and can contribute to all the tissues of the conceptus. Indeed, single cells from embryos at these stages have been demonstrated, under appropriate conditions, to form blastocysts and

Preimplantation Genetics, Edited by Y. Verlinsky and
A. Kuliev, Plenum Press, New York, 1991

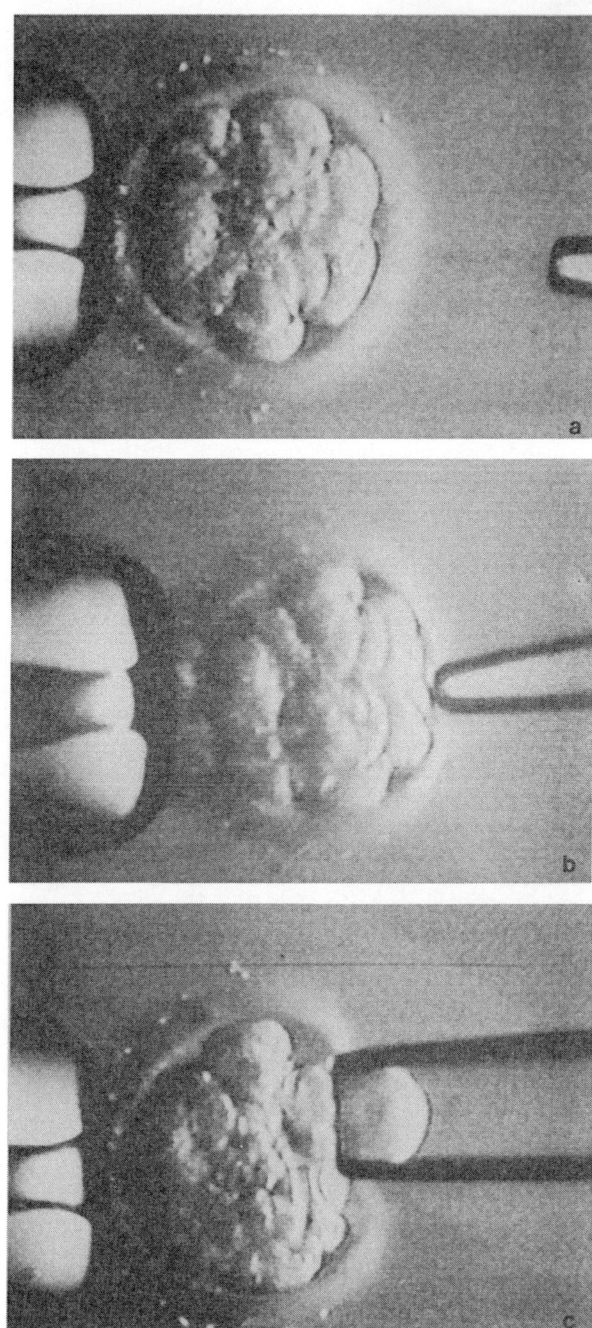

Figure 1 Cleavage stage biopsy of a 12-cell polyspermic human embryo on day 3. The embryo is immobilised on a flame-polished holding pipette (a) and a fine micropipette used to drill a hole in the zona pellucida with a stream of acid Tyrodes (b). Finally, a second larger micropipette is pushed through the hole in the zona to aspirate a single cell. Note that the nucleus of the cell biopsy is visible in this case. (Photographs from video recording).

to develop into normal offspring following transfer in a number of rodent and domestic species (Papaioannou and Ebert, 1986). The proportion developing successfully to term, however, declines rapidly with cells isolated at progressively later stages. This is because cleavage divisions simply subdivide the cytoplasm of the zygote into successively smaller cells and there appears to be a lower limit to the cellular mass of the embryo compatible with implantation and development.

The implications for cleavage stage biopsy of human embryos are that the reduction of cellular mass should be minimised to avoid an increase in biochemical pregnancies and implantation failure. We, therefore, decided to biopsy human embryos early on day 3, which is the latest day embryos have been routinely transferred without affecting pregnancy rate, at the 6- to 10-cell stage when the removal of a single cell at the appropriate stage only reduces the cellular mass by 1/8.

A simple biopsy method was developed in which normally fertilized embryos at the appropriate stage are placed in drops of medium under oil and transferred to a dissecting microscope for micromanipulation. The embryo is immobilised on a flame polished holding pipette (Figure 1a) and a hole drilled in the zona pellucida with a stream of acid Tyrodes from a fine micropipette (Figure 1b) (Gordon and Talansky, 1986). A second larger micropipette is then pushed through the hole in the zona to aspirate one or two cells (1/8 or 2/8 cells) (Figure 1c). Following biopsy, the corresponding 7/8 or 6/8 biopsied embryo is returned to culture and the cell biopsy prepared for DNA analysis.

PREIMPLANTATION DEVELOPMENT OF BIOPSIED EMBRYOS

Before attempting any diagnosis and transfer of biopsied cleavage stage embryos, we considered it was essential to examine their preimplantation development *in vitro* to find out how it had been affected by this process (Hardy et al., 1990). Biopsied embryos were examined for (1) development to the blastocyst stage *in vitro*, (2) the total cell number and numbers of cells in the TE and ICM at the blastocyst stage, and (3) the uptake of the energy substrates, pyruvate and glucose on successive days up to day 7.

The proportion of normally fertilized human embryos developing to the blastocyst stage in conventional media supplemented with heat inactivated maternal serum varies widely between patients but averages between about 40 and 70% (Hardy et al., 1989a,b and unpublished observations). After biopsy of one or two cells, the proportion developing to this stage (79 and 71%, respectively) was relatively high and even exceeded that of unmanipulated control embryos form the same patients (59%). Furthermore, many biopsied embryos (56%) hatched from the zona pellucida.

Cell numbers in biopsied embryos reaching the blastocyst stage were analysed by a modification of a method for differentially labelling the TE and ICM nuclei with two DNA specific fluorochromes which allows the nuclei to be distinguished by colour with fluorescence microscopy (Handyside and Hunter, 1986; Hardy et al., 1989a). Total numbers were reduced in both 7/8 and 6/8 blastocysts, especially on day 6, but only in proportion to the reduced cellular mass demonstrating that cleavage rate had not been affected (Figure 2). More importantly, since the fetus is derived from the ICM after implantation, the numbers of these cells as a proportion of the total was maintained.

Finally, viability of the embryos after biopsy was assessed by measuring the uptake of energy substrates. The uptake of pyruvate and glucose was measured using a non-invasive ultramicrofluorescence method (Leese and Barton, 1984) to assay the depletion of these substrates in droplets of culture medium over successive 24h incubations. The uptake of these substrates was reduced in both 7/8 and 6/8 blastocysts throughout their development after biopsy, but again only in proportion to the reduction in cellular mass (Figure 3). There was no indication of an additional affect on the viability of these

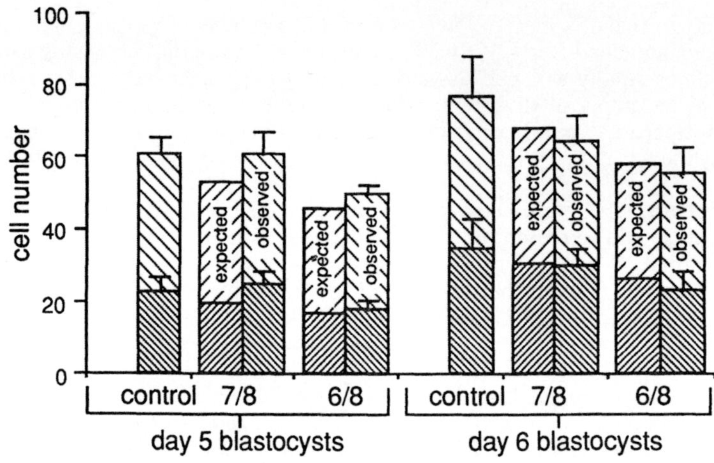

Figure 2 Comparison of expected and observed total (▨) and inner cell mass (▩) cell numbers in newly expanded blastocysts on day 5 and 6 derived from unmanipulated and 7/8 and 6/8 biopsied embryos. Expected values are calculated as proportions (7/8 and 6/8 of the control values). Error bars are standard errors. (From Hardy et al., 1990, with permission).

embryos. They also showed the characteristic switch in substrate preference from pyruvate to glucose at the blastocyst stage (Hardy et al., 1989b).

Preimplantation development of human embryos to the blastocyst stage *in vitro* does not, therefore, appear to be adversely affected by biopsy on day 3 at the 6- to 10-cell stage and effects on the allocation of cells to the ICM are limited to a proportionate reduction in numbers which is unlikely to increase significantly the incidence of biochemical pregnancies.

SEXING BY DNA AMPLIFICATION

Over 200 X-linked recessive diseases have been described which typically affect only males. Some of these, notably Duchenne muscular dystrophy, have been mapped and extensively characterised at the molecular level and prenatal diagnosis of normal or carrier females as well as normal or affected males is possible in many cases by DNA analysis after chorion villus sampling or amniocentesis. Other diseases are not so well characterised and all that can be offered is to diagnose the sex of the fetus giving the option of terminating a male with a 50:50 chance of being affected.

With a maximum of about 30 cells biopsied from each embryo, possible approaches for genetic analysis are very restricted. Amplification of a short fragment of the ß-globin gene using the polymerase chain reaction (PCR), however, had been used successfully for the prenatal diagnosis of sickle cell disease from as few as 75 fetal blood cells (Saiki et al., 1985). PCR also offers the advantage that short fragments of DNA can be amplified relatively quickly, within 3 to 5 hours, which is essential if embryos biopsied early on day 3 are to be transferred after analysis later the same day. More recently, a variety of unique sequences have been amplified from single cells including human sperm and fibroblasts (Li

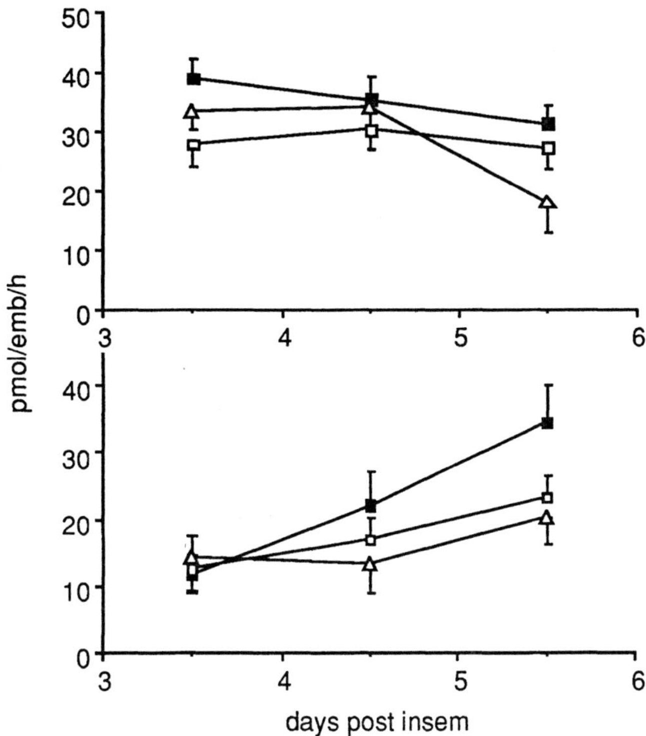

Figure 3 Pyruvate (upper panel) and glucose (lower panel) uptake by nineteen unmanipulated (—■—) embryos, twenty 7/8 (—□—) and ten 6/8 (—△—) biopsied embryos which had developed to the blastocyst stage by day 5 or 6. (From Hardy et al., 1990 with permission).

et al., 1988) and unfertilized eggs from which the first polar bodies had been removed (Coutelle et al., 1989). However, in most cases reamplification was necessary for simple detection of the fragment by gel electrophoresis and staining and amplification failure occured in 10-20% of single sperm and fibroblasts.

The identification of potentially affected male embryos requires amplification of a sequence specific for the Y chromosome. To minimise the possibility of amplification failure with single or small numbers of cells, which would lead to a false negative diagnosis and the transfer of a male embryo, we chose to amplify a sequence present in 800 to 5000 copies on the long arm of the Y chromosome (Kogan et al., 1987). Amplification of this Y-specific repeat fragment was consistently achieved from as little as 2pg of male DNA (about the amount present in a single cell) and in single cells biopsied from male embryos without reamplification (Figure 4). Also, the accuracy of sexing by DNA amplification from single biopsied cells was confirmed by cytogenetic analysis of the biopsied embryos after assessment of their preimplantation development *in vitro* (Handyside et al., 1989). Recently, as an alternative, we have amplified an alphoid repeat specific for the centromeric region of the Y chromosome which appears to be equally sensitive (Witt and Erickson, 1989).

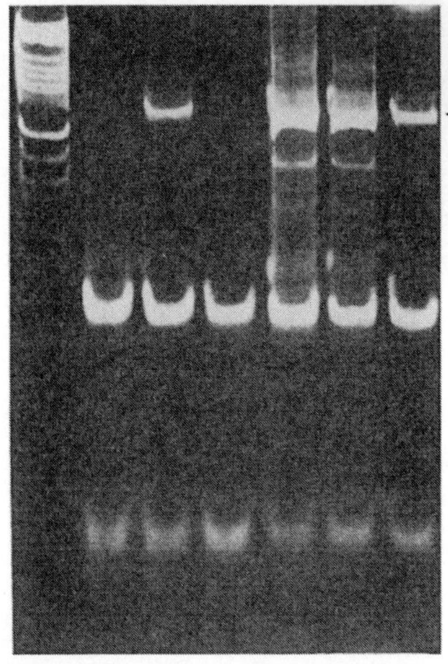

Figure 4 Analysis of polymerase chain reaction products with oligonucleotide primers for a 149 bp fragment (arrowed) of a 3.4 kb Y-specific repeat sequence by electrophoresis on a 12% polyacrylamide minigel and staining with ethidium bromide. Lane 1 (from left to right): DNA size markers; lane 2,3: female and male single cell biopsies; lane 4: control blank; lanes 5-7 dilutions of male genomic DNA - 20ng, 200pg and 2pg. (Modified rom Handyside, 1990, with permission)

PREVENTION OF X-LINKED DISEASE

On the basis of this preparatory work, we have attempted preimplantation diagnosis in an initial series of five (Handyside et al., 1990) and now a further three couples known to be at risk of transmitting various X-linked diseases. These included X-linked mental retardation, Lesch-Nyhan syndrome, adrenoleukodystrophy, retinitis pigmentosa and hereditary sensory motor neurone disease type II, and two of the couples at risk of Duchenne muscular dystrophy opted for this approach even though a specific DNA diagnosis by conventional methods may have been possible after CVS.

Initially, each of the couples underwent routine assessment for IVF. The women were then induced to superovulate using an established protocol involving an initial period of suppression of ovarian function with an LHRH agonist followed by stimulation of folliculogenesis and administration of human chorionic gonadotrophin (hCG) to trigger ovulation (Rutherford et al., 1988). The numbers of oocytes recovered and normally fertilized in these predominantly fertile women were similar to those obtained with infertile couples (Table 1). More critically, after biopsy, DNA amplification and diagnosis of sex, in a large proportion of treatment cycles, at least two female embryos were identified for transfer even though the probability of an embryo being female is only 50:50. This is important because pregnancy rates after the transfer of two embryos are significantly higher than after the transfer of one. On the other hand, we consider two the maximum number safe to transfer because of the risks of multiple pregnancies and the need to confirm the preimplantation diagnosis by CVS of each fetus which is feasible with twins but not triplets.

Table 1 Number and sex of normally fertilized human embryos biopsied at the 6- to 10-cell stage*

		Per cycle	(range)	Total
No. of oocytes collected		11.1	(7-21)	145
No. normally fertilized		6.2	(2-11)	81
No. biopsied on day 3		4.8	(2-7)	63
DNA analysis	Male	2.0	(1-4)	26
	Female	2.4	(1-4)	31
	Not sexed	0.5	(1-2)	6
No. transferred		1.7		22

*8 patients had a total of 13 treatment cycles.

Five out of the seven women became pregnant, three after one and one each after two and three treatment cycles. (The status of the most recent patient remains to be tested). The first two were twin pregnancies and these and the third singleton pregnancy have been confirmed by CVS to have normal female karyotypes. Both twins have gone to term and all four were normal females. The second of the second twins was stillborn at Caesarian delivery but detailed post-mortem examination failed to reveal any gross abnormality and the cause was probably intrapartum anoxia.

PROSPECTS FOR PREIMPLANTATION DIAGNOSIS

The implantation rate of biopsied cleavage stage embryos in these women was very much higher than commonly achieved following IVF with infertile couples. Ten out of 20 (50%) biopsied embryos implanted as detected by raised hCG levels around day 14. Although three of these embryos, including one 'vanishing' twin, failed to develop further, the remaining seven which established clinical pregnancies have resulted in a pregnancy success rate of 35% per embryo transferred. The use of IVF techniques for recovery of large numbers of preimplantation embryos for biopsy and genetic screening has, therefore, proved to be both reliable and efficient in terms of establishing pregnancies. The reason for this success may be that the majority of these couples are fertile. But it is also possible that zona drilling to enable cell biopsy assisted hatching and implantation. It has been suggested that zona hardening in culture contributes to implantation failure after IVF (Cohen et al., 1990).

Although two of the women had two and three cycles of treatment before becoming pregnant, this was achieved in a period of approximately six months. This indicates that a second major advantage of preimplantation diagnosis may be a substantial reduction in the time necessary to establish an unaffected pregnancy compared with what in many cases would be a series of spontaneous pregnancies followed by conventional postimplantation prenatal diagnosis.

To date, DNA amplification of a Y-specific repeat sequence has proved reliable for determining the sex of the biopsied embryos. Five of the seven fetuses, in which CVS has been carried out, have been confirmed to have normal female karyotypes. In all but two treatment cycles in which two cells were biopsied from each embryo, preimplantation diagnosis has been based on DNA amplification from single cells. We have recently demonstrated that amplification failure of chromosome-specific repeat sequences is as frequent as reported for unique sequences under these circumstances (Kontogianni and Handyside, this volume). Two or more biopsied cells may be necessary, therefore, for a sufficiently reliable diagnosis (Handyside, 1990). In general, however, because of the restrictions on the numbers of cells available from preimplantation embryos, genetic diagnosis at these early stages is unlikely to be as reliable as conventional postimplantation procedures and a follow up CVS or amniocentesis may continue to be necessary for the forseeable future.

ACKNOWLEDGEMENTS

I would like to thank my research students, Kate Hardy and Helen Kontogianni, also Karen Martin, Dr. Henry Leese and Professor Robert Winston for allowing me to discuss our joint work, and Karen Dawson, Joe Conaghan, Dr. Jacob Soussis for their help with clinical management and IVF. This work was supported by the Muscular Dystrophy Group of Great Britain and Northern Ireland.

REFERENCES

Bolton, V.N., Wren, M.E. and Parsons, J.H. (1991) Pregnancies following in vitro fertilization and transfer of human blastocysts. Fert. Steril. in press.

Cohen, J., Wright, G., Malter, H., Elsner, C., Kort, H., Massey, J., Mayer, M.P. and Wiemer, K.E. (1990) Impairment of the hatching process following in vitro fertilisation in the human and improvement of implantation by assisting hatching using micromanipulation. Hum. Reprod. 5, 7-13.

Coutelle, C., Williams, C., Handyside, A., Hardy, K., Winston, R and Williamson, R. (1989) Genetic analysis of DNA from single human oocytes - a model for pre-implantation diagnosis of cystic fibrosis. Brit. Med. J. 299, 22-24.

Dawson, K.J., Rutherford, A.J., Winston, N.J., Subak-Sharpe, R. and Winston, R.M.L. (1988) Human blastocyst transfer, is it a feasible proposition? Human Reprod. suppl 145, 44-45.

Dokras, A., Sargent, I.L., Ross, C., Gardner, R.L. and Barlow, D.H. (1990) Trophectoderm biopsy in human blastocysts. Human Reprod. 5, 821-825.

Gordon, J. and Talansky, B.E. (1986) Assisted fertilization by zona drilling: a mouse model for correction of oligozoospermia. J. exp. Zool. 239, 347-354.

Handyside, A.H. (1990) Preimplantation diagnosis by DNA amplification. In 'The embryo: normal and abnormal development and growth' (eds. Chapman, M., Grudzinskas, G., Chard, T. and Maxwell, D.) Springer-Verlag, London in press.

Handyside, A.H. and Hunter, S. (1986) Cell division and death in the mouse blastocyst before implantation. Roux's Arch. Dev. Biol. 195, 519-526.

Handyside, A.H., Pattinson, J.K., Penketh, R.J.A., Delhanty, J.D.A., Winston, R.M.L. and Tuddenham, E.D.G. (1989) Biopsy of human preimplantation embryos and sexing by DNA amplification. Lancet I. 347;-349.

Handyside, A.H., Kontogianni, E.H., Hardy, K. and Winston, R.M.L. (1990) Pregnancies from biopsied human preimplantation embryos sexed by Y-specific DNA amplification. Nature 344, 768-770.

Hardy, K., Handyside, A.H. and Winston, R.M.L. (1989a) The human blastocyst: cell number, death and allocation during late preimplantation development in vitro. Development 107, 597-604.

Hardy, K., Hooper, M.A.K., Handyside, A.H., Rutherford, A.J., Winston, R.M.L. and Leese, H.J. (1989b) Non-invasive measurement of glucose and pyruvate uptake by individual human oocytes and preimplantation embryos. Human Reprod. 4, 188-191.

Hardy, K., Martin, K.L., Leese, H.J., Winston, R.M.L. and Handyside, A.H. (1990) Human preimplantation development in vitro is not adversely affected by biopsy at the 8-cell stage. Hum. Reprod. 5, 708-714.

Kogan, S.C., Doherty, M. and Gitschier, J. (1987) An improved method for prenatal diagnosis of genetic disease by analysis of amplified DNA sequences. N. Eng. J. Med. 317, 985-990.

Leese, H.J. and Barton, A.M. (1984) Pyruvate and glucose uptake by mouse ova and preimplantation embryos. J. Reprod. Fert. 72, 9-13.

Li, A., Gyllensten, U.B., Cui, X., Saiki, R.K., Erlich, H.A. and Arnheim, N. (1988) Amplification and analysis of DNA sequences in single human sperm and diploid cells. Nature 335, 414-419.

Papaioannou, V.E. and Ebert, K.M. (1986) Comparative aspects of embryo manipulation in mammals. In 'Experimental approaches to mammalian embryonic development' (eds. Rossant, J. and Pedersen, R.) pp. 67-96. Cambridge University Press.

Rutherford, A.J., Subak-Sharpe, R.J., Dawson, K.J., Margara, R.A., Franks, S. and Winston, R.M.L. (1988) Improvement of in vitro fertilisation after treatment with buserelin, an agonist of luteinising hormone releasing hormone. Br. Med. J. 296, 1765-1768.

Saiki, R.K., Scharf, S., Faloona, F., Mullis, K.B., Horn, G.T., Erlich, H.A. and Arnhem, N. (1985) Enzymatic amplification of ß-globin sequences and restriction site analysis for diagnosis of sickle cell anaemia. Science 230, 1350-1354.

Witt, M. and Erickson, R.P. (1989) A rapid method for the detection of Y chromosomal DNA from dried blood specimens by the polymerase chain reaction. Hum. Genet. 82, 271-274.

BIOPSY OF BLASTOCYST

Sandra Carson

Department of Reproductive Endocrinology
University of Tennessee, College of Medicine
Memphis, TN 38163

INTRODUCTION

To date preimplantation diagnosis has been made in the two-eight cells embryos because these are available in IVF programs. However, most patients referred for IVF are infertile, very few undergoing IVF with the thought of preimplantation diagnosis. It may be possible that in the fertile patients, culturing to blastocysts may be a much more viable alternative.

The blastocyst offers a number of advantages for biopsy and preimplantation diagnosis. First of all, differentiation has started to occur, with the cells of the trophectoderm lining the periphery of the embryo. These cells can be removed without harming the inner cell mass. Because the blastocyst has undergone a number of divisions, there is a greater cell number available for diagnosis, making possible to overcome the problems of contamination in the PCR. In addition, the transfer efficiency of the blastocyst is much higher than the transfer efficiency of the embryo at the earlier stages.

However, if a 2-4 cell embryo is transferred after being retrieved from uterine lavage, the pregnancy rate for a single transfer is almost 10%. The pregnancy rate is 6 times higher if that embryo is a blastocyst, this pregnancy rate being almost identical throughout the mammalian kingdom.

Unfortunately, it is difficult to glean exactly the IVF pregnancy rate per single embryo transfer rate from the reports published, but it seems to be very close to the single embryo transfer rate, which is about 9-20% in lavage. J. Cohen (this volume), however, achieved a 22% per embryo pregnancy rate, by performing zona drilling.

The higher pregnancy rate from blastocysts may be explained by the selection of the healthiest embryos as not all 4 cell embryos will become blastocysts, i.e. a natural selection at the first few stages has already been bypassed.

Preimplantation Genetics, Edited by Y. Verlinsky and
A. Kuliev, Plenum Press, New York, 1991

The greatest advantage of the blastocyst biopsy is that the blastocyst can be obtained through uterine lavage, a much less expensive and easier procedure than IVF (Brambati, this volume): the human fertilized egg spends about 3 days in the fallopian tube and then floats freely inside the uterine cavity for another 3 days, during which it can be obtained through a lavage catheter.

There are, however, some <u>disadvantages</u> of the blastocyst biopsy. For example, the same discrepancy that exists between trophoblast tissue and the fetus in CVS may also exist between trophectoderm and inner cell mass or fetal cells. In addition, there may be limited incubation time available for genetic diagnosis. We are not sure exactly how long we can culture the biopsied blastocyst before hatching or what is the effect of keeping it in culture. J. Biggers (this volume) demonstrated that the embryo changes its environment considerably between the early embryo and the blastocyst. A lot of cell to cell communication happens in the developing blastocyst, the cells become very sticky, making their manipulation technically difficult, possibly leading to a higher premature hatching rate.

As N. First (this volume) mentioned the first embryo <u>microbiopsy techniques</u> were developed in the bovine system in the animal husbandry industry. This began with <u>bisection</u> and multiple section of the bovine embryo. Later, aspiration, particularly of trophectoderm cells, was developed. Finally, excision of herniating cells after spontaneous hatching and zona slitting were also used to biopsy the blastocysts. These <u>four techniques</u> are predominantly used to perform the biopsy of the blastocyst.

A micro scalpel, used in the bovine industry which is much larger than the instruments we use for early aspiration, is used to dissect the embryo and its inner cell mass in half. This is the way Ozyl (1983) bisected 22 blastocysts. First 11 bovine blastocysts were bisected through the inner cell mass, and all of them were transferred: 8 cows became pregnant, half of them with identical twins. From the second 11 embryos, only a single half of those bisected embryos were transferred with a lower pregnancy rate: only 4 cows became pregnant, one of the cows still bearing identical twins. Even though the blastocysts had begun differentiation, it is possible that one half of a blastocyst can go on to split into identical twins spontaneously.

The next technique to biopsy blastocysts is <u>aspiration</u>. A murine embryo from our own system is held and approximately 10 micron aspiration pipette is inserted through the zona into the trophectoderm. Cells are aspirated into the pipette. This technique is used on rabbits by Yang and Foote (1987). They found that aspirating trophectoderm really makes no difference in the live birth rate between aspirated blastocysts and their non-biopsied controls. However, aspiration of one cell from a two cell rabbit embryo led to a significantly lower development rate.

Trophectoderm herniation 4-6 hours after we have made a small hole in the zona with a finely drawn pipette results in a small mass, about 20 cells of trophectoderm, to herniate. Next, a finely drawn pipette excises the trophectoderm cells.

Such trophectoderm biopsy has been done in a number of species. Performing trophectoderm biopsy in a similar manner, Monk et al (1988) observed a slightly lower fetal development per embryo than in non-biopsied controls. Summers and his colleagues (1989) performed trophectoderm biopsy at the blastocyst stage in marmosets with no alteration of pregnancy or liveborn rate. Although the series studied was small, the data suggest that the technique is viable for use in the primate.

We have performed over 200 embryo biopsies using all the above 4 techniques. Using a knife to excise off a portion of the cells and the zona resulted in a significantly poorer embryo development than excision of trophectoderm after herniation either spontaneous or following zona slitting.

Using FITC BSA which stains the damaged cells rather than the living (the damaged cells highly fluoresce), we found that aspiration or excision yield a much greater number of damaged cells than just the removal of trophectoderm following its herniation. Therefore, according to our results, though all the four techniques yield a good number of trophectoderm cells for genetic analysis, bisection or excision as well as aspiration at the blastocyst stage result in significantly more damage to the embryo (leading to a poorer development and survival rate) than trophectoderm excision after herniation.

We found that it is possible to control the size of the slit and the number of cells which herniate; the number of the cells obtained also depends on the duration of the blastocyst in culture. According to our observation the location of the slit does not always control herniation.

As mentioned, the cells of the blastocyst are much stickier than in early stages, so we coated our pipettes with silicone prior to micromanipulation. We observed that the cells become more manageable if the embryo is incubated in calcium magnesium free media for about 20-30 minutes prior to the biopsy. We also found that increasing the media concentration of albumin to 15% significantly decreases the stickiness of the trophectoderm to the dish, the zona, and the aspirating pipette.

Our current interest is to make a single organ chimera (a trophectoderm chimera) using the blastocyst. So far, in the murine and bovine models, most chimeras that have been made were inner cell mass chimeras. To make single organ chimeras of trophectoderm in mice, we have taken murine blastocysts and stained them with FITC to identify the healthy cells before performing a trophectoderm biopsy. Then, we inserted the stained cells into the blastocoele cavity and watched the reexpansion of the embryos over the next 24 hours, with those trophectoderm cells spreading into the trophectoderm of the new non-stained embryo. The staining allows to look at the cell lineage, to observe stained cells living next to unstained trophectoderm and to evaluate the survival rate of the donor and the recipient embryos. We hope to report the results of these experiments at the next meetings.

REFERENCES

Monk, M., Muggleton-Harris, A.L., Rawlings, E., and Whittingham, D.G., 1988, Pre-implantation diagnosis of HPRT-deficient male and carrier female mouse embryos by trophectoderm biopsy, Hum. Reprod., 3(3):337-381.

Ozil, J.P., 1983, Production of identical twins by bisection of blastocysts in the cow, J. Reprod. Fertil., 69:463-468.

Smith, A.L., Gentry, W.L., Buster, J.E., and Carson, S.A., 1989, Murine blastocyst survival after blastomere excision: Comparison of 4 different methods, Abstract O-029, 45th Annual Meeting, The American Fertility Society, San Francisco, California, p.S12.

Summers, P.M., Campbell, J.M., and Miller, M.W., 1988, Normal in-vivo development of marmoset monkey embryos after trophectoderm biopsy, Hum. Reprod., 3(3):389-393.

Yang, X., and Foot, R.H., 1987, Production of identical twin rabbits by micromanipulation of embryos, Biol. Reprod., 37:1007-1014.

PART III

GENETIC ANALYSIS OF GAMETES AND PREEMBRYOS

CHROMOSOMAL ANALYSIS OF HUMAN SPERMATOZOA

Renee Martin

Division of Medical Genetics, Department of Pediatrics, University of Calgary and Genetics Clinic, Alberta Children's Hospital 1820 Richmond Rd., Calgary, Alberta, Canada T2T 5C7

INTRODUCTION

Human sperm chromosome complements have been studied for more than a decade since Edwina Rudak first published an ingenious technique in which hamster oocytes are used to reactivate human sperm (Rudak et al. 1978). This was a real breakthrough in the field of reproductive genetics because prior to this discovery information about the frequency, type and cause of chromosomal abnormalities had to be inferred from indirect studies of spontaneous abortions and livebirths. Since so many chromosomal abnormalities result in early pregnancy loss, these studies were subject to extreme ascertainment bias.

The analysis of human sperm chromosome complements has provided the first information available on the frequency and types of chromosomal abnormalities in human gametes. The majority of studies have been on normal males to obtain base-line data (Martin et al., 1982: 1983; 1987a; Brandriff et al., 1985; Sele et al., 1985; Kamiguchi and Mikamo, 1986; Jenderny et al., 1987; Pellestor and Sele, 1989; Navarro et al., 1990). There have also been studies on the effect of donor age on the frequency of chromosomal abnormalities in human sperm (Martin et al., 1987b), evaluation of sperm sex preselection techniques (Brandriff et al., 1986; Ueda and Yanigamachi, 1987; Beckett et al., 1989) and the effect of microinjection of human sperm into hamster oocytes (Martin et al., 1988). Men at increased risk of sperm chromosomal abnormalities have also been studied, such as men exposed to radiotherapy (Martin et al., 1986a, 1989; Jenderny and Rohrborn 1987) and men carrying constitutional chromosomal rearrangement such as translocations and inversions (Balkan and Martin, 1983a,b: Balkan et al., 1983; Martin, 1984; Brandriff et al., 1986; Martin, 1986a,b, Martin et al.,1986b, Burns et al., 1986 Pellestor et al., 1987; Benet and Martin, 1988; Martin 1989 a,b; Templado et al., 1988; Pellestor et al., 1989, 1990; Martin, 1990; Martin et al., 1990a,b; Templado et al.,1990).

Preimplantation Genetics, Edited by Y. Verlinsky and A. Kuliev, Plenum Press, New York, 1991

This report will review the information on sperm chromosomal abnormalities in normal men from the major studies, discuss how the data provide information on the etiology of numerical chromosomal abnormalities by studying the distribution of aneuploidy among the chromosomes, and summarize results of sperm chromosome studies in 28 men with constitutional chromosomal abnormalities.

MATERIALS AND METHODS

Human sperm chromosome complements were obtained after sperm penetration of hamster oocytes. This technique, which has been described in detail elsewhere (Martin, 1983, 1988b), allows analysis of human pronuclear sperm chromosomes which can be identified by banding techniques.

Sperm Chromosome Complements from Normal Men

Our laboratory has analyzed a total of 5629 sperm chromosome complements from 83 normal healthy men. The mean frequency of abnormalities in this population was 13.6%. Structural chromosomal abnormalities accounted for 9.4% and numerical for 4.2%. Of the numerical abnormalities, 0.6% were hyperhaploid, 3.5% were hypohaploid and 0.1% had multiple aneuploidy. If all of these numerical abnormalities were produced by meiotic nondisjunction, we would expect an equal frequency of hyperhaploid and hypohaploid complements. Since we observed a significant excess of hypohaploid complements, some of these complements must have arisen by other mechanisms such as anaphase lag or by artefactual loss of chromosomes during slide preparation. Since we used Tarkowski's technique (1966) for fixation of the fertilized eggs on slides and this technique is known to be associated with an excess of hypohaploidy, it is likely that our elevated frequency of hypohaploidy is a technical artefact. A conservative estimate of aneuploidy can be obtained by doubling the frequency of hyperhaploidy and multiple aneuploidy. This yields an aneuploidy frequency of 1.4%. Using this conservative estimate of aneuploidy, the

Table I. Chromosomal abnormalities in 83 normal men.

Total Sperm 5629

Abnormalities	Observed	Conservative Estimate
Structural	9.4%	9.4%
Numerical	4.2%	1.4%
Hyperhaploid	0.6%	
Hypohaploid	3.5%	
Multiple Aneuploidy	0.1%	
Total Abnormal	13.6%	10.8%

total mean of sperm chromosomal abnormalities was 10.8%. These results are summarized in Table 1.

Four other laboratories have reported on more than 5000 sperm karyotypes and these are outlined in Table II. Brandriff et al. (1985) in California reported on 2468 from 11 donors and had a very similar frequency of numerical and structural abnormalities as ours. Kamiguchi and Mikamo (1986) in Japan reported on 1091 from 4 donors and also had similar frequencies with structural abnormalities being a bit higher at 13%. Pellestor and Sele (1989) from France had a short summary on 787 karyotypes from 6 donors. They did not break down their numerical abnormalities into hyperhaploid and hypohaploid complements, so presumably their value of 7.3% numerical abnormalities is inflated by more hypohaploid complements. Navarro et al. (1990) from Spain summarized results on 503 sperm from an unknown number of donors with 4% numerical and 6.9% structural abnormalities. All of these studies had an overall frequency of approximately 10% abnormalities.

The three largest studies have reported the range in the frequencies of abnormalities and there has been considerable variation in the frequency of structural abnormalities among males (Table III). Brandriff et al.(1985) found a mean frequency of 7.7% structural abnormalities with a range of 2-16% in 11 donors, Kamiguchi and Mikamo (1986) reported 13.0% with a range of 11-17% in 4 donors and we found a mean frequency of 9.4% with a range of 0% to 22% in 83 donors. The frequencies of structural abnormalities appear to be donor-specific (Brandriff et al. 1985) and may reflect genetic differences, age differences (Martin et al. 1987b), or varying exposures to clastogens. A comparison of the mean frequency of hyperhaploid sperm complements in the individual men in the three largest studies demonstrates considerable similarities (Table IV). Brandriff et al. (1985) reported 1.7% numerical complements (range 0-3%), Kamiguchi and Mikamo (1986) reported to 0.9% (range 0-2%), and we have a mean of 1.4% (range 0-5%) in the individual men in our study. This suggests that an estimate of the mean frequency of aneuploidy in human sperm is approximately 1-2%. The range in the frequency of hyperhaploid sperm was quite small in all three studies.

Table II. Sperm chromosomal abnormalities observed in normal men from five large studies.

	Sperm	Donors	Numerical[a]	Structural	Total
Martin et al, this report	5629	83	1.4	9.4	10.8
Brandriff et al. 1985	2468	11	1.7	7.7	9.4
Kamiguchi & Mikamo, 1986	1091	4	0.9	13.0	13.9
Pellestor & Sele, 1989	787	6	7.3[b]	2.0	9.3
Navaro et al. 1990	503	?	4.0	6.9	10.3

[a] hyper/hypohaploid ratio not reported

[b] conservative estimate of aneuploidy (see text for explanation)

Table III. Structural chromosomal abnormalities in sperm

	Mean(%)	Range(%)
Martin et al. 1989	9.4	0 - 22
Brandriff et al. 1985	7.7	2 - 16
Kamiguchi & Mikamo, 1986	13.0	11 - 17

The frequencies of X- and Y-chromosome bearing sperm were not significantly different from 50% in our study or studies from the other laboratories, as theoretically expected.

Distribution of Aneuploidy among the Different Chromosomes

We have recently looked at the distribution and frequency of aneuploidy among the different chromosomes in sperm to determine whether all chromosomes are equally likely to be involved in aneuploid events or if some chromosomes are particularly susceptible to nondisjunction. It is important to determine the incidence of nondisjunction for different chromosomes because this information will give us clues about the mechanisms of nondisjunction and survival of chromosomally abnormal embryos. If all chromosome groups have the same frequency of nondisjunction, then the mechanism that causes nondisjunction must be common to all chromosomes, e.g. errors of spindle formation or attachment. If specific chromosomes have an increased frequency of nondisjunction, then we could focus on the specific characteristics of the chromosomes (e.g. small size, presence of heterochromatin, nucleolar organizing regions) to elucidate some of the factors which influence the rate of nondisjunction.

Results on the distribution of aneuploidy were pooled from a number of studies encompassing 8356 sperm chromosome complements from normal men and 3259 sperm complements from men carrying a constitutional chromosomal abnormality (Martin et al., 1990c). For the latter group, aneuploid complements unrelated to the constitutional chromosomal rearrangement were only included if no interchromosomal effect had been ascertained. The analysis was based on hyperhaploid complements since hypohaploid complements might be caused by technical artefact (as discussed above). The data were analyzed separately for the two groups of men and then pooled after statistical analysis demonstrated no difference in the two groups (chi square analysis). For statistical analysis it was assumed that all chromosomes had an equal probability of nondisjunction and two-tailed binomial tests using exact binomial probabilities were performed (Rosner, 1986) to determine if each chromosome (or chromosome group) had a frequency of hyperhaploidy that differed significantly from the expected frequency.

Table IV. Numerical chromosomal abnormalities in sperm

	Mean(%)[a]	Range(%)
Martin et al, this report	1.4	0 - 5
Brandriff et al. 1985	1.7	0 - 3
Kamiguchi & Makamo, 1986	0.9	0 - 2

[a]conservative estimate of aneuploidy (see text for explanation)

All chromosome groups were represented among the hyperhaploid sperm demonstrating that all groups are susceptible to nondisjunction, not just those commonly observed in studies of newborns and spontaneous abortions. There was a significantly increased frequency of hyperhaploidy for chromosome 1 (p=.02), 21 (p=.002) and the sex chromosomes (p=.0001). The increased frequency of hyperhaploidy for chromosome 1 is surprising since this is the only chromosome which has not been observed as a trisomy in spontaneous abortions. However, trisomy 1 has been observed in an eight-cell human pre-embryo (Watt et al. 1987) and it is possible that trisomy 1 is common in human conceptuses but is always lost in an early preimplantation stage. The increased frequency of hyperhaploidy for chromosome 21 and the sex chromosomes was even more dramatic. These results corroborate data from studies on human liveborn offspring and spontaneous abortions. A number of studies have demonstrated that the paternal contribution to trisomy 21 is approximately 20% (Juberg and Mowrey, 1983) whereas trisomies for other autosomes have a much lower frequency of paternal etiology (Hassold et al., 1984; Kupke and Muller, 1989). Therefore, it is not surprising that we found an increased frequency of hyperhaploidy for chromosome 21.

The highly significant increase in the frequency of hyperhaploidy for the sex chromosomes in human sperm is interesting since in many studies of meiosis I in human males, the X and Y chromosomes are often unpaired (Laurie and Hulten, 1985). It is possible that these univalents might predispose to nondisjunction of the sex chromosomes in males. There are four common aneuploidies for the sex chromosomes in humans: 45,X, 47,XYY, 47,XXY and 47,XXX. Recent evidence has demonstrated that the majority of these originate from a paternal error, unlike autosomal aneuploidies, in which maternal errors predominate (Hassold et al.,1988; Jacobs et al.,1988).

Sperm chromosome studies in men with a constitutional chromosomal abnormality

Men with a constitutional chromosomal abnormality have an increased risk of chromosomally abnormal children and spontaneous abortions in their partners. The ability to karyotype sperm for the first time has allowed detailed analysis of the segregation of chromosomes in men carrying a chromosomal rearrangement such as a translocation or inversion. Twenty-eight men have been analyzed from five different laboratories and this has provided our first glimpse of how these rearrangements affect meiosis (Table V). Most of the information has come from men heterozygous for a translocation, therefore I will discuss these results.

Table V. Chromosome sperm analysis in men with constitutional chromosomal abnormalities

Type of Abnormality		Laboratory(Reference)	% Unbalanced Sperm
Translocations			
Reciprocal	t(5;18)	Balkan and Martin, 1983	19
	t(5;13)	Pellestor et al, 1989	23
	t(5;11)	Burns et al, 1986	30
	t(6;14)	Balkan and Martin, 1983a	32
	t(2;5)	Templado et al, 1988	32
	t(4;7)	Pellestor et al, 1989	43
	t7;14)	Martin et al, 1990b	47
	t(6;7)	Pellestor et al, 1989	49
	t12;20)	Martin, 1990a	53
	t(4;6)	Martin et al, 1990b	54
	t(2;9)	Martin et al, 1990b	57
	t(1;2)	Templado et al, 1990	59
	t(9;10)	Martin, 1988a	60
	t(3;16)	Brandriff et al, 1986	63
	t(8;15)	Brandriff et al, 1986	63
	t(9;18)	Pellestor et al, 1989	66
	t(7;14)	Burns et al, 1986	70
	t(11;22)	Martin, 1984	77
Robertsonian	t(13;14)	Pellestor et al, 1987	8
	t(13;15)	Pellestor et al, 1990	10
	t(14;21)	Balkan and Martin, 1983b	13
	t(13;14)	Pellestor et al, 1987	27
Inversions			
Pericentric inv (3)		Balkan et alk, 1983	0
Paracentric inv (7)		Martin, 1986a	0
Pericentric inv (3)		Martin, 1990	31
Others			
Accessory marker chrom.		Martin et al, 1986b	0*
Fragile site 10q25		Martin, 1986b	Unknown
47,XYY		Benet and Martin, 1988	0

* 50% with marker chromosome

All theoretical segregations were observed among the reciprocal translocations, even though many of these are rarely observed in studies of human newborns and spontaneous abortions (Table VI). This demonstrates that these segregations are produced but lost early in embryonic development. The frequency of unbalanced sperm varied from 19% to 77% in the reciprocal translocations and 8% to 27% in the Robertsonian translocations. This decrease in the frequency of abnormalities in Robertsonian translocations compared to reciprocal translocations is interesting since males and females carrying a reciprocal translocation have an equivalent risk of producing chromosomally abnormal offspring whereas males carrying a Robertsonian translocation have a lower risk of producing an unbalanced child compared to female carriers (Boue and Gallano, 1984). Results from sperm chromosome studies suggest that this lower risk for male Robertsonian translocation carriers comes about by a meiotic mechanism which produces a low frequency of unbalanced sperm. It is possible that this pairing of Robertsonian translocations is more common during male meiosis as this type of pairing is more conducive to alternate segregation which yields chromosomally normal and balanced gametes.

Aneuploid offspring have often been observed in individuals who have a constitutional chromosomal anomaly and researchers have suggested that these individuals may have an increased risk of aneuploid children because of an "interchromosomal effect" (Aurias, 1978). It is thought that rearranged or extra chromosomes could disrupt the normal pairing and disjunction of homologous chromosomes. Human sperm chromosome analysis can be utilized to study this possibility of aneuploidy caused by an interchromosomal effect. Of 28 men with a constitutional chromosomal abnormality who have been studied by sperm karyotyping, two had an increased frequency of aneuploid sperm. All of the other men had a frequency of aneuploidy within the normal range. The one translocation carrier who had an increased frequency of aneuploidy was remarkable because he was heterozygous for two different reciprocal translocations (Burns et al., 1986). This man had an extremely high frequency of numerical chromosomal abnormalities, unrelated to the translocation. It is possible that the presence of two reciprocal translocations greatly disrupted pairing and crossing-over at meiosis, causing the anomalous segregations that were observed. The other man who had a significantly increases frequency of aneuploid sperm had a small bisatellited accessory marker chromosome (Martin et al., 1986b). It is interesting that in this case the chromosomes present as a numerical abnormality were all small chromosomes and thus consistent with a distributive pairing effect. Thus, it appears that an interchromosomal effect may occur for some constitutional chromosomal abnormalities but the data from sperm chromosome analysis suggest that it does not occur in most cases, at least not causing an enormous effect as in the double translocation carrier. However, the number of karyotypes analyzed have been small in most cases and there could still be an undetected doubling or tripling of the aneuploidy frequency.

Table VI. Segregation of chromosomes in sperm of reciprocal translocations

Segregations	Mean	Range
Alternate	49%	23-78
Adjacent 1	38%	16-23
Adjacent 2	8%	0-24
3:1	5%	0-25
4:0	0.2%	0-4

Sperm Selection

A central question to the utility of this technique in analysis of the segregation of chromosomes during meiosis is the possible selection of sperm by the system. Studies in mice and hamsters demonstrate that chromosome constitution does not influence the ability of sperm to affect fertilization (Epstein and Travis 1979; Sonta et al.,1984). The fact that we have aneuploidy in all chromosome groups, including small and big chromosomes (even chromosome number 1) argues against selection since one would expect sperm with less genomic imbalance if selection occurred. Similarly it is striking that in translocation carriers even sperm with a 4:0 chromosome segregation have been reported and the range in the frequency of abnormalities in the sperm of translocation heterozygotes is enormous. These studies demonstrate that chromosomally abnormal sperm are not at a disadvantage in fertilizing hamster eggs.

Subjecting sperm to different conditions also does not appear to affect the frequency of chromosomal abnormalities. We have recently shown that cryopreservation of sperm does not alter the sex ratio or the frequency of numerical and structural abnormalities (Chernos et al., 1989). Brandriff et al., (1986) have demonstrated that sperm selected for high motility (by the "swim-up" technique) do not have a statistically significant difference in the frequency of sperm chromosomal abnormalities and the sperm morphology in 30 donors (Martin and Rademaker, 1988). These studies, although indirect, all support the hypothesis of no sperm selection based on chromosomal content in this system.

ACKNOWLEDGMENTS

Sincere thanks go to Lynne Bell for preparing the manuscript and to Leona Barclay, Evelyn Ko and Kathy Hildebrand for expert technical assistance. Renee Martin is an Alberta Heritage Foundation for Medical Research Scientist, whose research is funded by the Medical Research Council of Canada and the Alberta Children's Hospital Foundation.

REFERENCES

Aurias, A., Peieur, M., Dutrillaux, B., Lejeune, J., 1978, Systematic analysis 95 reciprocal translocations of autosomes, Hum. Genet., 45:259-282.

Balkan, W., Burns, K., Martin, R.H., 1983, Sperm chromosome analysis of a man heterozygous for a pericentric inversion of chromosome number 3, Cytogenet. Cell Genet., 35:295-297.

Balkan, W., Martin, R.H., 1983a, Chromosome segregation into the spermatozoa of two men heterozygous for different reciprocal translocations, Hum. Genet., 63:345-348.

Balkan, W., Martin, R.H., 1983b, Segregation of chromosomes into the spermatozoa of a man heterozygous for a 14;21 Robertsonian translocation, Am. J. Med. Genet., 16:169-172.

Benet, J., Martin, R.H., 1988, Sperm chromosome complements in a 47,XYY man, Hum. Genet., 78:313-315.

Beckett, T.A., Martin, R.H. and Hoar, D.I., 1989, Assessment of the sephadex technique for selection of X chromosome-bearing human sperm by analysis of sperm chromosomes, deoxyribonucleic acid and Y-bodies, Fertil. Steril., 52:829-835.

Brandriff, B.F., Gordon, L.A., Aaendel, S., Ashworth, L.K., Carrano, A., 1986, The chromosomal constitution of human sperm selected from motility, Fertil. Steril., 46:686-690.

Brandriff, B., Gordon, L., Ashworth, L., Watchmaker, G., Moore, D., Wyrobek, A.J., Carrano, A.V., 1985, Chromosomes of human sperm: variability among normal individuals, Hum. Genet., 70:18-24.

Brandriff, B., Gordon, L., Ashworth, L.K., Littman, V., Watchmaker, G., Carrano, A.V., 1986, Cytogenetics of human sperm: meiotic segregation in two translocation carriers, Am. J. Hum. Genet., 38:197-208.

Brandriff, B.F., Gordon, L.A., Haendel, S., Singer, Moore DHII., Gledhill, B.L., 1986, Sex chromosome ratios determined by karyotypic analysis in albumin-isolated human sperm, Fertil. Steril., 46:678-685.

Boue, A., Gallano, P., 1984, A collaborative study of the segregation of inherited chromosome structural rearrangements in 1356 prenatal diagnosis, Prenat. Diagn., 4:45-67.

Burns, J.P., Koduru, P.R.K., Alonso, M.L., Chaganti, R.S.K., 1986, Analysis of meiotic segregation in a man heterozygous for two reciprocal translocations using hamster in vitro penetration system, Am.J. Hum. Genet., 38:467-478.

Chernos, J.E., and Martin, R.H., 1989, A cytogenetic investigation of the effects of cryopreservation on human sperm, Am. J. Hum. Genet., 45:766-777.

Epstein, C.J., and Travis, B., 1979, Preimplantation lethality of monosomy for mouse chromosome 19, Nature 280:144-145.

Hassold, T., Benham, F., Leppert, M., 1988, Cytogenetics and molecular analysis of sex chromosome monosomy, Am. J. Hum. Genet., 42:534-541.

Hassold, T., Chiu, D., Yamane, J.A., 1984, Parental origin of autosomal trisomies. Ann. Hum. Genet., 48:129-144.

Jacobs, P., Hassold, T., Whittington, E., Butler, G., Collier, S., Keston, M., Lee, M., 1988, Klinefelter's syndrome: an analysis of the origin of the additional sex chromosome using molecular probes, Ann. Hum. Genet., 52;93-109.

Jenderny, J., Rohrborn, G., 1987, Chromosome analysis of human sperm. 1. First results with a modified method, Hum. Genet., 76:385-388.

Juberg, R.C., Mowrey, P.N., 1983, Origin of non-disjunction in trisomy 21 syndrome: all studies compiled, parental age analysis and international comparisons, Am. J. Med. Genet., 16:111-116.

Kamiguchi, Y., Mikamo, K., 1986, An improved, efficient method for analysing human sperm chromosomes using zona-free hamster ova, Am. J. Hum. Genet., 38:724.

Kupke, K.G., Muller, U., 1989, Parental origin of the extra chromosome in Trisomy 18, Am. J. Hum. Genet., 45:599-605.

Laurie, D.A., Hulten, M.A., 1985, Further studies on bivalent chiasma frequency in human males with normal karyotypes, Ann Hum Genet., 49:189-201.

Martin, R.H., Balkan, W., Burns, K., and Lin, C.C., 1982, Direct chromosomal analysis of human spermatozoa: Results from 18 normal men, Am. J. Hum. Genet., 34(3):459-468.

Martin, R.H., 1983, A detailed method for obtaining preparations of human sperm chromosomes, Cytogenet. Cell Genet., 35:253-256.

Martin, R.H., Balkan, W., Burns, K., Rademaker, A.W., Lin, C.C., Rudd, N.L., 1983, The chromosome constitution of 1000 human spermatozoa, Hum. Genet., 63:305-309.

Martin, R.H., 1984, Analysis of human sperm chromosome complements from a male heterozygous for a reciprocal translocation t(11;22)(q23q11), Clin. Genet., 25:357-361.

Martin, R.H., 1986a, Sperm chromosome analysis in a man heterozygous a paracentric inversion of chromosome 7(q11q22), Hum Genet., 73:97-100.

Martin, R.H., 1986b, A fragile site 10q25 in human sperm chromosomes, J. Med. Genet., 23;279.

Martin, R.H., Hildebrand, K., Yamamoto, J., Rademaker, A., Barnes, M., Douglas, G., Arthur, K., Ringrose, T., Brown, I.S., 1986a, An increased frequency of human sperm chromosomal abnormalities after radiotherapy, Mut. Res., 174:219-225.

Martin, R.H., Hildebrand, K.A., Yamamoto, J., Peterson, D., Rademaker, A.W., Taylor, P., Lin, C.C., 1986b, The meiotic segregation of human sperm chromosomes in two men with accessory market chromosomes, Am. J. Med. Genet., 25;381-388.

Martin, R.H., Rademaker, A.W., Hildebrand, K., Long-Simpson, L., Peterson, D., Yamamoto, J., 1987a, Variation in the frequency and type of sperm chromosomal abnormalities among normal men, Hum. Genet., 77:108-114.

Martin, R.H., and Rademaker, A.W., 1987b, The effect of age on the frequency of sperm chromosomal abnormalities in normal men, Am. J. Hum. Genet., 41:484-492.

Martin, R.H., and Rademaker, A., 1988, The relationship between sperm chromosomal abnormalities and sperm morphology in humans, Mut. Res., 207:159-164.

Martin, R.H., Ko, E., and Rademaker, A., 1988, Human sperm chromosome complements after microinjection of hamster eggs, J. Reprod. Fertil., 84:179-186.
Martin, R.H., 1988a, Meiotic segregation of human sperm chromosomes in translocation heterozygotes: Report of a t(9;10)(q34q11) and a review of the literature, Cytogenet. Cell Genet., 47:48-51.
Martin, R.H., 1988b, Cytogenetic analysis of sperm from a male heterozygous for a 13;14 Robertsonian translocation, Hum. Genet., 80:357-361.
Martin, R.H., Rademaker, A., Hildebrand, K., Barnes, M., Arthur, K., Ringrose, T., Brown, I.S., and Douglas, G., 1989, A comparison of chromosomal aberrations induced by in vivo radiotherapy in human sperm and lymphocytes, Mut. Res., 226:21-30.
Martin, R.H., 1990, A cytogenetic study of the frequency of recombined chromosomes in sperm from a man heterozygous for an inversion of chromosome 3 (inv(3)(p25;q21)), Am. J. Hum. Genet., (in press).
Martin, R.H., McGillivray, B., Barclay, L., Hildebrand, K., Ko, E., Fowlow, S.B., 1990a, Sperm chromosome analysis in a man heterozygous for a reciprocal translocation 46,XYt(12;20)(q24.3;q11), Hum. Repro., 5:606-609.
Martin, R.H., Barclay, L., Hildebrand, K., Ko, E., Fowlow, S.B., 1990b, Cytogenetic analysis of 400 sperm from three translocation heterozygotes, Hum. Genet., (in press).
Martin, R.H., Ko, E., Rademaker, A., 1990c, The distribution of aneuploidy in human gametes: comparison between human sperm and oocytes, Am. J. Med. Genet., (in press).
Navarro, J., Templado, C., Benet, J., Lange, K., Rajmil, O., Egozcue, J., 1990, Sperm chromosome studies in an infertile man with partial complete asynapsis of meiotic bivalents, Hum. Repro., 5:227-229.
Pellestor, F., Sele, B., Jalbert, H., 1987, Chromosome analysis of spermatozoa from a male heterozygous for a 13:14 Robertsonian translocation, Hum. Genet., 76:116-120.
Pellestor, F., and Sele, B., 1989, Etude cytogenetique du sperme humain, Medecin & Sciences, 5:244-251.
Pellestor, F., Sele, B., Jalbert, H., Jalbert, P., 1989, Direct segregation analysis of reciprocal translocations: a study of 283 sperm karyotypes from four carriers, Am. J. Hum. Genet. 44: 464-473.
Pellestor, F., 1990, Analysis of meiotic segregation in a man heterozygous for a 13;14 Robertsonian translocation and a review of the literature, Hum. Genet., 85:49-54.
Rosner, B., 1986, "Fundamentals of Biostatistics," 2nd Ed, p 222, Duxbury Press, Boston.
Rudak, E., Jacobs, P.A., Yanagimachi, R., 1978, Direct analysis of the chromosome constitution of human spermatozoa, Nature, 274:911-913.
Sele, B., Pellestor, F., Jalbert, P., Estrade, C., 1985, Analyse cytogenetique des pronucleus a partir du modele de fecondation interspectifique homme hamster, Ann. Genet., 28:81.
Sonta, S., Fukui, K., Yamamura, H., 1989, Selective elimination of chromosomally unbalanced zygotes at the two-cell stage in the Chinese hamster, Cytogenet. Cell Genet., 38:5-13.

Tarkowski, A.K., 1966, An air-drying method for chromosome preparation from mouse eggs, Cytogenetics, 5;394-400.

Templado, C., Navarro, J., Benet, J., Genesca, A., Mar Penez, M., and Egozcue, J., 1988, Human sperm chromosome studies in a reciprocal translocation t(2;5), Hum. Genet., 79:24-28.

Templado, C., Navarro, J., Requena, K., Bene, J., Ballesta, F., Egozare, J., 1990, Meiotic and sperm chromosome studies in a reciprocal translocation t(1;2)(q32;q36), Hum. Genet., 84:159-162.

Ueda, K.U., Yanagimachi, R., 1987, Sperm chromosome analysis as a new system to test human X- and Y-sperm separation, Gamete Res., 17:221-228.

Watt, J.L., Templeton, A.A., Messinis, I., Bell, L., Cunningham, P., Dunca, R.O., 1987, Trisomy 1 in an eight cell human preembryo, J. Med. Genet., 24:60-64.

CHROMOSOME ANALYSIS OF OOCYTES AND EMBRYOS

Michelle Plachot

U. 173 INSERM
Hopital Necker
149, Rue de Sevres
75743 Paris Cedex 15, France

INTRODUCTION

As an extension of IVF techniques, preimplantation genetic diagnosis will make it possible to detect genetic and chromosomal diseases in embryos. With the use of molecular genetic technology, such as Polymerase Chain Reaction (PCR), several gene mutations have already been detected in polar bodies aspirated from mature oocytes or in blastomeres from biopsied embryos (Verlinsky, this volume). The same approach was used for gender determination in cases at high risk of X-linked diseases to avoid the birth of affected males (Handyside, this volume; Milayeva, this volume).

The diagnosis of chromosome abnormalities in the preimplantation stage is to date far from being as successful, although chromosome imbalance is the major cause of abnormal embryonic development and pre- and post-implantation embryonic loss. While the incidence of chromosomal disorders is 0.6% in liveborns (Nielsen, 1975), a hundred-fold increase is found in spontaneous abortions, (60%) (Boue and Boue, 1976) essentially as a result of meiotic non-disjunctions, fertilization anomalies, or mitotic non-disjunctions.

OOCYTE MEIOTIC NON-DISJUNCTIONS

Although proceeding from the same meiotic division process, oocytes are responsible for non-disjunctions much more often than spermatozoa. 62% of D and G trisomies observed in liveborns occur in metaphase I and 18% in metaphase II for oocytes, whereas only 12% in metaphase I and 8% in metaphase II for sperm.

Various techniques were employed for obtaining metaphase plates from human oocytes, the air-drying method of Tarkowski (1966) being the basic one: after a 6-10 min. hypotonic shock in 0.5% sodium citrate, oocytes are fixed on a slide with several drops of methanol/acetic acid (3:1), dried by gently blowing, and stained for 6 min. by 4% Giemsa in Sorensen's buffer.

Preimplantation Genetics, Edited by Y. Verlinsky and
A. Kuliev, Plenum Press, New York, 1991

The pooled data of cytogenetic studies in about 1200 mature human oocytes obtained from hyperstimulated ovaries show that the current estimates of chromosome abnormalities vary widely from as little as 4.5% to as high as 65%, with an average of 23.2% (Michelmann and Mettler, 1985; Plachot et al, 1986, 1988b; Martin et al, 1986; Wramsby et al, 1987; Pellestor and Sele, 1988; Bongso et al, 1988; Van Blerkom and Henry, 1988; Djalali et al, 1988; Papadopoulos et al, 1989a; Ma et al, 1989). These abnormalities include 14.8% nullisomies, 5.2% disomies, 3% diploid metaphase II oocytes, and 0.2% structural rearrangements (Table 1).

The most frequent anomaly is non-disjunction, with a great tendency for loss or gain of satellite chromosomes from groups D and G. While meiotic non-disjunction should produce an equal number of (n+1) (disomic) and (n-1) (nullisomic) gametes, the proportion of hypohaploid and hyperhaploid oocytes appeared to be not equal, hypohaploidy occurring approximately three times as often as hyperhaploidy. Since the excess of hypohaploid complements may be due to a technical loss of chromosomes during oocyte fixation, the incidence of non-disjunctions is generally evaluated by doubling the frequency of hyperhaploid oocytes (which is 5.2% in these pooled data), leading to a total of 13.6% of chromosome anomalies in oocytes. But this does not take into account the possibility of anaphase lag which is a non-artefactual loss of chromosomes. Using the heterospecific sperm penetration assay (zona-free hamster oocytes inseminated with human sperm), Martin (1984) demonstrated that male pronuclei show an equal frequency of hyperhaploid and hypohaploid complements, while a significant excess of hypohaploid complements is observed in female hamster pronuclei (Figure 1). As chromosomal preparations of both male and female pronuclei were performed under similar conditions within the same fertilized oocytes, it would be difficult to attribute the excess of female hypohaploidy to a technical artefact. Therefore, oogenesis seems to be more susceptible to chromosome loss than spermatogenesis.

The second most frequent anomaly due to non-disjunction is the presence of diploid metaphase II oocytes which may lead to triploidy after fertilization. In fact, about 6% of fertilized oocytes after IVF are triploid, although 83% of these are a result of dispermy involving an impairment of the cortical granule reaction rather than a meiotic disorder (Plachot et al, 1988).

The fact that structural rearrangements were observed rarely may be due to the difficulty of detecting minor structural anomalies in meiotic chromosomes, which are usually short, curly and contracted, presenting difficulties for obtaining banding patterns.

Although a selected population, oocytes unfertilized after insemination show no higher incidence of chromosome anomalies than freshly recovered uninseminated oocytes, the range being from 4 to 50% (Martin et al, 1986; Wramsby et al, 1987; Van Blerkom and Henry, 1988).

Table 1. Total incidence and conservative estimation of chromosome abnormalities in 1187 human oocytes.

	Total incidence		Conservative estimation
Nullisomies	14.8 %		
Disomies	5.2 %	x2	10.4 %
Diploid metaphase II	3.0 %		3.0 %
Structural anomalies	0.2 %		0.2 %
	23.2 % (a)		13.6 % (b)

Effect of clinical and biological parameters on chromosome anomalies in oocytes

Among the parameters possibly involved in the occurrence of chromosome imbalance in human, or more generally in mammalian oocytes, the following are discussed below: maternal age, stimulation treatment, and oocyte aging.

Effect of maternal age. The influence of maternal age on meiotic non-disjunctions, leading to increased incidence of chromosome anomalies in spontaneous abortions (Boue and Boue, 1973) or in newborns (Hook 1981) has been clearly demonstrated for most trisomies, while no effect was noted for monosomy-X, triploidy, tetraploidy, and structural aberrations. Paternal age is not involved in the occurrence of chromosome anomalies.

However, the data available are conflicting with regard to the rate of aneuploidy in oocytes observed, which could be due to a small size of a series impairing statistical significance. In our study we found an increased rate of aneuploidy in patients over 35 (38%) compared to younger patients (24%) (Plachot et al, 1988b, Bongso et al, 1988), whereas no effect of maternal age was reported by Pellestor and Sele (1988) and Djalali et al (1988).

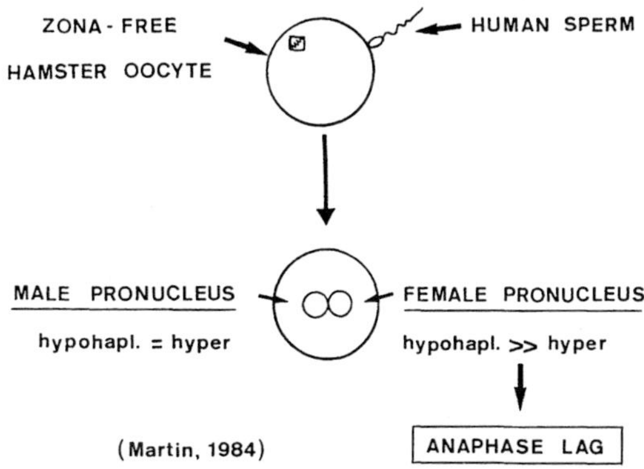

Figure 1. The heterospecific sperm penetration assay shows an excess of hypohaploid complement in female pronucleus.

In fact, non-disjunction during meiosis is facilitated when the number of chiasmata is low, as in small chromosomes. According to the "production line" hypothesis, non-disjunction seems to be predetermined by the order of formation of oocytes during fetal development: the oocytes which are formed last with fewer chiasmata and more univalent pairs are the last to be ovulated, late in life (Gindoff et Jewelewicz, 1986).

Effect of stimulation treatment. It has been reported in humans that exogenous gonadotropins increase the frequency of chromosomal anomalies in spontaneous abortions (83%), compared to patients having no induced ovulation (61%), such adverse effect disappearing after 2 months of discontinuation of gonadotropins (Boue and Boue, 1973). Unfortunately, no information is available on the chromosome complements of human oocytes recovered in spontaneous cycles to appraise the real advantages or eventual disadvantages of stimulated cycles for IVF. Our comparison of different types of stimulation treatment, namely the administration of luteinizing hormone releasing hormone (LH-RH) analogs in combination with human menopausal gonadotropin (hMG) and the ovulation induction without LH-RH analogs, i.e. Clomid/hMG or hMG alone, did not reveal significant differences in meiotic disorders: 33% and 26% accordingly.

Effect of oocyte aging. Several lines of evidence indicate the importance of a synchronized maturation and fertilization of the oocytes. Fertilization is spontaneously delayed in 1% of the cases of IVF, pronuclei being first observed 42 hours after gamete mixing, followed by cleavage 24 hours later. This delayed fertilization of mature oocytes induces oocyte aging in vitro, which is known to induce chromosome anomalies in rats and mice. We observed a drastically high rate of chromosome abnormalities after fertilization delay (87%) compared to undelayed fertilization (29.2%) (Plachot et al, 1988a). Such anomalies could be the consequence of the changes in the cytoskeleton and in the organization of microtubules leading to abnormal spindles (Webb et al, 1986). Thus, it may be recommended that oocytes in which fertilization is delayed should not be transferred to avoid the possible increase in the incidence of chromosomally abnormal fetuses.

FERTILIZATION ABNORMALITIES

Cytogenetic analysis of so called unfertilized oocytes and the observation of pronuclei the day after insemination revealed several fertilization abnormalities such as premature chromosome condensation (PCC), parthenogenesis and triploidy.

Premature chromosome condensations. Our cytogenetic observation of so-called unfertilized oocytes revealed the presence of paternal chromosomes in some of the oocytes (Plachot et al, 1987b), suggesting that they have obviously been fertilized (Figure 2). These chromosomes were single stranded since condensation occurred in G_1 phase, i.e. before the S phase which takes place in fully grown pronuclei. The overall incidence of paternal PCC among 301 "unfertilized"

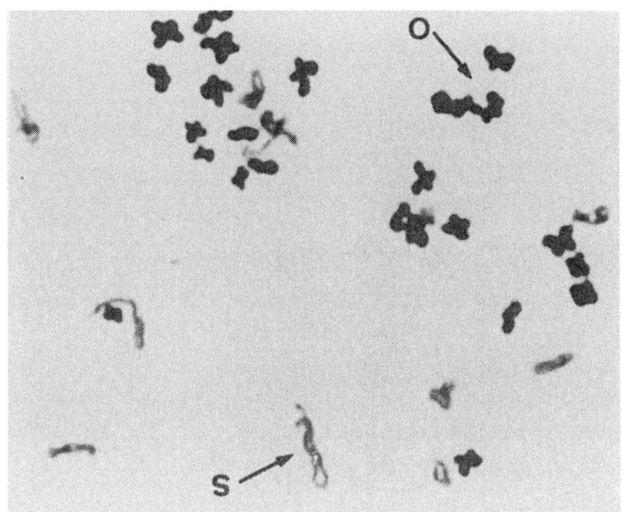

Figure 2. Premature chromosome condensation of sperm chromosomes in supposed unfertilized oocytes.

O: Oocyte chromosomes
S: Sperm chromatids

oocytes was as high as 18%, 1 out of 5 having more than one sperm chromosome complement (up to 8), thus pointing to a high rate of polyspermy. The incidence of PCC in metaphase I oocytes was higher (34%) than in metaphase II (14%), probably due to an abnormal persistence of condensation factors normally present at the time of germinal vesicle breakdown.

This abnormality is the result of the consecutive failures both in oocyte activation following sperm penetration and in full pronuclei growth and DNA synthesis. Regarding metaphase II oocytes, we observed neither the effect of IVF indication nor of maternal age, but a slight increase following ovarian stimulation with FSH (30%) compared to LH-RH agonists/hMG (9%) (the latter stimulation protocol routinely have 9% false fertilization failures).

Parthenogenesis. The fact that 1.6% of human oocytes display a single pronucleus the day after insemination provides the possibility for analysis of parthenogenesis in humans (Plachot et al, 1988b; 1989b). The chromosome complements of embryos resulting from such oocytes are not homogeneous: 46% are haploid (as expected), 29% are diploid as a result of non- extrusion of the second polar body or early diploidisation, and 25% are mosaic N/2N in case of late diploidisation (Table 2). When cultured for 5 days, 77% of parthenogenetically activated oocytes were able to divide, 20% of them reaching the blastocyst stage. Their morphology, the rate of division, and the capacity of in vitro development were identical to those of normally fertilized embryos. However, this anomaly was lethal during or following implantation.

Table 2. Chromosome complement of embryos resulting from parthenogenetic oocytes.

Haploid	46 %
Diploid	29%
Mosaics N/2N	25 %

Triploidy. 1.5% of all conceptions occurring after *in vivo* fertilization, and 20% of chromosomally abnormal spontaneous abortions are triploid. The incidence of triploid zygotes is about 4 times higher after IVF than after *in vivo* fertilization.

The karyotype of embryos resulting from tripronucleate eggs was rather unexpected: 34% were triploid, 29% -diploid/triploid mosaics, 5% -haploid and 32% -diploid (Table 3). The haploid and diploid embryos might appear due to the exclusion of 2 or 1 pronuclei, respectively, from the first mitotic division, probably induced by a synchronous pronuclei growth and migration.

In contrast to bipronucleate eggs which first divide into 2 cells and then into 4 cells through a bipolar spindle, half of tripronucleate ova first divide into 3 cells and then into 6 cells probably through a tripolar spindle.

These observations allowed to propose a model for the development of tripronucleate eggs, in accordance with different karyotypes and different way of cleavage (Kola et al, 1987, Plachot et al, 1989b). Four types of events appear to occur:

- First, all 69 chromosomes gather on the metaphase plate and proceed to divide regularly, producing daughter cells with a full triploid complement. Such embryos could develop to the first trimester of pregnancy and exceptionally to birth.

- Second, spindle formation at the first cleavage division may be tripolar leading directly to a three-cell embryo. In this situation, movement of the chromosomes to the poles would be disorganized, leading to nuclei containing various chromosome complements and therefore to mosaicism.

- Third, one of the 3 pronuclei could be entirely excluded from the metaphase plate at the first cleavage. In case of exclusion of an extra male pronucleus or an extra female pronucleus, a normal diploid two-cell embryo could be produced, probably able to develop normally.

- Conversely, in case of exclusion of a female pronucleus from a dispermic egg, androgenetic embryo will result, leading to an early cleavage arrest or to a hydatidiform mole because of a lack of maternal contribution.

CHROMOSOME ANALYSIS OF PREIMPLANTATION EMBRYOS

Throughout the world, 363 embryos from 2 to 8 cells, have been karyotyped, the overall incidence of cytogenetic

Table 3. Chromosome complement of embryos resulting from tripronucleate eggs.

Triploid : 34 %
Mosaic 2N/3N : 29 %
Diploid : 32 %
Haploid : 5 %

disorders ranging from 23 to 71% (Angell et al, 1986; Plachot et al, 1987a, 1988b; Veiga et al, 1987; Wimmers et Van der Merwe, 1988; Papadopoulos et al, 1989b).

This wide variation is mainly due to a small size of samples analyzed, as well as due to the different sources from which the spare embryos were obtained for studies: normal embryos or fragmented ones rejected from transfer or freezing.

A high rate of mosaicism appeared to be the most important feature in early embryos (7%) (Plachot et al, 1989b) meaning that the karyotype of a single cell might not always reflect the karyotype of the embryo, presenting the source of a potential misinterpretation.

CHROMOSOME ANALYSIS OF POST IMPLANTATION EMBRYOS

A European survey, carried out under ESHRE's auspices, made it possible to undertake karyotyping of 34 spontaneous abortions after IVF (Plachot, 1989a). Twenty one (62%) conceptuses were chromosomally abnormal: 20 had de novo anomalies and 1 was transmitted by the mother who had a t(9;14) translocation. The types of chromosomal abnormalities were as follows: 14 trisomies (chromosomes from A, D, E and G groups were involved), 1 double trisomy (48,XX, +17, +22), 3 monosomy X, 1 triploidy and 1 tetraploidy.

No differences were observed either in the frequency or in the types of anomalies between _in vivo_ and _in vitro_ fertilization, except for a decrease in the incidence of triploidy after _in vitro_ fertilization, probably due to the exclusion of detected tripronucleate zygotes from the transfer (Table 4).

CONCLUSION

Several lines of evidence indicate that meiotic nondisjunction is a major factor responsible for the high rate of chromosome imbalance in human embryos. Since the incidence of chromosome anomalies in liveborns after IVF (0.6%) is not augmented compared to _in vivo_ fertilized conceptuses, a drastic selection must occur, probably either during embryo cleavage or during the first trimester fetal development (Figure 3).

With the improvement of cytogenetic techniques preimplantation diagnosis of chromosomal disturbances will certainly increase the efficiency of IVF by preventing the transfer of affected embryos.

Table 4. Chromosomally abnormally spontaneous abortions occurring after in-vitro or in-vivo conceptual cycles (Plachot et al., 1989a).

		Fertilization in vitro			Boue and Boue (1976), fertilization in vivo
		No.		%	%
Monosomy	45,X	3		14.2	15.3
Trisomy	A +	1	14	66.6	52
	D +	6			
	E +	2			
	G +	5			
Double trisomy		1		4.8	1.7
Triploidy	69,XXX	1		4.8	19.9
Tetraploidy		1		4.8	6.2
Translocation		1		4.8	3.8
Mosaicism		0		0	1.1

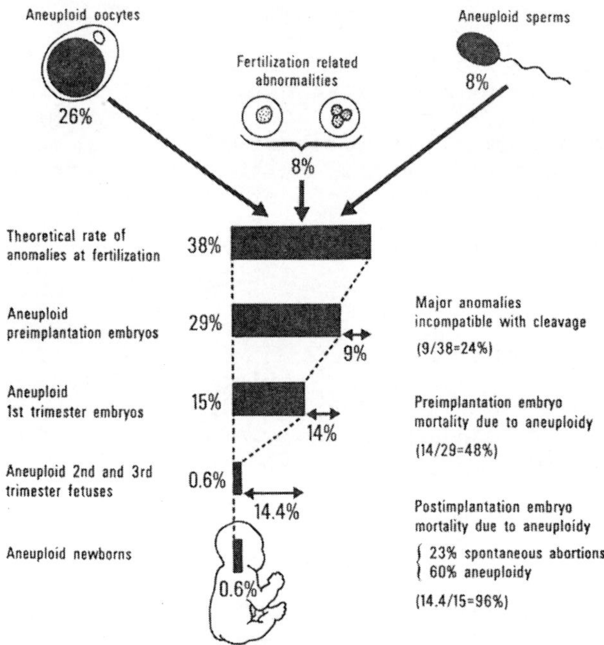

Figure 3. Model-deduced from IVF-of Natural Selection Against Chromosome Anomalies

REFERENCES

Angell R., Templeton A.A., Aitken R.J. (1986). Chromosome studies in human in vitro fertilization. Hum. Genet., 72, 333 -339.

Bongso A., Chye N.S., Ratnam S., Sathananthan H., Wong P.C. (1988). Chromosome anomalies in human oocytes failing to fertilize after insemination in vitro. Hum. Reprod., 3, 645 -649.

Boue J.G., Boue A. (1973). Increased frequency of chromosomal anomalies in abortions after induced ovulation. Lancet, 7804, 679 - 680.

Boue J.G. and Boue A. (1976). Chromosomal anomalies in early spontaneous abortion. (Their consequences on early embryogenesis and in vitro growth of embryonic cells). Curr. Top. Pathol., 62, 193 - 208.

Djalali M., Rosenbusch B., Wolf M., Sterzik K. (1988). Cytogenetics of unfertilized human oocytes. J. Reprod. Fert., 84, 647 - 652.

Gindoff P.R., Jewelewicz R. (1986). Reproductive potential in the older woman. Fertil. Steril., 46, 989 - 1001.

Hook E.B. (1981). Rates of chromosome abnormalities at different maternal ages. Obstet. Gynecol., 58, 282.

Kola I., Trounson A., Dawson G., Rogers P. (1987). Tripronuclear human oocytes: altered cleavage patterns and subsequent karyotypic analysis of embryos. Biol. Reprod., 37, 395 - 401.

Ma S., Yuen B.H., Kalousek D.K., Gomel V., Zouves C., Moon Y.S. (1989). Chromosome analysis of human oocytes failing to fertilize in vitro. Fertil. Steril., 51, 992 - 997.

Martin R.H. (1984). Comparison of chromosomal abnormalities in hamster egg and human sperm pronuclei. Biol. Reprod., 31, 819 - 825.

Martin R.H., Mahadevan M.M., Taylor P.J., Hildebrand K., Long-Simpson L., Peterson D., Yamamoto J., Fleetham J. (1986). Chromosomal analysis of unfertilized human oocytes. J. Reprod. Fert., 78, 673 - 678.

Michelmann H.W., Mettler L. (1985). Cytogenetic investigations on human oocytes and early human embryonic stages. Fertil. Steril., 43, 320 - 322.

Nielsen J. (1975). Chromosome examination of newborn children. Purpose and ethical aspects. Humangenetik, 26, 215 - 222.

Papadopoulos G., Randall J., Templeton A.A. (1989a). The frequency of chromosome anomalies in human unfertilized oocytes and uncleaved zygotes after insemination in vitro. Hum. Reprod., 4, 568 - 573.

Papadopoulos G., Templeton A.A., Risk N., Randall J. (1989b). The frequency of chromosome anomalies in human preimplantation embryos after in vitro fertilization. Hum. Reprod., 4, 91 - 98.

Pellestor F., Sele B. (1988). Assessment of aneuploidy in the human female by using cytogenetics of IVF failures. Am. J. Hum. Genet., 42, 274 - 283.

Plachot M., Junca A.M., Mandelbaum J., Grouchy J. De, Salat-Baroux, J., Cohen J. (1986). Chromosome investigations in early life. I. Human oocytes recovered in an IVF programme. Hum. Reprod., 1, 547 - 551.

Plachot M., Junca A.M., Mandelbaum J., Grouchy J. de, Salat-Baroux J., Cohen J. (1987a). Chromosome investigations in early life. II. Human preimplantation embryos. Hum. Reprod., 2, 29 - 35.

Plachot M., Grouchy J. de, Junca A.M., Mandelbaum J., Turleau C., Couillin P., Cohen J., Salat-Baroux J. (1987). From oocyte to embryo: a model, deduced from in vitro fertilization, for natural selection against chromosome abnormalities. Ann. Genet., 30, 22 - 32,

Plachot M., Grouchy J. de, Junca A.M., Mandelbaum J., Cohen J. (1988a). Chromosome analysis of human oocytes and embryos: does delayed fertilization increase chromosome imbalance? Hum. Reprod., 3, 125 - 127.

Plachot M., Veiga A., Montagut J., Grouchy J. de, Calderon G., Lepretre S., Junca A.M., Santalo J., Carles E., Mandelbaum J., Barri P., Degoy J., Cohen J., Egozcue J., Sabatier J.C., Salat-Baroux J., (1988b). Are clinical and biological IVF parameters correlated with chromosomal disorders in early life: a multicenter study. Hum. Reprod., 3, 627 - 635.

Plachot M. (1989a). Chromosome analysis of spontaneous abortions after IVF. A European survey. Hum. Reprod., 4, 425 -429.

Plachot M., Mandelbaum J., Junca A.M., Grouchy J. de, Salat-Baroux J., Cohen J. (1989b). Cytogenetic analysis and developmental capacity of normal and abnormal embryos after IVF. Hum. Reprod., 4 supp., 99 - 103.

Tarkowski A.K. (1966). An air - drying method for chromosome preparation from mouse eggs. Cytogenetics, 5, 394 - 400.

Van Blerkom J., Henry G. (1988). Cytogenetic analysis of living human oocytes: cellular basis and developmental consequences of perturbations in chromosomal organization and complement. Hum. Reprod., 3, 777 - 790.

Veiga A., Calderon G., Santalo J., Barri P.N., Egozcue J. (1987). Chromosome studies in oocytes and zygotes from an IVF programme Hum. Repro., 2, 425 - 430.

Webb M., Howlett S.K., Maro B. (1986). Parthenogenesis and cytosqueletal organization in ageing mouse eggs. J. of Embryol. Exp. Morph., 95, 131 - 145.

Wimmers M.S.E., Van der Merwe J.V. (1988). Chromosome studies on early human embryos fertilized in vitro. Hum. Reprod., 3, 894 - 900.

Wramsby H., Fredga K., Liedholm P. (1987). Chromosome analysis of human oocytes recovered from preovulatory follicles in stimulated cycles. New Engl. J. of Med., 316, 121 - 124.

PREIMPLANTATION DIAGNOSIS OF GENETIC DISEASE USING ENZYME ASSAYS

Peter Braude

Department of Obstetrics and Gynecology, United Medical and Dental Schools of St. Thomas's and Guy's Hospitals, Lambeth Palace Road London SE1 7EH

INTRODUCTION

A number of severe abnormalities are caused by a genetically inherited enzyme defect (table 1)[1]. In many cases, the inherited disease manifests early in childhood with severe physical and mental handicap, and often results in the demise of the child before its teens. Many couples who carry such genetic diseases already have one affected child and thus are eager to take advantage of methods which will allow antenatal diagnosis of the condition during a subsequent pregnancy. However, a number of these women, whether for religious or personal motives, cannot accept termination of an affected pregnancy and would avail themselves willingly of methods to diagnose the genetic disease prior to conception.

Table 1

Genetic diseases caused by enzyme abnormalities

	syndrome	deficient enzyme
GM1 Gangliosidoses		β galactosidase
GM2 Ganglisidoses	Tay-Sachs disease	hexosaminidase
	Fabry disease	galactosidase A
	Gaucher disease	β glucosidase
	M Leucodystrophy	aryl sulfatase A
	Globoid leucodystrophy	gal-cerase
	Nieman-Pick disease	sphyngomyelinase
Mucopolysaccharidoses	Hurler syndrome	α -L-iduronidase
	Hunter syndrome	iduronate sulfatase
Purine salvage diseases	Lesch Nyhan syndrome	hypoxanthine-prt'ferase
	Severe Combined Immunodeficiency Disease	adenine deaminase

Preimplantation Genetics, Edited by Y. Verlinsky and A. Kuliev, Plenum Press, New York, 1991

For preimplantation diagnosis of genetic disease to be effective, three important criteria need to be fulfilled. First, methods of embryo biopsy must be efficient and safe. In mice, single blastomeres have been removed from 4- and 8-cell embryos[2], and biopsies have been taken by micro-manipulation from blastocyst stages[3], with encouraging success rates for pregnancy following transfer of the biopsied embryos. In the human it has been demonstrated that removal of cells from early cleavage stages interferes minimally with blastocyst development[4], and is compatible with survival of the embryo after replacement[5]. Second, there should be a good chance of pregnancy following embryo replacement. Currently, "take home baby" rates following IVF and embryo replacement are low, in the order of 8%-12% per treatment cycle[6,7]. However, pregnancy success rates for infertile couples may not necessarily be comparable with the chances of pregnancy for fertile couples embarking on IVF solely for the purposes of preimplantation diagnosis of genetic disease. In the single therapeutic series reported so far, results are encouraging with 4 out of 17 embryos replaced after biopsy resulting in an ongoing pregnancy[5]. Third, appropriate methods for diagnosing the genetic disease must be available. Genetic disease can be diagnosed at one of two levels (Fig.1): directly from DNA, or from the synthesized product of the gene. For diagnosis at a DNA level, an appropriate probe must be available for the defective gene to be identified by in situ hybridization, or by restriction fragment length polymorphism analysis[8]. In addition to this, the amount of DNA that is available for analysis from the tiny biopsy that can be taken safely from the cleavage stage embryo is miniscule. Although amplification of the DNA using the polymerase chain reaction[9-11] can overcome this difficulty[12], faithfulness of amplification and contamination are serious confounding problems. Alternatively, the genetic disease may be diagnosed at a post-transcriptional level by measuring the product of the messenger RNA translation. In general, the product is synthesized in such minute amounts that it cannot be identified directly. However, if the product is an enzyme, microassays are available for their detection, provided that the appropriate substrate for the enzyme reaction is known[13]. These enzyme assays are only useful in preimplantation diagnosis of genetic disease if the embryonic enzyme activity can be distinguished clearly from the activity of any maternal enzyme carried over in the egg.

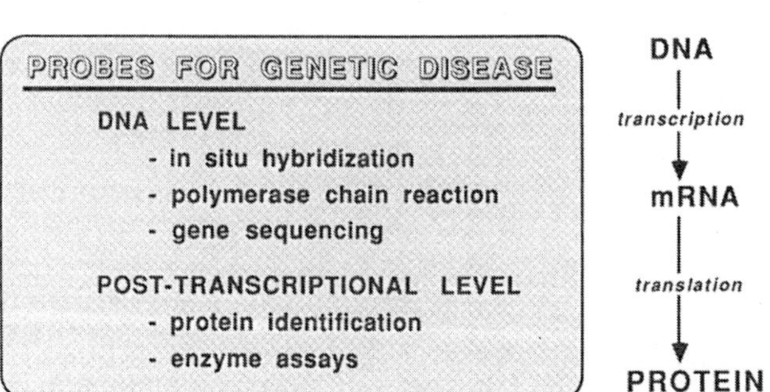

Figure 1. Levels of diagnosis of genetic disease.

The embryonic genome has been shown to activate during early cleavage, although the exact stage at which this activation occurs varies between species[14]. In the mouse[15,16] and the gerbil[17] it occurs at the mid 2-cell stage, and in the cow[18] and the goat[19] between the 8 and 16 cell stages. As in the sheep[20], activation of the human embryonic genome seems to occur between the 4- and 8-cell stage of development[21,22]. Prior to this time, cleavage is dependent on mRNA synthesized during oogenesis and inherited in the egg. This maternal mRNA seems to be degraded simultaneously with activation of the embryonic genome in the mouse, but no data are available about the persistence of maternal mRNA in the human.

We can postulate two different patterns of expression of enzyme activity operating in the early human embryo. Firstly, maternal enzyme could be present and active only in very small amounts in the egg. Thus, until the stage of embryonic genome activation, measurable activity in early cleavage stages would be constant and low (Fig.2a). If the enzyme to be measured is synthesized on embryonic templates during the first wave of embryonic gene activity, enzyme activity might be expected to increase after the time of genome activation, and would exceed greatly the activity of any maternal enzyme present. Thus, measurement of enzyme activity during this post transcriptional phase of development should be able to differentiate cells with abnormally low amounts of enzyme activity due to a defective gene. Alternatively, enzyme activity in the oocyte may be high. Hence, during the early cleavage stages, this post-transcriptional maternally inherited enzyme activity may overshadow newly synthesized embryonically coded enzyme activity (Fig.2b). If the half life of the enzyme is long, it could even exceed embryonically coded activity throughout the preimplantation period. Our experience in measuring hypoxanthine phosphoribosyl transferase (HPRT) in early human embryos serves to highlight these alternative mechanisms.

The Lesch Nyhan syndrome is a recessively inherited sex linked disease caused by a deficiency of HPRT. The syndrome manifests with spastic cerebral palsy, mental retardation and severe self mutilation[23,24]. Boys affected with this disease usually die before the age of ten. In the mouse, measurable HPRT activity increases substantially after the 8-cell stage, approximately 2 cleavage divisions following gene activation[25]. Prevention of this rise in activity by the transcriptional inhibitor α-amanitin, demonstrates that the increase is due to transcription and new enzyme synthesis. Thus, in the mouse, HPRT activity measured from the 8-cell onwards, will reflect embryonic rather than maternally coded activity. However, HPRT activity in the human embryo seems to be different. Measurement of HPRT activity in 46 preimplantation human embryos cultured <u>in vitro</u> up to the blastocyst stage failed to reveal any significant rise[26] (Fig.3a). Indeed, the average levels of activity for 4-cell, morula and blastocyst stages were similar to those found in unfertilized eggs (Fig.3b). This steady state of activity could indicate that measurable enzyme activity was wholly maternal in origin, or that any new synthesis that was occurring, was matched by an equivalent breakdown in

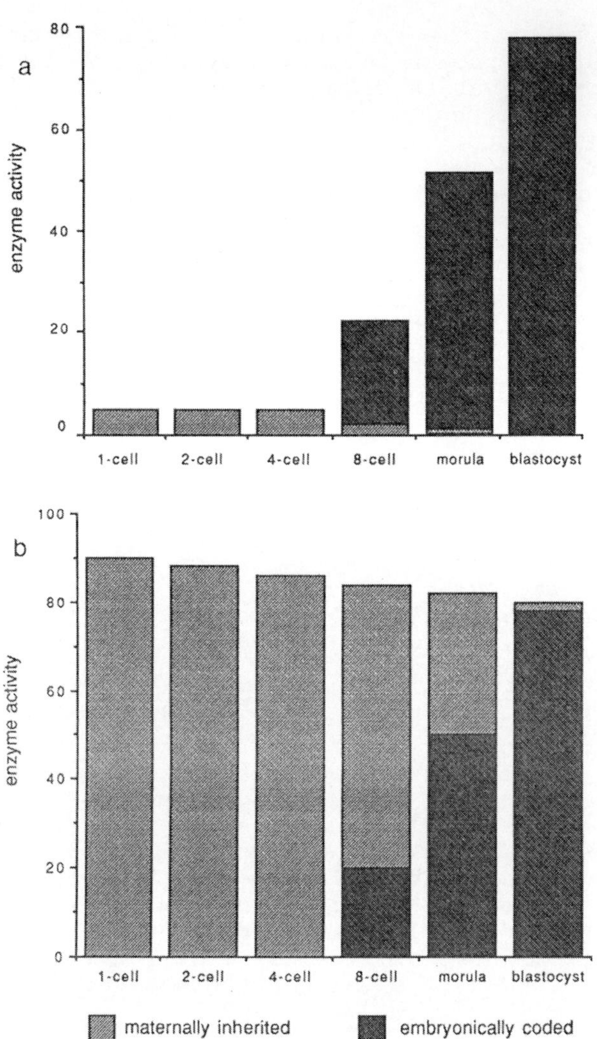

Figure 2. Changes in measurable enzyme activity during early cleavage for an enzyme (a) with low endogenous activity in the egg and coded for mainly by the embryonic genome; and (b) with high maternally inherited activity.

maternally derived enzyme. Since exposure of 4-cell or morula stage embryos to α-amanitin failed to produce any significant decrease in enzyme activity, and the levels of activity in the early cleavage stages are not significantly different from those in the oocyte, it would be appropriate to conclude that measured enzyme activity during the *in vitro* period is wholly maternal in origin. In addition to these disappointing findings, levels of enzyme activity were found to vary substantially between batches of embryos, individual embryos, and blastomeres taken from the same embryo[26]. Although this might have been due in part to variation in this sensitive micro-assay, it is now clear that many cleavage stage human embryos cultured *in vitro* are highly abnormal, and have deficient karyokinesis and cytokinesis[27]. Thus, even if embryonic activity could be distinguished from maternal activity, it would be difficult

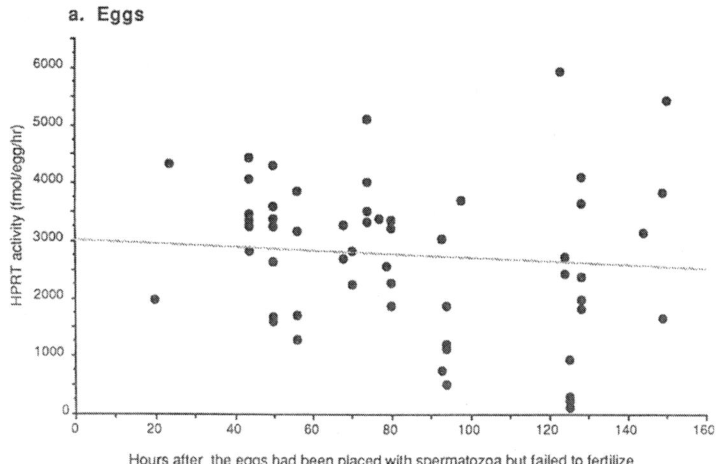

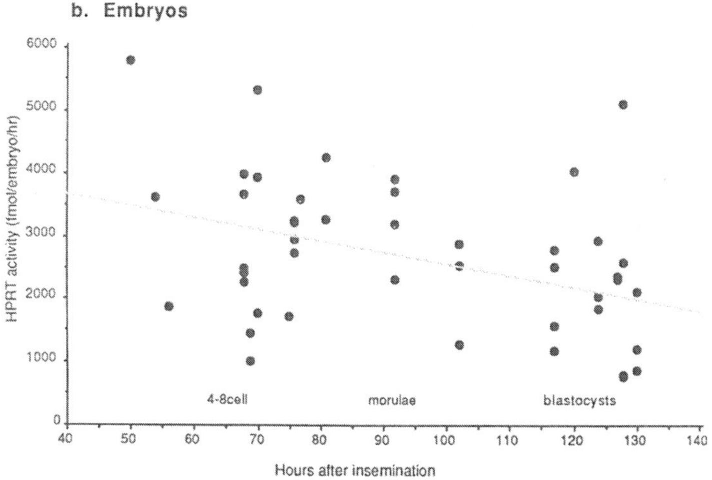

Figure 3. HPRT activity measured in single human (a) unfertilized oocytes and (b) cleavage stage preimplantation embryos cultured in vitro. The mean regression line for each graph is shown. For experimental details see (26).

to be sure what an appropriate "normal" level for an embryo or biopsy should be. It would appear, therefore, that preimplantation diagnosis of Lesch Nyhan syndrome is unlikely to be successful using these methods. Sadly, diagnosis of Tay Sachs disease by microassay of hexosaminidase is proving equally disappointing (I.Liebaers & A.Van Steirteghem, personal communication).

These results serve to highlight important general points about preimplantation diagnosis of genetic disease. Firstly, results obtained using animal models may not necessarily be applicable to the human. Although it is essential that new methods are tried initially on the preimplantation embryos of other mammalian species, direct extrapolation of results is dangerous, and intermediate experiments on "spare" human preimplantation embryos should be undertaken before the techniques are considered for clinical application[28].

Secondly, although the technology for genetic assay may be available[29], and results can be obtained from single cell biopsies[30], this feasibility does not necessarily mean that the methods are accurate enough or reliable enough to be applied clinically. Meticulous experiments need to be carried out to demonstrate reproducibility and accuracy before these methods are applied to a rather vulnerable group of patients.

REFERENCES

1. A. Milunsky, "Genetic disorders and the fetus," Plenum, New York, (1986).
2. M. Monk, A. Handyside, K. Hardy and D. Whittingham, Preimplantation diagnosis of deficiency of hypoxanthine phophoribosyl transferase in a mouse model for Lesch-Nyhan syndrome, Lancet, ii:423 (1987).
3. M. Monk, A.L. Muggleton-Harris, E. Rawlings and D.G. Whittingham, Pre-implantation diagnosis of HPRT-deficient male and carrier female mouse embryos by trophectoderm biopsy., Human Reproduction, 3:377 (1988).
4. K. Hardy, A.H. Handyside, K. Martin, H. Leese and R.M.L. Winston, Effects of pre-implantation biopsy on the human blastocysts, Development, (1989).
5. A. H. Handyside, E.H. Kontogianni, K. Hardy and R.M.L. Winston, Pregnancies from human preimplantation embryos sexed by Y-specific DNA amplification, Nature, 344:768 (1990).
6. ILA, The Fifth Report of the Interim Licensing Authority For Human In Vitro Fertilisation and Embryology, (1990).
7. American Fertility Society, In vitro fertilization-embryo transfer in the United States: 1988 results from the IVF-ET Registry, Fertil. Steril., 53:13 (1990).
8. R. J. Akhurst, The use of gene probes in studying human reproduction and embryology, Human Reproduction, 1:213 (1986).
9. H. A. Erlich, D.H. Gelfand and R.K. Saiki, Specific DNA amplification, Nature, 331:461 (1988).
10. R. K. Saiki, D.H. Gelfand, S. Stoffel, S.J. Scharf, R. Higuchi, G.T. Horn, K.B. Mullis and H.A. Erlich, Primer-directed enzymatic amplification of DNA with a thermostable DNA polymerase, Science, 239:487 (1988).
11. H. Li, U.B. Gyllensten, X. Cui, R.K. Saiki, H.A., Erlich and N. Arnheim, Amplification and analysis of DNA sequences in single human sperm and diploid cells, Nature, 335:414 (1988).
12. A. H. Handyside, J.K. Pattinson, R.J.A. Penketh, J.D.A. Delhanty, R.M.L. Winston and E.G.D. Tuddenham, Biopsy of human pre-embryos and sexing by DNA amplification, Lancet i:347 (1989).
13. M. Monk, in "Mammalian Development. A Practical Approach," M. Monk, ed., IRL Press, Oxford (1987).
14. J. Tesarik, Developmental control of human preimplantation embryos: a comparative approach, J. In Vitro Fertil. Embryo Transfer, 5:347 (1988).

15. G. Flach, M. Johnson, P. Braude, R. Taylor and V. Bolton, The transition from maternal to embryonic control in the 2-cell mouse embryo, EMBO.J., 1:681 (1982).
16. P. R. Braude, H.R.B. Pelham, G. Flach and R. Lobatto, Post transcriptional control in the early mouse embryo, Nature 282:102 (1979).
17. M. L. Norris, S.C. Barton and M.A.H. Surani, Changes in protein synthesis during early cleavage of the Mongolian gerbil embryo, J. Exp. Zool., 236:149 (1985).
18. R. E. Frei, G.A. Schulz and R.B. Church, Qualitative and quantitative changes in protein synthesis occur at the 8-16 cell stage of embryogenesis in the cow, J. Reprod. Fert., 86:637 (1989).
19. D. Sakkas, P.A. Batt and A.W.N. Cameron, Development of preimplantation goat embryos in vivo and in vitro, J. Reprod. Fert., 87:359 (1989).
20. I. M. Crosby, F. Gandolfi and R.M. Moor, Control of protein synthesis during early cleavage of sheep embryos, J. Reprod. Fert., 82:769 (1988).
21. P. R. Braude, V.N. Bolton and S. Moore, Human gene expression first occurs between the four- and eight-cell stages of preimplantation development, Nature, 332:459 (1988).
22. J. Tesarik, V. Kopecny, M. Plachot and J. Mandelbaum, Early morphological signs of embryonic genome expression in human preimplantation development as revealed by quantitative electron microscopy, Devl. Biol., 128:15 (1988).
23. W. L. Nyhan, The Lesch-Nyhan syndrome, Ann. Rev. Med., 24:41 (1973).
24. M. Lesch and W.L. Nyhan, A familial disorder of uric acid metabolism and central nervous system function, Am. J. Med., 36:561 (1984).
25. M. Harper and M. Monk, Evidence for translation of HPRT enzyme on maternal mRNA in early mouse embryos, J. Embryol. Exp. Morphol., 74:15 (1983).
26. P. R. Braude, M. Monk, S.J. Pickering, A. Cant and M.H. Johnson, Measurement of HPRT activity in the human unfertilized oocyte and pre-embryo, Prenat. Diag., 9:839 (1989).
27. N. J. Winston, P.R. Braude, S.J. Pickering, M.A. George, A. Cant, J. Currie and M.H. Johnson, The incidence of abnormal morphology and nucleo-cytoplasmic ratios in 2,3 and 5 day human pre-embryos., Hum. Reprod., (in press): (1991).
28. P. R. Braude and M.H. Johnson, Embryo Research - yes or no, Brit. Med. J., 299:1349 (1989).
29. Y. Verlinsky, E. Pergament and C. Strom, The preimplantation genetic diagnosis of genetic diseases, J. in Vitro Fertil. Embryo Transfer, 7:1 (1990).
30. M. Monk and C. Holding, Amplification of a ß-haemoglobin sequence in individual human oocytes and polar bodies, Lancet, 335:985 (1990).

SINGLE SPERM PCR ANALYSIS - IMPLICATIONS FOR PREIMPLANTATION GENETIC DISEASE DIAGNOSIS

Norman Arnheim[1], Honghua Li[1], Xiangfeng Cui[1] and William Navidi[2]

Departments of Molecular Biology[1] and Mathematics[2]
University of Southern California
Los Angeles, California 90089-1340

INTRODUCTION

The polymerase chain reaction (PCR; Saiki et al., 1985, 1988; Mullis and Faloona, 1987) is a method of selective _in vitro_ gene amplification. The principle of the PCR method is shown in Figure 1. Two small stretches of DNA of known sequence that flank the target region to be amplified (Figure 1a) are used to design two oligonucleotide primers. The synthetically made primers are chosen so that one is complementary to one flanking sequence while the other is complementary to the other flanking sequence. The 3' hydroxyl ends of the primers face the target sequence. Following DNA denaturation of the target, the single stranded primers hybridize to their complementary flanking sequences (Figure 1b). In the presence of a DNA polymerase the primers will be extended through the target sequence (Figure 1c). DNA denaturation, primer hybridization and DNA polymerase extension represent one PCR cycle. If the first cycle is followed by a second one (Figure 1d), more copies of the target sequence will be made. The major product of PCR is a DNA fragment which is exactly equal in length to the sum of the lengths of the two primers and the target DNA. Production of copies of the target sequence is exponential with respect to cycle number. Since the amount of target approximately doubles with each cycle, as few as 20 cycles will generate about a million times more target sequence than is present initially.

The ability of PCR to make hundreds of millions to billions of copies of a specific DNA segment makes it especially valuable for analyzing DNA sequences from sources where only very small amounts of cells or tissues can be obtained. In fact, the first application of PCR was to the prenatal diagnosis of sickle cell anemia (Saiki et al., 1985). Soon after its original development, PCR was shown to have the ultimate sensitivity in that a single molecule of DNA present in a single sperm cell was shown to be

Preimplantation Genetics, Edited by Y. Verlinsky and A. Kuliev, Plenum Press, New York, 1991

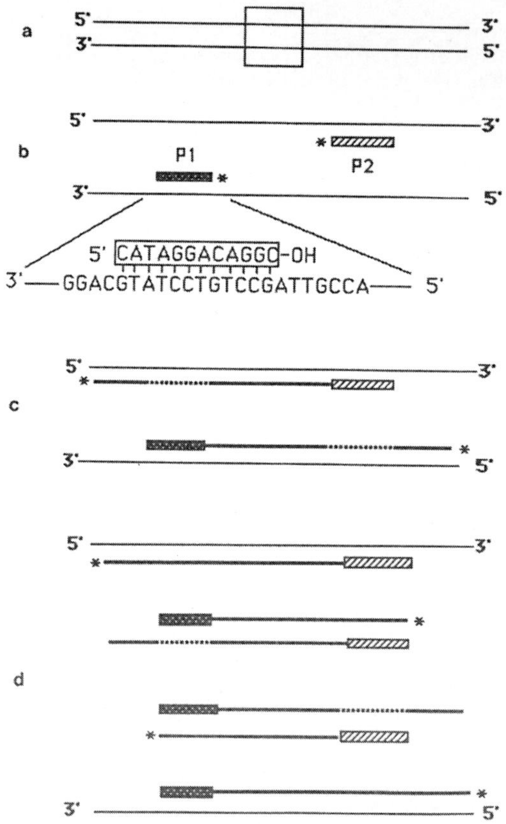

Figure 1. Principle of the Polymerase Chain Reaction. a). The boxed target sequence is shown within a doubled stranded DNA molecule. The 5' and 3' orientations of the two single strands is indicated. b) The two PCR primers, P1 and P2, are shown annealed to the sequences flanking the target after the DNA has been denatured. The 3' end of the primer undergoing elongation by DNA polymerase is denoted by a *. Shown below primer P1 are the details of the base pairing between the primer (boxed) and the DNA strand. c). Result of the extension of the two primers by DNA polymerase. The region of the extension product of each primer which is complementary to the other primer is shown by a broken line. 1.c). The second cycle of PCR. Each of the four DNA strands shown in part 1.c, above, can anneal to a primer which is subsequently extended. Note that at the completion of this second cycle there are four double-stranded copies of the target that was originally present (part a) as a single double-stranded molecule. Note also that two of the eight single stranded products are equal in length to the length of the two primers and the intervening target. Products of this size accumulate exponentially as additional cycles are carried out. [Reprinted with permission of the American Institute of Biological Sciences from Arnheim et al., 1990b].

amplified (Li et al., 1988). This same study and that of Jeffreys et al. (1988) also demonstrated that unique gene sequences in a single diploid cell could also be analyzed using PCR. This ability to study DNA sequences in a single haploid or diploid cell led to the idea that genetic disease diagnosis could be carried out in a human embryo produced by _in vitro_ fertilization prior to implantation (Li et al., 1988; Handyside et al., 1989; Coutelle et al., 1989). Handyside et al. (1989) showed that single blastomeres can be removed from human embryos at the 6 or 8 cell stage and highly repeated Y chromosome sequences can be detected by PCR (Handyside et al., 1989). In their report, a large fraction of the sampled human embryos developed into later stages _in vitro_, suggesting the possibility that the diagnosed embryos could be implanted in the mother's uterus for further development. Recently, pregnancies from biopsied embryos sexed by Y chromosome-specific DNA amplification have been reported and appear to be normal by ultrasound examination (Handyside et al., 1990). Studies similar to these have also been carried out in mice (Holding and Monk, 1989; Gomez et al., 1990; Bradbury et al., 1990).

Preimplantation genetic disease diagnosis could be made even before fertilization by analyzing DNA sequences in the first polar body accompanying the oocyte (Strom et al., 1990; Monk and Holding, 1990). If the polar body from a woman heterozygous for a disease gene is found to contain the mutant allele, the oocyte can be assumed to contain the normal allele and thus fertilized and implanted. However, this approach will only be useful for disease genes that are within 50 cM of the centromere. If a gene is 50 cM or more away from the centromere a diagnosis is not possible since both oocyte and polar body will always be heterozygous.

SINGLE CELL PCR ANALYSIS

Successful preimplantation diagnosis depends upon reliably being able to determine the genotype of a single cell using PCR. Over the past few years we have been developing methods to carry out this procedure with high efficiency and minimal error (Cui et al., 1989; Li et al., 1990; Arnheim et al., 1990). Our most recent system allows us to amplify DNA in a single sperm and to determine the allelic state of the chromosome based upon the size of the PCR product itself using gel electrophoresis (Li et al., 1990). These experiments are carried out in two stages. A preliminary amplification step amplifies the target sequences from each locus of interest in a single cell. The second step involves allele-specific amplification for each locus separately. Considering a locus with two alleles which differ by a single base substitution, two PCR primers can be designed where each primer is identical to one allele and differs from the other allele by a single base substitution at its 3' end (Figure 2). Under the appropriate PCR conditions the extension by DNA polymerase of the completely matched primer will be more efficient than the extension of the primer with a mismatch at the 3' end. To distinguish between the PCR products from the two alleles, the allele-specific primers are constructed so that they differ in length by 15 base pairs and therefore their products are of a different size and can be distinguished by gel electrophoresis. Using this method we have analyzed three independent genetic loci (parathyroid hormone, PTH; G-gamma globin, $^{G}\gamma$; Low density lipoprotein receptor, LDLr) simultaneously from single sperm samples (Li et al., 1990).

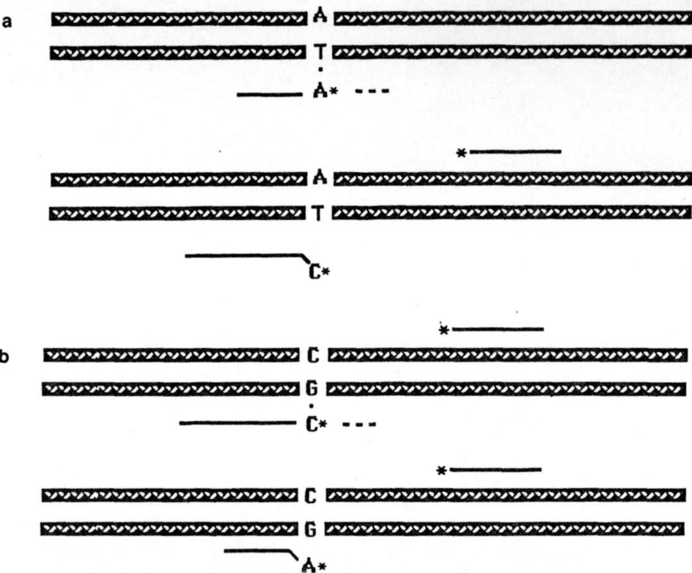

Figure 2. Two possible PCR targets are shown which are identical in sequence except for an A/T (a) to G/C (b) substitution. PCR allele-specific primers are used. Under the appropriate conditions only the primer with the perfectly matched 3' end is extended by DNA polymerase and incorporated into the PCR product (dashed lines). Since the lengths of the two primers are different (———A* is shorter than ———————C*), the PCR products also differ in size and can be distinguished by gel electrophoresis.

Data from Li et al. (1990) are shown in Figure 3. Because we are using a sperm system where each cell, as a result of meiosis, is expected to contain only one of two alleles present in the donors somatic tissue, our data provide the strongest kind of evidence that we are in fact analyzing the amplification products from a single DNA molecule.

RISK ASSESSMENT FOR PREIMPLANTATION DIAGNOSIS

PCR, like any other technique, is subject to experimental error. The application of PCR to preimplantation genetic disease diagnosis therefore should be accompanied by some estimate of the risk that the ultimate diagnosis will be incorrect. Our experience with PCR analysis of single sperm cells has given us an indication of the levels of errors that can be expected when analyzing single cells. Using estimates of these errors we can make risk assessments.

Our best estimates of errors associated with single cell PCR comes from a study on 708 single sperm samples (Cui et al., 1989). Statistical analysis of the data reveals the following error rates: (1) on the average there is about a 5-10% chance that a locus present in a tube will not be amplified to a detectable level; (2) every sample has a less than 5% probability of being contaminated.

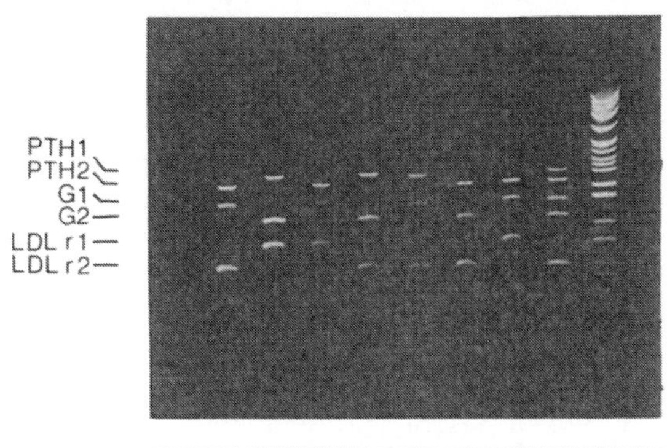

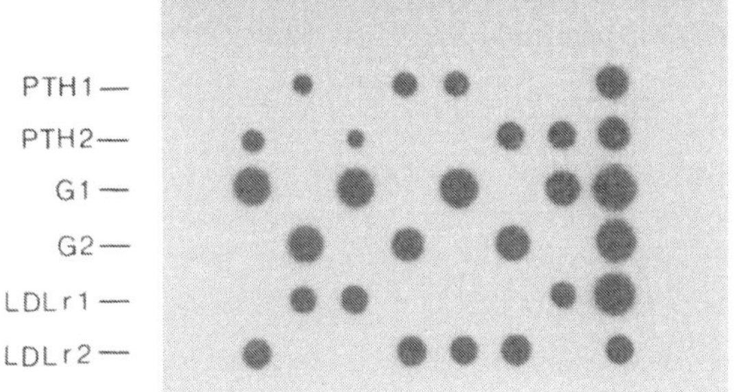

Figure 3. Determination of the allelic state at three loci in single sperm. The initial rounds of amplification of each sperm contained three primer pairs; one pair each for each locus. Following this, 3 aliquots of each single sperm PCR reaction were taken. Each aliquot was further amplified using the allele-specific primers. Upper panel: Electrophoresis after PCR using allele-specific primers. Lane 1: Negative PCR control which received all reagents except a sperm. Lanes 2-8, Single sperm samples. Lane 9, products from a 3 μl sample of semen from the triply heterozygous sperm donor. Lane 10, pBR322 digested with Msp I. The sizes of the allele-specific products are: 1) Alleles PTH1 and PTH2, 172 and 157 base pairs respectively. 2) G-$^{G}\gamma$ globin alleles G1 and G2, 139 and 124 base pairs respectively. 3) LDLr1 and LDLr2 alleles, 106 and 91 base pairs respectively. Lower panel: Determination of the allelic state using the allele-specific hybridization approach to confirm the electrophoretic results. The aliquots from each sample are arranged in a column below the electrophoretic results and the genotype of each sample is determined by the presence or absence of the radioactive dot after hybridization. (Reprinted with permission of the National Academy of Sciences from Li et al., 1990).

These PCR typing errors can lead to errors in determining the genotype of a sample. We have considered the effect of these errors on preimplantation genetic disease diagnosis involving embryo biopsy and polar body analysis. Errors in typing can be divided into three categories, reflecting varying degrees of severity. The least serious error ("acceptable") is one which results in using an embryo or oocyte that should have been used anyway or not using one which should have rejected. Thus, in the case of a recessive disease an example of such an error is implanting an embryo of genotype **Aa** which was typed **AA** by mistake. Given the limited number of embryos or oocytes available a somewhat more undesirable error is that which would lead to not using an embryo or oocyte which could have been used. Thus, for recessive diseases, no serious consequences would result by not using an **AA** embryo which was typed as **aa** by mistake. We call this kind of errors "tolerable". The most serious error, which we consider "unacceptable", is that resulting in the implantation of an embryo which should not be implanted.

In Tables 1-3 we summarize the errors in embryo biopsy and polar body analysis for autosomal recessive diseases. The calculations are based upon the efficiency of an allele being amplified to a detectable level (r), the probability of contamination (c) and, in the case of polar body typing, the recombination fraction Θ between the disease gene and the centromere. Table 1 categorizes the possible blastomere typing errors and gives the conditional probabilities for their occurrence in the case of an autosomal recessive disease when both parents are heterozygous for the disease gene **(a)**. Calculations are made for several values of r and c. The column labeled "Primary Cause of Error" shows which kind of PCR error, efficiency or contamination, is primarily responsible for each incorrect diagnosis. The probabilities are conditional on the predicted genotype. Thus, the table tells us that if r=0.9 and c=0.05, a blastomere which has been typed **Aa** has a probability of about 1.5% of in fact being of genotype **aa** which, if implanted, would produce an unacceptable result.

Table 2 categorizes the various polar body typing errors in the case of an autosomal recessive disease. The errors in the second and fourth rows are classified as "potentially unacceptable". In these cases, an oocyte containing the disease allele **a** is used and whether, after fertilization, the resulting embryo is the unacceptable genotype **aa** is a matter of chance and depends upon whether the fertilizing sperm carried an **a** or **A**. Table 3 gives conditional probabilities for unacceptable errors for various values of r, c and Θ.

Table 1

ERRORS IN BLASTOMERE TYPING FOR AUTOSOMAL RECESSIVE DISEASES								
					CONDITIONAL PROBABILITY OF ERROR			
TRUE GENOTYPE	OBSERVED GENOTYPE	USE?	ERROR CATEGORY	PRIMARY CAUSE OF ERROR	(1) r=.8 c=0	(2) r=.9 c=.05	(3) r=.9 c=0	(4) r=1 c=0
AA	Aa	YES	ACCEPTABLE	c	.00	.015	.00	.00
AA	aa	NO	TOLERATED	c	.00	.00022	.00	.00
Aa	AA	YES	ACCEPTABLE	r	.25	.15	.15	.00
Aa	aa	NO	TOLERATED	r	.25	.15	.15	.00
aa	AA	YES	UNACCEPTABLE	c	.00	.00022	.00	.00
aa	Aa	YES	UNACCEPTABLE	c	.00	.015	.00	.00

Table 2

ERRORS IN POLAR BODY TYPING FOR AUTOSOMAL RECESSIVE DISEASES						
POLAR BODY		OOCYTE				
TRUE GENOTYPE	OBSERVED GENOTYPE	TRUE GENOTYPE	DEDUCED GENOTYPE	USE?	ERROR CATEGORY	PRIMARY CAUSE OF ERROR
AA	Aa	aa	Aa	NO	ACCEPTABLE	c
AA	aa	aa	AA	YES	POTENTIALLY UNACCEPTABLE	c
Aa	AA	Aa	aa	NO	ACCEPTABLE	r
Aa	aa	Aa	AA	YES	POTENTIALLY UNACCEPTABLE	r
aa	AA	AA	aa	NO	TOLERATED	c
aa	Aa	AA	Aa	NO	TOLERATED	c

Table 3

CONDITIONAL PROBABILITIES OF UNACCEPTABLE ERRORS IN POLAR BODY TYPING FOR AUTOSOMAL RECESSIVE DISEASES					
	PRIMARY CAUSE OF ERROR	(1) r=.8 c=0	(2) r=.9 c=.05	(3) r=.9 c=0	(4) r=1 c=0
Polar Body, aa observed, θ=0	c	.00	.00013	.00	.00
Polar Body, aa observed, θ=.10	r	.019	.011	.011	.00
Polar Body, aa observed, θ=.20	r	.045	.027	.027	.00
Polar Body, aa observed, θ=.25	r	.063	.039	.038	.00
Polar Body, aa observed, θ=.30	r	.083	.054	.054	.00
Polar Body, aa observed, θ=.40	r	.14	.11	.11	.00

Comparing the probabilities in the last two rows of table 1 with the probabilities in table 3 shows that for autosomal recessive diseases, and for the same values of r and c, unacceptable results occur less frequently with blastomere typing than with polar body typing unless θ is quite small or unless r and c are both quite high.

Tables 4-6 consider the risks associated with preimplantation diagnosis of autosomal dominant diseases when one parent is affected **(Aa)** and the other normal **(aa)**. Table 4 gives the conditional probabilities for blastomere analysis. Comparison with table 1 shows that unacceptable errors are more likely with dominant diseases. Tables 5 and 6 categorize and give conditional probabilities of unacceptable error for polar body analysis when the mother is **Aa**. With polar body typing, errors are twice as likely when the disease is dominant than when the disease is recessive. In general (compare the first row of table 4 with table 6) blastomere typing errors are less frequent than polar body typing errors if the recombination fraction θ is greater than about 0.25. Otherwise polar body typing errors are less frequent.

Table 4

ERRORS IN BLASTOMERE TYPING FOR AUTOSOMAL DOMINANT DISEASES

TRUE GENOTYPE	OBSERVED GENOTYPE	USE?	ERROR CATEGORY	PRIMARY CAUSE OF ERROR	CONDITIONAL PROBABILITY OF ERROR (1) r=.8 c=0	(2) r=.9 c=.05	(3) r=.9 c=0	(4) r=1 c=0
Aa	aa	YES	UNACCEPTABLE	r	.14	.084	.083	.00
aa	Aa	NO	TOLERATED	c	.00	.027	.00	.00

We have recently carried out more extensive calculations of the errors associated with embryo biopsy and polar body analysis (Navidi and Arnheim, manuscript submitted). This new work takes into consideration a third type of error in which the probability that a polar body or blastomere is actually deposited into the PCR reaction tube is estimated. This study also considers X-linked genetic disorders typed by PCR of Y chromosome markers in addition to autosomal diseases. The number of embryos or oocytes that have to be examined to be 95% sure that one of the correct genotypes will be found is also calculated. This paper provides a better understanding of the method of calculating the conditional probabilities and more realistic risk assessments than those provided here and could prove informative when an application to preimplantation diagnosis is being considered.

Table 5

ERRORS IN POLAR BODY TYPING FOR AUTOSOMAL DOMINANT DISEASES						
POLAR BODY		OOCYTE				
TRUE GENOTYPE	OBSERVED GENOTYPE	TRUE GENOTYPE	DEDUCED GENOTYPE	USE?	ERROR CATEGORY	PRIMARY CAUSE OF ERROR
AA	Aa	aa	Aa	NO	TOLERATED	c
AA	aa	aa	AA	NO	TOLERATED	c
Aa	AA	Aa	aa	YES	POTENTIALLY UNACCEPTABLE	r
Aa	aa	Aa	AA	NO	ACCEPTABLE	r
aa	AA	AA	aa	YES	UNACCEPTABLE	c
aa	Aa	AA	Aa	NO	ACCEPTABLE	c

Table 6

CONDITIONAL PROBABILITIES OF UNACCEPTABLE ERRORS IN POLAR BODY TYPING FOR AUTOSOMAL DOMINANT DISEASES					
	PRIMARY CAUSE OF ERROR	(1) r=.8 c=0	(2) r=.9 c=.05	(3) r=.9 c=0	(4) r=1 c=0
Polar Body, AA observed, θ=0	c	.00	.00025	.00	.00
Polar Body, AA observed, θ=.10	r	.038	.022	.022	.00
Polar Body, AA observed, θ=.20	r	.091	.054	.054	.00
Polar Body, AA observed, θ=.25	r	.13	.077	.077	.00
Polar Body, AA observed, θ=.30	r	.17	.11	.11	.00
Polar Body, AA observed, θ=.40	r	.29	.21	.21	.00

REFERENCES

Arnheim, N., Li, H., and Cui, X. (1990a) PCR analysis of DNA sequences in single cells: single sperm gene mapping and genetic disease diagnosis. Genomics 8:415-419.

Arnheim, N., White, T. and Rainey, W.E. (1990b) Application of PCR: Organismal and Population Biology. Bioscience 40:174-182.

Bradbury, M.W., Isola, L.M. and Gordon, J.W. (1990) Enzymatic amplification of a Y chromosome repeat in a single blastomere allows identification of the sex of preimplantation mouse embryos. Proc. Natl. Acad. Sci.87:4053-4057.

Coutelle, C., Williams, C., Handyside, A., Hardy, K., Winston, R. and Williamson, R. (1989). Genetic Analysis from DNA from Single Human Oocytes: A Model for Preimplantation Diagnosis of Cystic Fibrosis. Br. Med. J. 299:22-24.

Cui, X., Li, H., Goradia, T.M., Lange, K., Kazazian, H.H., Galas, D.J. and Arnheim, N. (1989) Single Sperm Typing: Determination of Genetic Distance Between the G-Gamma Globin and Parathyroid Hormone Loci. Proc. Natl. Acad. Sci. USA. 86:9389-9393.

Gomez, C.M., Muggleton-Harris, A.L., Whittingham, D.G., Hood, L.E. and Readhead, C. (1990) Rapid preimplantation detection of mutant (shiverer) and normal alleles of the mouse myelin basic protein gene allowing selective implantation and birth of live young. Proc. Natl. Acad. Sci. 87:4481-4484.
Handyside, A.H., Pattinson, J.K., Penketh, R.J.A., Delhanty, J.D.A., Winston, R.M.L., and Tuddenham, E.G.D. (1989). Biopsy of Human Preimplantation Embryos and Sexing by DNA Amplification. Lancet i:347-349.
Holding, C. and Monk, M. (1989) Diagnosis of Beta-Thalassemia by DNA Amplification in Single Blastomeres from Mouse Preimplantation Embryos. Lancet September 2,:532-535.
Jeffreys, A.J., Wilson, V., Neumann, R., and Keyte, J. (1988) Amplification of Human Minisatellites by the Polymerase Chain Reaction: Towards DNA Fingerprinting of Single Cells. Nucleic Acids Research 16:10953-10971.
Li, H., Gyllensten, U., Cui, X., Saiki, R., Erlich, H. and Arnheim, N. (1988) Amplification and Analysis of DNA Sequences in Single Human Sperm. Nature 335:414-417.
Li, H., Cui, X. and Arnheim, N. (1990) Direct Electrophoretic Detection of the Allelic State of Single DNA Molecules in Human Sperm Using PCR. Proc. Natl. Acad. Sci. 87:4580-4584.
Monk, M. and Holding, C. (1990) Amplification of a β- hemoglobin sequence in individual human oocytes and polar bodies. Lancet 335:985-988.
Mullis, K.B., and Faloona, F.A. (1987) Specific Synthesis of DNA in vitro via a polymerase catalyzed chain reaction. Methods Enzymol. 155:335-351.
Navidi, W. and Arnheim, N. (1991) Using PCR in Preimplantation Genetic Disease Diagnosis. Submitted to Human Reproduction.
Saiki, R., Scharf, S., Faloona, F., Mullis, K., Horn, G., Erlich, H., and Arnheim, N., (1985) Enzymatic Amplification of Beta-Globin Genomic Sequences and Restriction Site Analysis for Diagnosis of Sickle Cell Anemia. Science 230: 1350-1354.
Saiki, R.K., Gelfand, D.H., Stoffel, S., Scharf, S.J., Higuchi, R., Horn, G.T., Mullis, K.B., and Erlich, H.A. (1988) Primer Directed Amplification of DNA with a Thermostable DNA Polymerase. Science 239:487-491.
Strom, C.M., Verlinsky, Y., Milayeva, S., Evsikov, S., Cieslak, J., Lifchez, A., Valle, J., Moise, J., Ginsberg, N. and Applebaum, M. (1990) Preconception genetic diagnosis of cystic fibrosis. Lancet 336:306-307.

PREIMPLANTATION GENETIC ANALYSIS USING PCR

Charles M. Strom, Gloria Enriquez
and Svetlana Rechitsky*

Reproductive Genetics Institute, Illinois
Masonic Medical Center, Chicago, IL 60657

INTRODUCTION

Preimplantation genetic diagnosis is designed to offer couples at high risk for having children with genetic diseases an alternative to prenatal diagnosis with termination of affected fetuses. The technique, first proposed in 1987 (Verlinsky et al, 1987) involves performing genetic analysis of material derived from zygotes or embryos using _in vitro_ fertilization (IVF) protocols and the transfer of only unaffected pre-embryos (Coutelle et al, 1989, Handyside et al, 1989, 1990, Strom et al, 1990, Verlinsky et al, 1990a,b). Pre-embryos determined to be affected or potentially affected are frozen for the possibility of future diagnosis or gene therapy.

In preimplantation genetic analysis, diagnosis must be accomplished within 72 hours of oocyte harvest in order to optimally transfer the pre-embryos during the initial cycle, thereby precluding freezing of the embryos and the resultant reduction of the implantation rate (Handyside et al, 1989, 1990, Verlinsky et al, 1990a, 1990b).

The advent of the Polymerase Chain Reaction (PCR) enables genetic diagnosis to be performed on the DNA from a single cell in less than 72 hours (Li et al, 1988, Handyside et al, 1989, Strom et al, 1990, Verlinsky et al, 1990b). The PCR is a technique that amplifies specific genetic sequences for subsequent genetic analysis. The ability of PCR to amplify minute quantities of DNA is simultaneously its greatest strength and its greatest weakness. Any contaminating DNA sequences present in buffers, reagents, enzymes or tubes will be amplified simultaneously with the material to be diagnosed thus preventing accurate genetic analysis.

In the usual clinical or research application 0.1 - 1.0 mcg of DNA is subjected to PCR for genetic analysis (Saikai et al, 1985, 1988, Cheehab et al, 1987, Myerowitz, 1988,

* In previous publications Svetlana Milayeva

Preimplantation Genetics, Edited by Y. Verlinsky and
A. Kuliev, Plenum Press, New York, 1991

Myerowitz and Costigan, 1988, Dermer and Johnson, 1988, Speer et al, 1989, Kerem et al, 1989). This is equivalent to the DNA from 15,000 - 150,000 cells. Under these circumstances contamination of the PCR reaction with minute amounts of DNA does not cause diagnostic errors because several fold more DNA from the sample to be analyzed is present than any contaminating sequences. Techniques have been developed to prevent large scale contamination which have been called "carry-over" problems. These include the use of positive displacement pipettes, separate rooms for oligonucleotide manipulations and PCR reactions, extensive purification of oligonucleotides, and filtering liquids through 0.2 micron filters. These techniques are successful in preventing large scale contamination and "carry over" problems for standard clinical analysis. In analyzing the DNA from a single cell, however, even minute amounts of DNA contamination are sufficient to prevent successful DNA diagnosis.

When PCR analysis of single cells is used for scientific purposes intermittent problems with contamination have been reported (Li et al, 1988). In this setting, detection of experiments that have suffered from contamination and statistical corrections are appropriate. In contrast, inaccurate analysis and/or failed analysis due to contaminating DNA sequences is unacceptable when PCR is used for preimplantation genetic analysis. This report contains our experience on PCR analysis of single cells and on the application of this analysis for preimplantation genetic diagnosis.

APPLICATION OF PCR FOR GENETIC ANALYSIS IN SINGLE CELLS FOR PREIMPLANTATION DIAGNOSIS OF GENETIC DISORDERS

As the usual techniques of preventing "carry over" are ineffective in preventing DNA contamination or eliminating DNA contamination once it has occurred when analyzing single cells, we have developed a technique of "decontamination". Decontaminated PCR reaction mixtures can reliably be used for the genetic analysis in single cells, as demonstrated by our experience in applying this technique for preimplantation genetic analysis for the presence Y specific long arm repeats, and X alphoid repeats, the Delta-F508 mutation of cystic fibrosis (CF), the XbaI polymorphism in the Factor VIII gene for Hemophilia A and the PI type Z mutation of alpha-1-antitrypsin deficiency (Verlinsky, et al, 1990a, Strom et al, 1990).

Initially we used primers Y1.1 and Y1.2 described by Kogan et al (1987) to amplify a 149 base pair repeated sequence specific to the human Y chromosome (Nakahori et al, 1986, Kogan et al, 1987). Using these primers we were able to establish the sex of 0.1 mcg of DNA. Lanes containing female DNA have no Y specific band at 149 base pairs whereas the lanes containing male DNA have an intense band at 149 base pairs (Kogan et al, 1987).

Figure 1 shows the application of this technique to the analysis of single amniocytes obtained by micromanipulation.

As can be seen from this experiment all lanes, including those containing distilled water and female cells demonstrated the male specific 149 base pair band. Since even the distilled water contained amplified Y specific sequences it was clear that extraneous DNA sequences were contaminating one or more of the reagents. It is important to note that this contamination is not a major problem when analyzing 0.1 mcg or more of DNA. Under these circumstances only 20-30 cycles of PCR are performed, and the assay reliably determines the sex of the individual from whom the DNA sample was obtained. It is only when analyzing single cells that the low level contamination becomes a problem.

To overcome this problem, a systematic decontamination of all reagents used in the PCR using restriction enzyme digestion has been initiated as described in technical part of this volume (Rechitsky et al, this volume).

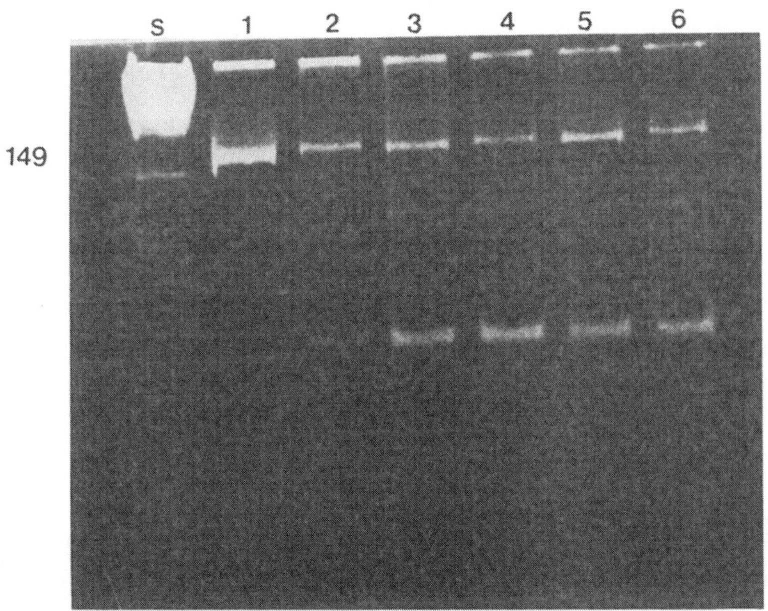

Figure 1. <u>PCR amplification of Single Cells without Decontamination</u>

40 rounds of PCR performed with primers Y1.1 and Y1.2 as described in methods.

Lane S: Size standards
Lane 1: Control male 0.1 mcg DNA
Lane 2: Distilled water control (No added DNA)
Lane 3: Female single amniocyte
Lane 4: Female single amniocyte
Lane 5: Male single amniocyte
Lane 6: Male single amniocyte

This decontamination procedure will only eliminate double stranded DNA sequences (Porter-Jordan & Garrett, 1990; Bianchi et al, 1990). Rarely, a reagent has become contaminated with sequences resistant to decontamination procedures (presumably single stranded DNA). In such occasions the reagent must be discarded and new reagents made. All reagents are aliquotted so that if contamination occurs uncontaminated reagents will be immediately available.

Following decontamination the tubes are only opened once, under sterile techniques, when the single cells are added to reaction mixture. The tubes are then placed in the thermal cycler and incubated at 95° for 7 minutes followed by 40 cycles of PCR.

To avoid any possible contamination and to perform systematic decontamination procedure, a separate laboratory was constructed and designed for preimplantation genetics diagnosis. Design of the Preimplantation Genetics Laboratory is presented in the technical part of this volume (Rechitsky et al, this volume).

Another important problem in preimplantation gender determination is false negative results which we observed in 20% of the cases due to a failure of PCR. We have proposed a method of simultaneous amplification of X and Y specific sequences which make it possible to sort out the failure of PCR from the absence of the Y specific sequence (Fig. 2).

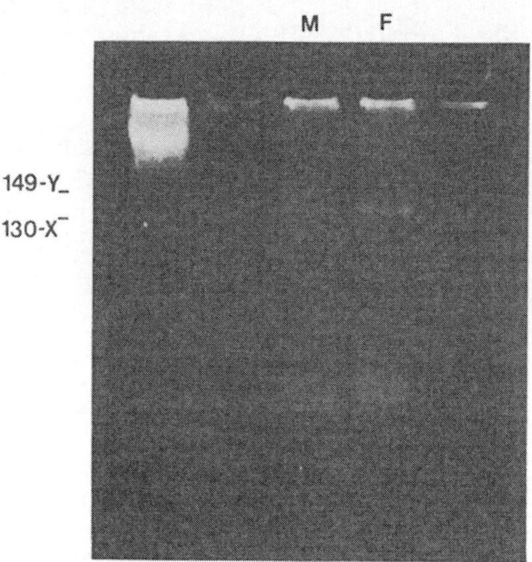

Figure 2. Simultaneous amplification of X and Y specific sequences.

25 cycles of PCR using primers Y1.1, Y1.2, X1, and X2 (see methods).

Lane M: 0.1 mcg Male DNA
Lane F: 0.1 mcg Female DNA

However, preimplantation gender determination will eliminate 50% of male embryos that are not affected by X-linked disorders. Therefore, it is of a great practical importance to develop methods for preconception and preimplantation diagnosis of X-linked disorders based on gene specific probes or linkage analysis. We have applied the latter approach to the analysis of the XbaI polymorphism in the coagulation Factor VIII gene for the linkage analysis for Hemophilia A. In this technique slot blots and allele specific oligonucleotide hybridization are performed. Preimplantation genetic analysis has been performed on 21 oocytes and blastomeres in a couple informative for the XbaI polymorphism (Fig. 3).

Although many problems are still left to be solved regarding a reliability and accuracy of the test, the method seems promising as an extention or a combination with gender determination using X and Y specific primers.

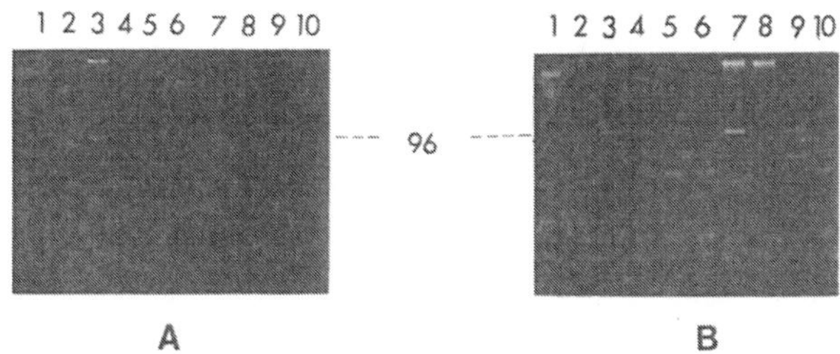

Figure 3. Analysis of Polar Bodies for Xba Polymorphism in Factor VIII gene.

Ethidium bromide stained 8% polyacrylaimide gels of PCR products from polar bodies following 45 cycles of PCR. Polar body numbers correspond to the slot shown in figure 4.

A:
Lane 1: Size standard (pBR-322/HaeIII)
Lane 2: Buffer control (slot 6b)
Lane 3: 10 ng control DNA hemizygous DNA Xba - (slot 7c)
Lane 4: Polar body (not shown on slot blot)
Lane 5: Polar body (slot 3a)
Lane 6: Polar body (slot 7b)
Lane 7: Polar body (slot 3b)
Lane 8: Polar body (slot 4b)
Lane 9: Polar body (slot 4c)
Lane 10: Polar body (slot 6c)

B:
Lane 1: Size standard (pBR-322/HaeIII)
Lane 2: Polar body (slot 5b)
Lane 3: Polar body (slot 5a)
Lane 4: Polar body (slot 6a)
Lane 5: Buffer control (slot 1a)
Lane 6: Buffer control (slot 2a)
Lane 7: 10 ng Maternal heterozygous DNA (slot 5c)
Lane 8: 10 ng Genomic DNA hemizygous DNA Xba - (slot 1c)
Lane 9: Polar body (slot 7a)
Lane 10: Polar body (slot 1b)

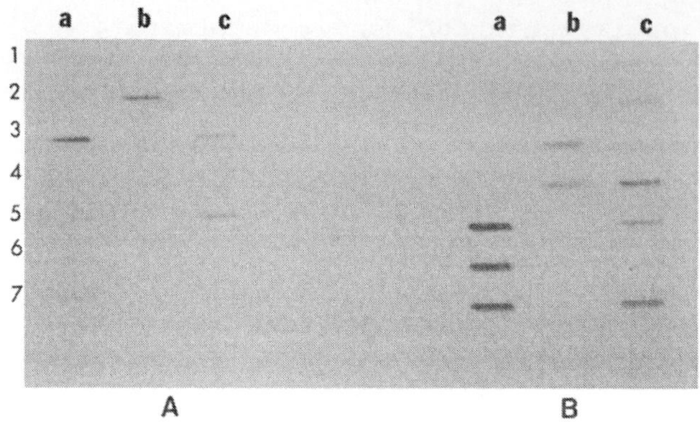

Figure 4. Slot blot analysis of PCR products shown in figure 3:

A: Hybridized with probe 7.3 (Xba +)
B: Hybridized with probe 7.4 (Xba -)

Preconception and preimplantation diagnosis of autosomal recessive diseases have the greatest potential for prevention of autosomal recessive disorders. Our experience regarding preimplantation diagnosis of the CF delta-F508 mutation shows that the test is highly reliable both for polar body analysis and blastomeres. We performed genetic analysis in 22 polar bodies and 15 blastomeres, which was successful in 18 polar bodies and in 13 blastomeres (Verlinsky et al, this volume).

In most cases, the diagnosis of predicted CF was confirmed in subsequent stages of development. For example, one oocyte, predicted to be homozygous normal by polar body analysis did not fertilize and was analyzed and found to be hemizygous for the normal allele, thus confirming the diagnosis. An embryo developed from an oocyte predicted to contain the Delta-F508 gene by polar body analysis resulted in a pre-embryo whose genotype was determined to be homozygous Delta-F508 by blastomere biopsy (Strom et al, 1990), thus confirming the preconception diagnosis.

CONCLUSION

We have found that contamination of PCR reagents and/or embryology reagents to be the most significant obstacle to successful PCR genotyping of single cells for preimplantation diagnosis. We have developed a method to minimize contamination, to detect contamination when it occurs prior to preimplantation genetic analysis, and to decontaminate reagents after contamination has been discovered. Using these techniques we are able to reliably determine the genotype of single human cells for the purposes of preimplantation and preconception genetic diagnosis. This was applied to preconception and preimplantation DNA analysis for gender determination for diagnosis of Haemophilia, α-1-antitrypsin deficiency and CF delta-F508 mutation.

REFERENCES

Bianchi DW, Flint AF, Pizzimenti MF, Knoll JF, Latt SA (1990) Isolation of fetal DNA from nucleated erythrocytes in maternal blood. Proc Natl Acad Sci USA 87(9):3279-83.

Chehab FF, Doherty M, Cai S, Kan Y-W, Cooper S, Rubin EM (1987) Detection of sickle cell anaemia and thalassaemias. Science 329:293-294.

Coutelle C, Williams C, Handyside A, Hardy K, Winston R, Williamson R (1989) Genetic analysis of DNA from single human oocytes: a model for preimplantation diagnosis of cystic fibrosis. BMJ 299:22-24.

Dermer SJ, Johnson EM (1988) Rapid DNA analysis of alpha-1-antitrypsin deficiency: application of an improved method for amplifying mutated gene sequences. Lab Invest 59:403-408.

Handyside AH, Penketh RJA, Winston RML, Pattinson JK, Delhanty JDA, Tudenham EGD (1989) Biopsy of human preimplantation embryos and sexing by DNA amplification. Lancet i(8634), 347-349.

Handyside AH, Kontogianni EH, Hardy K, Winston RML (1990) Pregnancies from biopsied human preimplantation embryos sexed by Y-specific DNA amplification. Nature 344:768-770.

Kerem B-E, Rommens JM, Buchanan JA, Markiewicz D, Cox TK, Chakravarti A, Buchwald M, Tsiu L-C (1989) Identification of the cystic fibrosis gene: genetic analysis Science 245:1073-1080.

Kogan SC, Doherty M, Gitschier J (1987) An improved method for prenatal diagnosis of genetic diseases by analysis of amplified DNA sequences. New Eng J Med 317(16):985-990.

Li HH, Gyllensten UB, Cui XF, Saikai RK, Ehrlich HA, Arnheim N (1988) Amplification and analysis of DNA sequences in single human sperm and diploid cells. Nature 335(6189):414-7.

Myerowitz R (1988) Splice junction mutation in some Ashkenazi Jews with Tay-Sachs disease: evidence against a single defect within this ethnic group. Proc Nat Acad Sci USA 85(11):3955-9.

Myerowitz R, Costigan FC (1988) The major defect in Ashkenazi Jews with Tay-Sachs disease is an insertion in the gene for the alpha-chain of beta-hexosaminidase. J Biol Chem 263(35):18587-9.

Nakahori Y, Mitani K, Yamada M, Nakagome Y (1986) A human Y-chromosome specific repeated DNA family (DYZ1) consists of a tandem array of pentanucleotides. Nuc Acids Res 14(19):7569-79.

Porter-Jordan K, Garrett CT (1990) Source of contamination in polymerase chain assay. Lancet 235(May 19, 1990);1220.

Saikai RK, Chang C-A, Levensen CH, Warren TC, Boehm CD, Kazazian HH, Erlich HA (1988) Diagnosis of sickle cell anemia and beta-thalassemia with enzymatically amplified DNA and nonradioactive allele-specific oligonucleotide probes. New Eng J Med 319:537-541.

Saikai RK, Scharf S, Faloona F, Mullis KB, Horn GT, Erlich HA, Arnheim N (1985) Enzymatic amplification of beta-globin genomic sequences and restriction site analysis of diagnosis of sickle cell anemia. Science 230:1350-1354.

Speer A, Rosenthal A, Billwitz H, Hanke R (1989) DNA amplification of a further exon of DuChenne muscular dystrophy locus increase possibilities for deletion screening. Nucleic Acids Research 17(16):6774-6784.

Strom CM, Verlinsky Y, Rechitsky S, Evsikov S, Cieslak J, Lifchez A, Valle J, Moise J, Ginsberg N, Applebaum M (1990) Preconception Genetic Diagnosis for Cystic Fibrosis By Polar Body Removal and DNA Analysis. Lancet, 336:306-308.

Verlinsky Y, Pergament E, Binor Z and Rawlins R (1987) Genetic Analysis of Human Embryos Prior to Implantation: Future Applications of In-Vitro Fertilization in Treatment and Prevention of Human Genetic Diseases. In: Ferchlinger and Kemeter (Ed). "Future Aspects in Human In-Vitro Fertilization". Springer-Verlag, Berlin, pp 262-266.

Verlinsky Y, Pergament E, Strom C (1990a) The Preimplantation Genetic Diagnosis of Genetic Diseases. J In Vitro Fert and Embryo Trans, 7:1-5.

Verlinsky Y, Ginsberg N, Lifchez A, Valle J, Moise J, Strom CM (1990b) Analysis of the First Polar Body: Preconceptual Genetic Diagnosis. Hum Reproduction, 5(7):826-829.

Waye JS, Willard HF (1985) Chromosome specific alpha satellite DNA: nucleotide sequence analysis of the 2.0 kilobasepair repeat from the human X chromosome. Nuc Acids Res 13(8):2731-2743.

Witt M, Erickson RP (1989) A rapid method for detection of Y-chromosomal DNA from dried blood specimens by the polymerase chain reaction. Hum Genet 82:271-274.

CO-AMPLIFICATION OF X- AND Y-SPECIFIC SEQUENCES FOR SEXING PREIMPLANTATION HUMAN EMBRYOS

Eleni H. Kontogianni, Kate Hardy and Alan H. Handyside

Institute of Obstetrics and Gynaecology
Royal Postgraduate Medical School
Hammersmith Hospital
Du Cane Road
London W12 0NN, UK

INTRODUCTION

For couples at risk of X-linked recessive diseases, which typically affect only males, *in vitro* fertilization followed by the identification and transfer of female preimplantation embryos provides an alternative to prenatal diagnosis at later stages of pregnancy and the possibility of terminating an affected pregnancy. DNA amplification of a Y chromosome-specific repeat sequence by the polymerase chain reaction (PCR) from single cells biopsied at the 6- to 10-cell stage allows the identification of male, and absence of amplification, female embryos, within a few hours (Handyside et al., 1989). Several pregnancies have now been established using this approach with these couples (Handyside et al., 1990; Handyside, this volume) and, to date, all five fetuses in two sets of twins and a singleton pregnancy have been confirmed, by chorion villus sampling and cytogenetic analysis, to have normal female karyotypes.

AMPLIFICATION OF REPEAT SEQUENCES

The identification of potentially affected male embryos by PCR requires amplification of a sequence specific for the Y chromosome, if possible from one or two cells, since this number of cells can be biopsied from cleavage stage embryos without adversely affecting preimplantation development (Hardy et al., 1990). Unique sequences have been successfully amplified from single cells including human sperm and fibroblasts (Li et al., 1988) and unfertilized eggs from which the first polar bodies had been removed (Coutelle et al., 1989). However, reamplification is often necessary for the simple and rapid detection of the fragment by gel electrophoresis and staining, or otherwise, detection requires hybridisation with labelled probes which would take longer than the time currently available between biopsy and the optimal time for transfer. An additional problem when amplifying unique sequences from single cells is that in about 10 to 20% of cases amplification fails for reasons which are not understood (Li et al., 1988; Boehnke et al., 1989).

Preimplantation Genetics, Edited by Y. Verlinsky and A. Kuliev, Plenum Press, New York, 1991

Table 1 Sex chromosome specific repeat sequences detected by DNA amplification

Y long arm repeat primers (Kogan et al., 1987):

5'-ATTACACTACATTCCCTTCCA-3'	94°C	30 sec
5'-AGTGAAATTGTATGCAGTAGA-3'	65°C	1 min 30 sec

149 bp amplified from a 3.4 kb Hae III fragment 800 - 5000 copies (Cooke, 1976)

Y alphoid repeat primers (Witt and Erickson, 1989):

5'-ATGATAGAACGGAAATATG-3'	94°C	30 sec
5'-AGTAGAATGCAAAGGGCTCC-3'	55°C	1 min 30 sec

170 bp amplified from a 5.5 kb EcoR1 fragment (Wolfe et al., 1985) 100 copies

X alphoid repeat primers (Witt and Erickson, 1989):

5'-AATCATCAAATGGAGATTTG-3'	94°C	30 sec
5'-GTTCAGCTCTGTGAGTGAAA-3'	55°C	1 min 30 sec

130 bp amplified from a 2.0 kb Bam H1 fragment (Waye and Willard, 1985) 5000 copies

In an attempt to overcome these problems, we decided to amplify repeated sequences specific for the sex chromosomes to increase the number of target sequences present in single or small numbers of biopsied cells. Initially, we amplified a 149 base pair (bp) fragment of a sequence repeated 800 to 5000 times on the long arm of the Y chromosome (Cooke, 1976; Kogan et al., 1987) but recently, we have also amplified alphoid sequences specific for either the X or Y chromosomes (Willard and Waye, 1985; Witt and Erickson, 1989) (Table 1). The alphoid sequences have the advantage that they are centromeric, whereas the Y-specific long arm sequence can, in rare cases, be deleted in normal men or translocated to autosomes and inherited by women, so that each couple has to be screened beforehand (Bobrow et al., 1971; Cooke and Noel, 1979).

With repeat sequences, limited cross-homology between different repeats and minor variations in the target sequence within the overall repeat could result in amplification of non-specific fragments or multiple specific bands, respectively. Thus, after optimal annealing temperatures and segment times had been found for specific amplification from male and/or female genomic DNA (Table 1), the identity of each of the amplified fragments was confirmed by digestion with appropriate endonucleases for known restriction sites. For each of these three repeats, forty cycles of amplification generated enough of the fragment, from dilutions of genomic DNA down to 2pg or single embryo cells, to be easily visualised on ethidium bromide stained polyacrylamide gels. Sporadic contamination among control blanks was rare. However, despite rigid precautions to separate sample preparation and product handling (Kwok and Higuchi, 1989) carry-over contamination of the amplified product was sometimes detectable in all female samples and control blanks. Nevertheless, the level of amplification from this form of contamination was generally very low and easily distinguished from that generated from single cells.

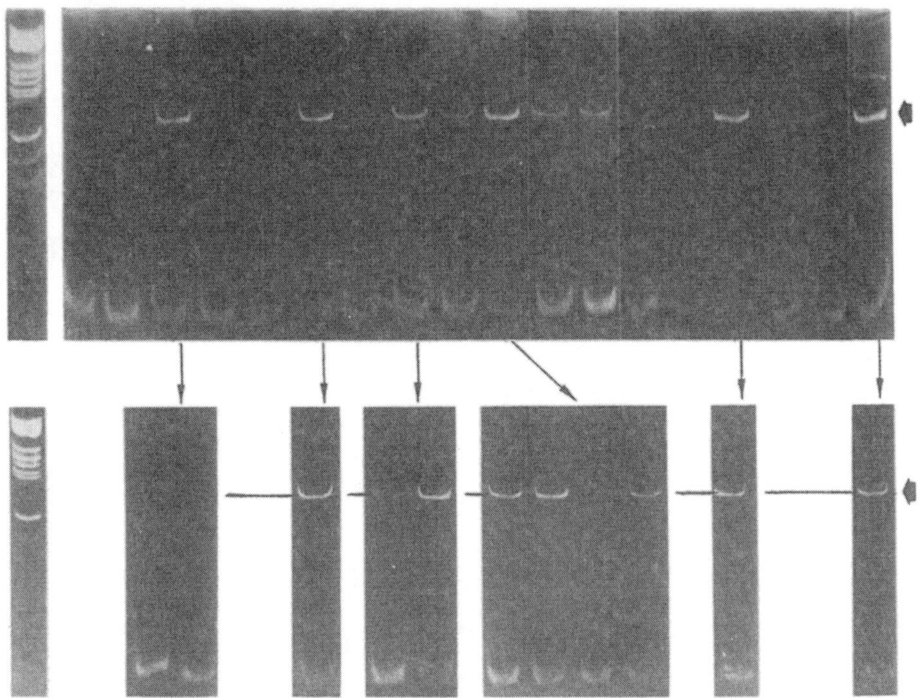

Figure 1 Amplification of a 149 bp fragment (arrowheads) of the Y-specific long arm repeat (Kogan et al., 1987). Cleavage stage embryos on day 3 post insemination were partially disaggregated after complete removal of the zona pellucida with acid Tyrode and about half (2-4) the cells from each embryo used to identify the sex by PCR. Single cells from selected male embryos were then checked for nuclei and amplified individually. Upper panel: nine out of eighteen embryos were identified as male by the detection of the Y-specific repeat fragment. Lower panels: three out of eleven single cells from six of the male embryos (correspondence indicated by the arrows) failed to amplify. Far left lanes: DNA size markers.

CO-AMPLIFICATION OF X- AND Y-SPECIFIC REPEAT SEQUENCES

Identifying the sex of an embryo on the basis of the presence or absence of a Y-specific fragment alone is not ideal since PCR is susceptible to contamination and conversely at the single cell level to amplification failure. In the context of X-linked disease, the former would generate false positives which would only reduce the number of female embryos available for transfer. On the other hand, amplification failure would be more serious and could result in the transfer of a potentially affected male embryo. The accuracy of sexing by PCR with the Y long arm repeat primers has been examined in a limited series of biopsied embryos by cytogenetic analysis using *in situ* hybridization with a Y-specific probe or fluorescent labelling of the Y chromosome in metaphase spreads (Handyside et al., 1989). The sex of 15 (4 male and 11 female) normally fertilized embryos was accurately identified in all cases. However, in two polypronucleate embryos in which the Y-specific fragment had been amplified from the single cell biopsy, the presence of a Y chromosome in the biopsied embryos could not be confirmed. These false positives may have been caused by contamination, but alternatively, polypronucleate embryos, which typically arise by fertilization with more than one sperm, are often genetic mosaics.

To investigate the incidence of amplification failure from single cells for different repeat sequences, we have begun attempts to amplify from each or several cells of individual disaggregated embryos. Since between 5% of cells from embryos with good morphology, and 15% from embryos with poor morphology, are anucleate at the 5- to 8-cell stage (Hardy et al., 1991), each cell was checked for the presence of a single nucleus after disaggregation either by phase contrast microscopy or by vital labelling with a DNA fluorochrome before preparation for PCR (Handyside and Kontogianni, this volume). With both the Y long arm repeat (Figure 1) and Y alphoid repeat fragments (Figure 2), amplification from the majority of single male cells was easily detected. However, amplification failed in some cases with both repeats and, in the case of the Y alphoid

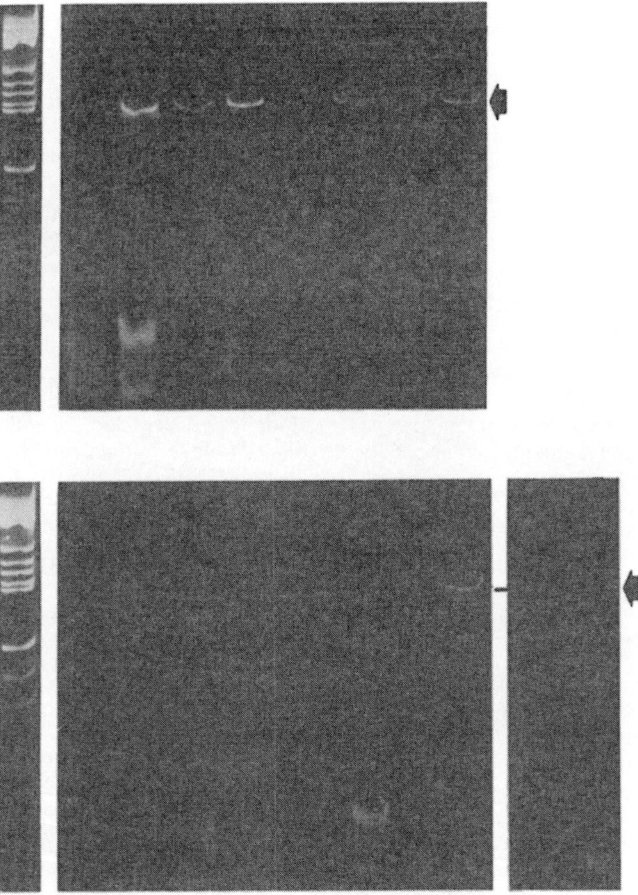

Figure 2 Amplification of a 170 bp fragment (arrowheads) of the Y-specifc alphoid repeat (Witt and Erickson, 1989) from single cells disaggregated from an 8-cell male (upper panel) and female embryo (lower panel) after confirming the presence of nuclei.Two out of eight male cells failed to amplify and in the female embryo there was non-specific amplification in two cells and Y-specific amplification in a third presumably as a result of contamination. One of the blanks was also contaminated (bottom panel, far right two lanes) Far left lanes: DNA size markers.

repeat, sporadic contamination was also encountered in a female cell (Figure 2, lower panel). Insufficient cells have been analysed to obtain an accurate estimate of the frequencies with different target sequences, but initial indications suggest they may be similar to the 10 to 20% observed with unique sequences. If this is the case and the initial number of target sequences present does not affect the frequency with which amplification fails it suggests that sample preparation may be more critical than factors related to PCR itself.

Multiplex PCR with an additional set of primers to a constant target sequence has been suggested as an internal control for amplification failure. We have, therefore, investigated the possibility of co-amplifying X- and Y-specific repeat sequences. Using a modification of a multiplex buffer (Chamberlain et al., 1989) and a reduced combined annealing and extension temperature (60°C), it was possible to co-amplify the Y long arm and X alphoid repeat fragments. However, the sensitivity of amplification, especially of the X alphoid repeat fragment was significantly reduced and co-amplification was not successful at the level of a single cell. In contrast, co-amplification of X- and Y-specific alphoid fragments was possible, again at a reduced annealing/extension temperature (50°C), with as little as 2 pg of male DNA (Figure 3).

CONCLUSIONS

Amplification from each or several single cells of disaggregated male embryos has revealed that amplification failure can occur from a repeated target sequence, even after checking for the presence of a nucleus, and the frequency appears to be similar to that observed with unique sequences. Improved sample preparation for PCR, possibly using a lysis buffer, may help to reduce the frequency of these failures. Co-amplification of X- and Y-specific alphoid repeat sequences is possible down to the single cell level. However, further work is necessary to examine whether this will be an effective control since there is evidence in other systems that amplification from different sets of primers

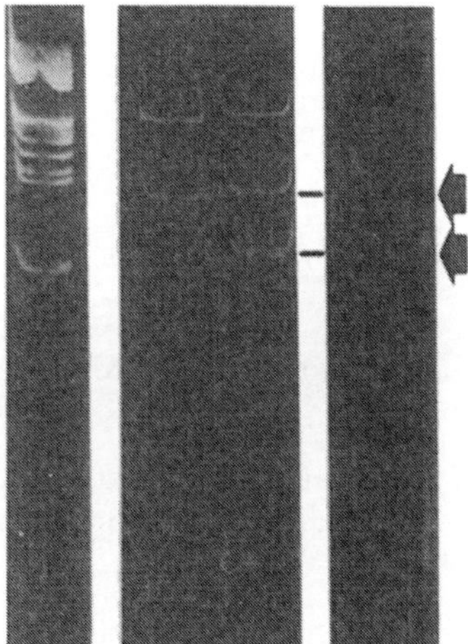

Figure 3 Co-amplification of 130 and 170 bp fragments (lower and upper arrowheads) of X- and Y-specific alphoid repeat sequences (Witt and Erickson, 1989) from male DNA. Lane 1 (from left to right): DNA size markers; lanes 2 and 3: 20 and 2pg male DNA; lane 4: control blank with no DNA. Note the additional X-specific dimer band (approx 300 bp) and low level amplification of the X-specific fragment from contaminant DNA in the control blank.

can fail independently of each other (Li et al., 1988; Boehnke et al., 1989). If the problem of amplification failure cannot be sufficiently reduced in these ways, an alternative may be to biopsy two cells from each cleavage stage embryo (Hardy et al., 1990; Handyside, 1990), carry out duplicate amplifications and avoid selecting embryos in which the results conflict.

ACKNOWLEDGEMENTS

We would like to thank Professor Robert Winston, Karen Dawson and the IVF team for their help. This work was approved by the Research Ethics Committee of the Royal Postgraduate Medical School and the Interim Licensing Authority for Human Fertilization and Embryology, and was supported by the Muscular Dystrophy Group of Great Britain and Northern Ireland.

REFERENCES

Bobrow, M., Pearson , P.L., Pike, M.C. and El-Alfi, 0.S. (1971) Length variation in the quinacrine-binding segment of human Y chromosomes of different sizes. Cytogenetics 10:190-198.

Boehnke, M., Arnhem, N., Li, H. and Collins, F.S. (1989) Fine structure genetic mapping of human chromosomes using the polymerase chain reaction on single sperm: experimental design considerations. Am. J. Hum. Genet. 45:21-32.

Chamberlain, J.S., Gibbs, R.A., Ranier, J.E. and Caskey, C.T. (1990) Multiplex PCR for the diagnosis of Duchenne muscular dystrophy. In 'PCR protocols: a guide to methods and applications' (eds. Innis, M.,Gelfand, D.H., Sninsky, J.J. and White, T.J.) pp.272-281 Academic Press, San Diego.

Cooke, H.J. (1976) Repeated sequences specific to human males. Nature 262:182-186.

Cooke, H.J. and Noel, B. (1979) Confirmation of Y/autosome translocations using recombinant DNA. Hum. Genet. 67:222-224.

Coutelle, C., Williams, C., Handyside, A., Hardy, K., Winston, R and Williamson, R. (1989) Genetic analysis of DNA from single human oocytes - a model for pre-implantation diagnosis of cystic fibrosis. Brit. Med. J. 299, 22-24.

Handyside, A.H. (1990) Preimplantation diagnosis by DNA amplification. In 'The embryo: normal and abnormal development and growth' (eds. Chapman, M., Grudzinskas, G., Chard, T. and Maxwell, D.) Springer-Verlag, London in press.

Handyside, A.H., Pattinson, J.K., Penketh, R.J.A., Delhanty, J.D., Winston, R.M.L. and Tuddenham, E.D.G. (1989) Biopsy of human preimplantation embryos and sexing by DNA amplification. Lancet: 347-349.

Handyside, A.H., Kontogianni, E.H., Hardy, K. and Winston, R.M.L. (1990) Pregnancies from biopsied human preimplantation embryos sexed by Y-specific DNA amplification. Nature 344, 768-770.

Hardy, K., Martin, K.L., Leese, H.J., Winston, R.M.L. and Handyside, A.H. (1990) Human preimplantation development *in vitro* is not adversely affected by biopsy at the 8-cell stage. Hum. Reprod. 5, 708-714.

Hardy, K., Winston, R.M.L. and Handyside, A.H. (1991) Binucleate cells in normally fertilized human preimplantation embryos *in vitro*: failure of cytokinesis during early cleavage. Development, submitted.

Kogan, S.C., Doherty, M. and Gitschier, J. (1987) An improved method for prenatal diagnosis of genetic disease by analysis of amplified DNA sequences. N. Eng. J. Med. 317:985-990.

Kwok, S. and Higuchi, R. (1989) Avoiding false positives with PCR. Nature 339:237-238.

Li, A., Gyllensten, U.B., Cui, X., Saiki, R.K., Erlich, H.A. and Arnheim, N. (1988) Amplification and analysis of DNA sequences in single human sperm and diploid cells. Nature 335, 414-419.

Willard, H.F. and Waye, J.S. (1987) Hierarchical order in chromosome-specific human alpha satellite DNA. Trends Genet. 3:192-198.

Witt , M. and Erickson, R.P. (1989) A rapid method for the detection of Y chromosomal DNA from dried blood specimens by the polymerase chain reaction. Hum. Genet. 82:271-274.

IN SITU HYBRIDIZATION OF BLASTOMERES FROM EMBRYO BIOPSY

Jamie A. Grifo, David C. Ward and Ann Boyle

The Center for Reproductive Medicine and Infertility
The New York Hospital-Cornell Medical Center
New York, NY 10021

INTRODUCTION

Recent advances in molecular genetics have improved the sensitivity of methods utilized for genetic diagnosis. Two techniques in particular, the polymerase chain reaction (PCR) and in situ hybridization permit the analysis of the genetic material of a single cell. The sensitivity of PCR is achieved via enzymatic amplification of target DNA sequences by a factor of 10^5 fold or more.[1] In contrast in situ hybridization allows the direct visualization of single genes in metaphase chromosomes or interphase nuclei.[2] The micromanipulation of preembryos could allow specific preimplantation diagnosis to be made. Currently, genetic diagnoses are limited to the post implantation period and are performed by chorionic villus sampling or amniocentesis. Embryo biopsy has been performed in animals and humans with resultant liveborn offspring[3]. Since a broad range of diseases can be diagnosed by PCR or in situ hybridization the technique of *in vitro* fertilization coupled with preembryo biopsy could provide a mechanism for preimplantation diagnosis.

Gardner and Edwards sexed embryos by sampling trophoblast cells removed from blastocysts in 1968.[4] Bovine preembryos have been sexed by chromosome analysis[5,6] or Y chromosome specific DNA probes.[7] Monk et al. have diagnosed a deficiency of the X-linked enzyme hypoxanthine phosphoribosyl transferase in mouse preembryos prior to transfer.[8] Trophectoderm biopsy has been successfully carried out in marmoset monkey. The cells were cultured for analysis and the embryos were transferred with resultant liveborn offspring.[9] The human preembryo has been successfully sexed using chromosome specific probes and in situ hybridization. However, this utilized a tritiated Y chromosome specific probe which used the whole embryo in a destructive fashion.[10] This methodology as described has no clinical application. Recently, Handyside et al. have sexed the human preembryo in an indestructible fashion utilizing embryo biopsy in conjunction with PCR.[11] They were able to remove a single blastomere, amplify Y-specific DNA repeat sequences and culture biopsied embryos to the blastocyst stage at the same rate as nonbiopsied embryos. Holding and Monk diagnosed ß thalassaemia by PCR in single blastomeres from mouse preimplantation embryos.[12] The unique feature of this study is that they were able to amplify a region of a gene in a single diploid blastomere by a method that is theoretically applicable to any genetic disorder in which a mutation or a deletion exists. Wilton and Trounsen have recently described a method of mouse preembryo biopsy with successful cryopreservation.[13] In addition, they have developed a method for culturing blastomeres which after 6 days resulted in 20 cell nuclei per cultured blastomere.[14] While in situ hybridization can be applied to a single cell, the ability to culture blastomeres would facilitate a reliable diagnosis. In situ hybridization allows numerical chromosome analysis and is not limited by contamination. These are distinct advantages over PCR for the analysis of single cells.

MICROMANIPULATION AND ANALYSIS OF BLASTOMERES

A single blastomere was removed from four to eight cell mouse and human embryos using an inverted microscope and Narishige micromanipulators. A holding pipette of approximately 80 μM outside diameter with a 30 μM inside diameter was fire polished using a Nikon microforge. A sharpened dissection pipette was used for zona dissection and a microbiopsy pipette was pulled to a 10 to 20 μM outside diameter with a 5 to 15 μM inside diameter. The preembryo was stabilized with gentle suction through the holding pipette and a slit was made through the zona pellucida using a sharpened dissection pipette similar to the method of Malter and Cohen.[15] The dissection pipette was removed and replaced by a biopsy pipette which was placed into the zona slit to aspirate a blastomere partially into the pipette under controlled suction. Later experiments used acid Tyrodes to drill a hole in the zona as described by Gordon and Talansky.[16] The blastomere was removed from the zona and the preembryo was placed into the Hoppe-Pitts media for continued culture after three passes into fresh media droplets. Embryo culture was done in microdrops under paraffin oil equilibrated with media.

Blastomeres obtained in media droplets under oil were fixed directly on the tissue culture slides under a Zeiss dissecting microscope in the following fashion. Blastomeres were placed in cold methanol: acetic acid (3:1) in a watch glass under a dissecting microscope. When they turned brown (after 30 sec) they were removed, placed on a glass slide and air dried. Human oocytes and polyploid embryos were processed in a manner similar to mouse preembryos, however, biopsied polyploid preembryos were not cultured after biopsy as all blastomeres were fixed to glass slides.

The X repeat probe 60-36, and 4.2 kb insert was kindly provided by Christine Disteche.[17] It is a 4.2 kb sequence repeated approximately 50 times which maps just below the centromere. Probe MαG-6, a cosmid containing the genomic sequence for the sodium/potassium ATPase gene was provided by David Houseman. It is a 40 kb insert which hybridizes the alpha·subunit single copy gene.[18] The human Y repeat probe hybridizes to pHY2.1, a 2.1 kb sequence that is repeated approximately 2000 times on the long arm of the Y chromosome.[19] Probes were labelled with biotinylated-11-dUTP via nick translation.[2] The probes were nick translated intact with both the vector and the insert and used in this form for in situ hybridization.

The X repeat probes and MαG-6 probes were used at a concentration of 6 μg/ml which are the standard amounts for in situ hybridization performed in our laboratory. In addition to salmon sperm DNA, mouse genomic DNA was added to suppress repetitive sequences in the probe DNA. The probe mixtures contained 20% dextran sulphate and 2X sterile saline citrate (SSC). Slides and probe were denatured separately. After denaturation for 2 minutes at 70°C in 20% formamide/2X SSC, slides were dehydrated through a cold ethanol series and allowed to air dry. Probe mixtures were denatured at 75°C for 5 minutes followed by ice (for repeat probes) or a 37°C incubation for those probes requiring suppression.[2] Ten microliters of denatured probe was applied to the slide, coverslips applied and sealed and incubated overnight at 37°C. Slides were incubated for 5 minutes 3 times in 0.1X SSC. They were blocked for 30 minutes at 37°C in 3% BSA, 4X SSC and 0.1% Tween 20. Biotinylated probes were detected by incubation with FITC:avidin (Vector Labs) diluted to 50 μg/ml in 1% BSA, 4X SSC and 0.1% Tween 20 at 37°C for 30 minutes. After washing 3 times for 5 minutes in 4X SSC, 0.1% Tween 20 at 42°C, slides were mounted with glycerol containing propidium iodide as counterstain and DABCO to reduce photobleaching. The total time required for preembryo biopsy, fixation and in situ hybridization of blastomeres is approximately 48 hours.

APPLICATION OF IN SITU HYBRIDIZATION TO BLASTOMERES FROM HUMAN AND MOUSE EMBRYOS

The majority of embryos (84%) cultured in our system proceeded to the blastocyst stage. Manipulations required for partial zona dissection did not inhibit embryo growth as approximately 90% of embryos treated in this fashion proceeded to blastocyst. The additional steps required to remove a blastomere has a minimal negative impact on the ability of preembryos to develop into blastocysts as 74% continued to divide. Blastocysts from biopsied preembryos were morphologically indistinguishable from control blastocysts except that the slit in the zona was sometimes apparent. Occasionally, the total size of the blastocyst was somewhat smaller with biopsied preembryos but this was not a consistent finding. Later embryo biopsy experiments were performed with a modified biopsy medium which contained bovine serum albumin (0.2%) and sucrose (0.1M). A series of 31 biopsies performed in this media resulted in an 84% blastocyst rate. More recently, a series of mouse embryo biopsies utilizing acid Tyrodes solution to make a hole in the zona have provided a more efficient system for embryo biopsy. Data on preimplantation and postimplantation development following embryo biopsy are presented in Table I.

Table I. Preimplantation and postimplantation development following biopsy of mouse preembryos.

	Embryo Biopsy	Control
# Embryos	219	54
# Blastocysts	191	54
% Blastocysts	87%	100%
# Embryos Transferred	102	33
# Implanted	51	19
% Implanted	50%	30%
# Liveborn	22	9
% Liveborn	22%	14%

Figure 1, Panel A demonstrates a metaphase spread of mouse chromosomes which had been subjected to in situ hybridization with a probe which hybridizes to the sodium and potassium ATPase gene on chromosome 3. This probe had been nick translated in the presence of biotinylated dUTP which specifically incorporated biotin into the probe. After in situ hybridization with the probe FITC linked to avidin was incubated with the slide allowing the specific avidin-biotin interaction to label the DNA sequence of interest with a fluorescent probe. Signal was then visualized using a laser scanning confocal microscope and an image was constructed by an image processor utilizing data recorded by a photomultiplier detector. The image obtained demonstrates a signal on both chromatids of each copy of chromosome 3 (4 total signals). Panel 1B demonstrates the same probe hybridized to a blastomere obtained by preembryo manipulation. Two signal domains are noted in this interphase nucleus. This cell is from the G2 phase in which the synthesis phase has duplicated the chromatids, thus each signal domain has a bilobed configuration. Comparison of metaphase chromosomes to interphase nuclei demonstrates the same signal domains.

Panel 1C demonstrates a blastomere which has been hybridized with a probe which specifically labels the centromeric portion of the X chromosome. There are two clear signals consistent with a female karyotype. Panel 1D demonstrates a blastomere hybridized with the X probe which happens to be in metaphase. There is one clear signal consistent with a male karyotype. Even though this cell is in metaphase, in situ hybridization allows one to analyze the DNA for the signal of interest. Finally, Panel 1E demonstrates a human male metaphase spread hybridized with a chromosome Y specific probe. One clear signal domain is

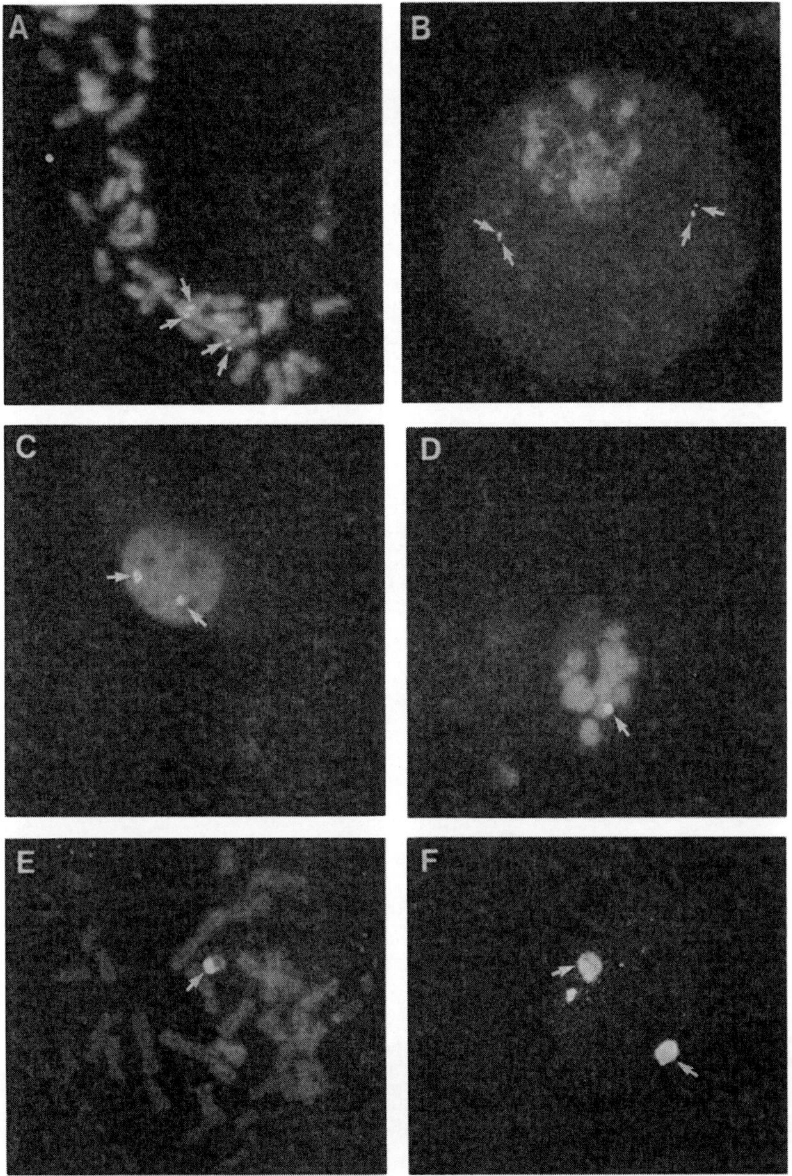

Figure 1. In situ hybridization. Panel A. In situ hybridization to a metaphase spread with a cosmid clone which maps to chromosome 3 (sodium/potassium ATPase gene). Panel B. In situ hybridization to a mouse blastomere with a cosmid clone which maps to chromosome 3 (sodium/potassium ATPase gene). Panel C. In situ hybridization to a mouse consistent with an XX karyotype. Panel D. In situ hybridization to a mouse blastomere with an X specific repeat probe, 1 signal is noted. Panel F. In situ hybridization with a Y specific repeat probe to a blastomere from a polyspermic human preembryo which has two major signals consistent with an YY karyotype. Two other cells from this preembryo also had two signals each. Used with permission of Publisher of American Journal of Obstetrics & Gynecology.

apparent. In situ hybridization to a blastomere from a human polyspermic embryo is demonstrated in Panel 1F. Two clear major signal domains are noted. Two other cells from this same preembryo had two clear signals as well (data not shown) and presumably this represents an XYY karyotype.

CONCLUSION

In situ hybridization is possible for analysis of blastomeres obtained via embryo biopsy. It offers a method for numerical chromosome analysis as well as for sexing by a rapid analysis. Contamination appears not to be as significant a problem as it is with PCR. Further technical development is required to make this a more efficient procedure. In addition, the problem of mosaicism must be addressed. In situ hybridization provides a complementary technique to PCR and has promise for preimplantation genetic diagnosis.

Acknowledgements: We would like to thank Christine Disteche and David Houseman for providing probe sets, Jacques Cohen and Henry Malter for teaching us partial zona dissection and Beth Talansky for teaching us zona drilling. The work was supported by the ACOG-Ortho Fellowship Award and the AFS Serono Fellowship Award.

REFERENCES

1. R. K. Saiki, D.H. Gelfand, S. Stoffel, et. al Primer directed enzymatic amplification of DNA with a thermostable DNA polymerase. Science, 239:487-491 (1988).
2. P. L. Lichter, T. Cremer, J. Borden, L. Manuelidis and D.C. Ward, Delineation of individual human chromosomes in metaphase and interphase cells by in situ suppression hybridization using recombinant DNA libraries. Hum. Genet., 80:224-234 (1988).
3. A. H. Handyside, E.H. Kontogianni, K. Hardy, and R.M.L. Winston, Using pregnancies from biopsied human preimplantation embryos sexed by Y specific DNA amplification, Nature, 344:768-770 (1990).
4. R. L. Gardner, R.G. Edwards, Control of sex ratio at full term in the rabbit by transferring sexed blastocysts. Nature, 218:346-348 (1968).
5. E. L. Signh, W.C.D. Hare, The feasibility of sexing bovine morula state embryos prior to embryo transfer. Theriogenology, 14:421-427 (1980).
6. L. Picard, W.A. King, K.J. Betterridge, Production of sexed calves from frozen thawed embryos. Bet. Rec., 117:603-608 (1985).
7. M. Leonard, M. Kirszenbuam, C. Cotinot, P. Chesne, Y. Heyman, M.G. Stinnakre, C. Bishop, C. Delouis, M. Vaiman, M. Fellows, Sexing bovine embryos using y chromosome specific DNA probes. Theriogenology, 27:248 (1987).
8. M. Monk, A. Handyside, K. Hardy, D. Whittingham, Preimplantation diagnosis of deficiency of hypoxanthine phosphoribosyl transferase in a mouse model for Lesch-Nyhan Syndrome. Lancet, ii:423-425 (1987).

9. P. Summers, J.M. Campbell, M.W. Miller, Normal in vivo development of marmoset monkey embryos after trophectoderm biopsy. Human Reproduction, 3:389-393 (1988).
10. J. D. West, J.R. Gosden, R.R. Angell, et al, Sexing the human preimplantation embryo by DNA:DNA in situ hybridization. Lancet, ii:1345-1347 (1987).
11. A. H. Handyside, R.J.A. Penketh, R.M.L. Winston, J.K. Pattinson, J.D.A. Delhanty, E.G.D. Tuddenham, Biopsy of human preimplantation embryos and sexing by DNA amplification. Lancet, i:347-349 (1989).
12. C. Holding, M. Monk, Diagnosis of beta thalassaemia by DNA amplification in single blastomeres from mouse preimplantation embryos. Lancet, ii:532-535 (1989).
13. L. J. Wilton, J.M. Shaw. A.O. Trounsen, Successful single-cell biopsy and cryopreservation of preimplantation mouse embryos. Fertil. Steril., (1989) 51:513-517.
14. L. J. Wilton, A.O. Trounsen, Biopsy of preimplantation mouse embryos: development of micromanipulation embryos and proliferation of single blastomeres in vitro. Biol. Reprod., 40:145-152 (1989).
15. H. Malter, and J. Cohen, Partial zona dissection of the human ooctye:a nontraumatic method using micromanipulation to assist zona pellucida penetration. Fertil. Steril., 51:139-148 (1989).
16. J. Gordon, B.E. Talansky, Assisted fertilization by zona drilling: A mouse model for correction of oligospermia. J. Exp. Zool., 239:347-54 (1986).
17. C. M. Disteche, U. Trantravahi, S. Gandy, N. Eisenhard, D. Adler, and L.M. Kunkel, Isolation and characterization of two repetitive DNA fragments located near the centromere of the mouse X chromosome. Cytogenet. Cell. Genet., 39:262-268 (1985).
18. R. B. Kent, D.A. Fallows, E. Geissler, T. Glaser, J. Rettig-Emanuel, P.A. Lalley, R. Levenson, and D.E. Houseman, Genes encoding the alpha and beta subunits of sodium/potassium ATPase are located on three different chromosomes in the mouse. PNAS, 84:5369-5373 (1984).
19. J. Burns, V. Chan, J.A. Jonasson, K.A. Fleming, S. Taylor, and J.O.D. McGee, Sensitive system for visualising biotinylated DNA probes hybridised in situ: rapid sex determination of intact cells. J. Clin. Pathol., 38:1085-1092 (1985).

PART IV

CLINICAL ASPECTS OF PREIMPLANTATION GENETICS

IN VITRO FERTILIZATION AND PREIMPLANTATION DIAGNOSIS

Andre Van Steirteghem, Jiaen Liu, Etienne Van den Abbeel, Inge Liebaers[1] and Paul Devroey

Centre for Reproductive Medicine and [1]Department of Medical Genetics, Academisch Ziekenhuis, Vrije Universiteit Brussel, Belgium

INTRODUCTION

The preimplantation diagnosis of genetic disorders became possible since the availability of in vitro fertilization technology, the advances made in the micromanipulation of gametes or embryos and the ultra-sensitive procedures to detect gene mutations. The purpose of this report is to analyze the actual performance and possibilities of three important features influencing the clinical application of preimplantation diagnosis. In sequence we shall discuss the results of assisted procreation procedures, the cryopreservation of excess human embryos and the outcome of an experimental mouse study that combined embryo biopsy and cryopreservation of the biopsied embryos.

ASSISTED PROCREATION

The clinical application of preimplantation diagnostic procedures requires a robust and successful assisted procreation program. In the United Kingdom the Fifth Report of the Interim Licensing Authority for Human In Vitro Fertilization and Embryology[1] reported that for the period between 1985 to 1988 the percentage of live births per treatment cycle ranged from 8.6 to 10.0. For 1988 the Report mentioned differences in the live birth rate in 10 small (5.7%), 19 medium (8.7%) and 6 large (9.7%) Centers. This difference could indicate that the success of IVF is related to the experience of the clinical and scientific staff of a Centre.

The performance of medically assisted procreation procedures will be illustrated by the results in the Center for Reproductive Medicine of the Vrije Universiteit Brussel since 1985. Between January 1985 and December 1989, we started 4667 treatment cycles for in vitro fertilization and embryo transfer (IVF-ET), gamete or zygote intrafallopian transfer (GIFT-ZIFT). The details of these procedures have been reported extensively in several publications[2-9].

Preimplantation Genetics, Edited by Y. Verlinsky and A. Kuliev, Plenum Press, New York, 1991

The couples were suffering from longstanding infertility that was mainly due to tubal disease (one third of cycles), andrological subfertility (20%) and idiopathic infertility (20%). In one out of every five couples multiple causes of infertility were detected. Among the less frequent indications we noticed endometriosis, immunologic infertility, polycystic ovarian disease and dysovulation. An overview of the relevant data of the treatment cycles is summarized in Table 1. The number of treatment cycles in our Centre increased substantially during this five year period; the number of trials in 1989 (n=1243) was more than three times higher than in 1985 (n=390). From 1985 to 1987 between 11 and 16% of the started cycles had to be cancelled prior to oocyte retrieval. In 1988 and 1989 less than 6% of the cycles were stopped. This reduction in the number of cancelled cycles coincided with the introduction in late 1987 of the ovarian stimulation protocol combining the gonadotrophin releasing hormone analog buserelin (Suprefact®, Hoechst, Belgium) with hMG. In 1988 and 1989 this stimulation protocol was used in almost all cycles. The type of ovarian stimulation protocol that was used was also reflected in the mean number of oocytes that could be retrieved per pick-up: from 5.6 and 6.5 in 1985 and 1986 to 8.8 and 9.9 in 1988 and 1989. The reduction in the number of GIFT procedures in 1988 and 1989 was due to the fact that in case where a tubal transfer was possible, we selected more often to replace fertilized oocytes. The fertilization rate of the inseminated oocytes was rather constant over the five year period (from 55% to 62%). Even so, the cleavage rate of the zygotes that were left in culture did not vary a lot. As the number of cycles increased, more fertilized oocytes and early embryos were cryopreserved.

The pregnancy rate per transfer during this five year period is summarized in Table 2. No valid comparisons can be made between the three procedures. This would require a prospective controlled comparative study. For the total period the pregnancy rate per started cycle was 20.7% (967/4667) and the conception rate per oocyte retrieval was 22.8% (967/4244). The chance for a transferred embryo or zygote to generate a fetal sac was 14.6% per replaced embryo (1905 embryo transfers) and 20.5% per replaced zygote (420 ZIFT procedures). The analysis of the pregnancy outcome revealed that fetal loss occurred in one third of the pregnancies. In the children born after assisted procreation the incidence of congenital abnormalities did not differ from such incidence in those born after natural conception.

In 1990 the number of oocyte retrievals was even higher than in the previous years (1459). The number of cancelled cycles was very low (66) and the pregnancy rate per started cycles was 23.9%. If we include the pregnancies obtained after replacement of frozen and later thawed supernumerary embryos the conception rate per cycle was 26.2%. Taking into account a fetal loss of about one third of the conceptions, a take home baby rate of 18% per cycle can be expected.

Table 1. Overview of Assisted Procreation Procedures from 1985 to 1989.

	1985	1986	1987	1988	1989	1985-1989
No. of started cycles	390	777	1111	1146	1243	4667
No. of oocyte retrievals	333	655	984	1100	1172	4244
(Percent of started cycles)	(85.4%)	(84.3%)	(88.6%)	(96.0%)	(94.3%)	(90.9%)
No. of oocytes retrieved	1857	4252	7655	9665	11619	35048
(Mean per pick-up)	(5.6)	(6.5)	(7.8)	(8.8)	(9.9)	(8.3)
No. of oocytes replaced / No. of GIFT	45/15	561/192	455/155	146/49	54/17	1257/428
(Mean per GIFT)	(3.0)	(2.9)	(2.9)	(2.9)	(3.1)	(2.9)
No. of oocytes inseminated	1812	3677	7041	9208	11268	33006
No. of oocytes fertilized	995	2269	4193	5240	6176	18873
(Percent fertilization)	(55%)	(61.7%)	(59.6%)	(56.9%)	(54.8%)	(57.2%)
No. of zygotes replaced / no. of ZIFT	-	-	115/42	470/181	549/204	1134/427
(Mean per ZIFT)	-	-	(2.7)	(2.6)	(2.7)	(2.7)
No. of zygotes frozen	-	363	260	287	272	1182
No. of zygotes cultured in vitro	930	1717	3510	4259	5063	15479
No. of zygotes further cleaved as						
type 1 + 2 embryos (%)	589 (63%)	1082 (63%)	2088 (60%)	2265 (53%)	2784 (55%)	8808 (57%)
type 1 + 2 + 3 embryos (%)	-	1396 (81%)	2927 (83%)	2958 (70%)	3277 (65%)	-
No. of embryos transferred/ no. of transfers	768/273	823/352	1486/624	1537/620	1871/716	6485/2585
(Mean per transfer)	(2.8)	(2.3)	(2.4)	(2.5)	(2.6)	(2.5)
No. of embryos frozen	82	479	810	907	1056	3334

Table 2. Pregnancies after GIFT, ZIFT and IVF-ET

	1985	1986	1987	1988	1989	Total
No. of pregnancies/no. of GIFT	4/15	57/192	34/155	16/49	13/17	124/428
(Percent)	(27%)	(28%)	(22%)	(33%)	(76%)	(29%)
No. of pregnancies/no. of ZIFT	-	-	18/42	73/181	80/204	171/427
(Percent)	-	-	(43%)	(40%)	(39%)	(40%)
No. of pregnancies/no. of ET	43/273	93/352	141/624	181/620	214/716	672/2585
(Percent)	(16%)	(26%)	(23%)	(29%)	(30%)	(26%)

The performance and outcome of medically assisted procreation in a particular Centre can certainly be positively influenced by several factors such as the number of patients treated, the presence of a specialized and qualified multidisciplinary clinical, laboratory and research staff, the use of standardized protocols for the different procedures including stringent quality control procedures.

CRYOPRESERVATION OF HUMAN EMBRYOS

The main application of cryopreservation of human embryos obtained after IVF is the freezing of supernumerary embryos. To reduce the risk of multiple pregnancy without compromising the success rate after IVF, it is now a generally accepted practice to limit the number of transferred concepti (oocytes, zygotes or embryos) to a maximum of three. Excess embryos have been cryopreserved for eventual later use[10-12]. Freezing of human embryos is also very useful in oocyte and embryo donation by circumventing the difficult synchronization of the ovarian cycles of the donor and the acceptor[13-15]. Cryopreservation of embryos might also be necessary for preimplantation diagnosis if the time required for the diagnostic procedure would require more time than the biopsied embryo can be kept in culture. Two world surveys on human embryo cryopreservation indicated that after freezing slightly more than half of the embryos survived the thawing procedure and could be transferred[16,17]. The results of the human cryopreservation of the Dutch-speaking Free University of Brussels (VUB) Centre for Reproductive Medicine are summarized in Table 3. These results report the experience since the beginning of our human embryo freezing in 1985 until the end of 1990.

The chance for a cryopreserved embryo to implant in the uterus is markedly lower than for a "fresh" one (about 5% for a cryopreserved embryo versus about 15% for a fresh embryo). In our program, out of 1428 pregnancies obtained until December 1990 after assisted procreation, 116 pregnancies (8.1% of the total) were obtained after the transfer of frozen-thawed embryos. Among the factors influencing the outcome of the cryopreservation of excess human embryos, we could mention the ovarian stimulation protocol,the morphological quality and the development stage of the supernumerary embryos at the time of freezing as well as the methodology used to freeze and thaw the embryos[11,12,18-31].

Table 3. Results of Human Embryo Cryopreservation at the Brussels' University.

Total no. of frozen embryos	6281
No. of embryos thawed to be transferred	3522
No. of embryos of sufficient quality after thawing to be replaced	1484 (42.1 %)
No. of transfers of frozen-thawed embryos	800
No. of pregnancies	116
Pregnancy rate per transfer	14.5 %

We investigated the influence of blastomere loss of frozen-thawed embryos on their implantations after transfer. In a series of 927 frozen-thawed embryos that were replaced in 598 transfers we obtained 90 pregnancies and 101 implantations (fetal sacs on ultrasound). Fetal loss occurred in 26% of the implanted embryos. Of the embryos transferred, 64% (593) were fully intact and 36% (334) had undergone some blastomere loss not exceeding 50% of the initial number of blastomeres present. The pregnancy rate per transfer was 15.9% for intact embryos and 10.9% for embryos who had undergone blastomere loss (NS). The implantation rate per embryo transferred was respectively 13.8% and 7.7% ($P<0.05$) and the fetal loss was 18.4% and 50.0% ($P<0.03$). Prior to freezing the morphological quality was similar in the embryos remaining intact after freezing and thawing and those exhibiting some loss of blastomeres during cryopreservation. Sixteen implantations after transfer of cryopreserved embryos with blastomere loss resulted in eight (pre)clinical abortions, six healthy children and two ongoing pregnancies. Thirty-eight implantations after transfer of intact cryopreserved embryos resulted in seven (pre)clinical abortions and thirty-one healthy children. The results clearly indicated that blastomere loss after cryopreservation impaired the implantation and further development of the embryo. Whether blastomere loss is an indication of less intrinsic embryonic quality inducing a decreased implantation or whether the blastomere loss itself was the reason for the reduced implantation requires further clarification.

BIOPSY AND CRYOPRESERVATION OF MOUSE EIGHT-CELL EMBRYOS

In the preimplantation diagnostic procedure, it might be necessary to combine the biopsy of one or two blastomeres and the cryopreservation of the biopsied embryo. In 1989 Alan Trounson's group at Monash University reported that sampling a single blastomere from the four-cell mouse embryo does not compromise continued development *in vitro* or *in vivo*. They also demonstrated that the biopsied embryo can be successfully cryopreserved by ultrarapid freezing even though they had a punctured zona pellucida[32].

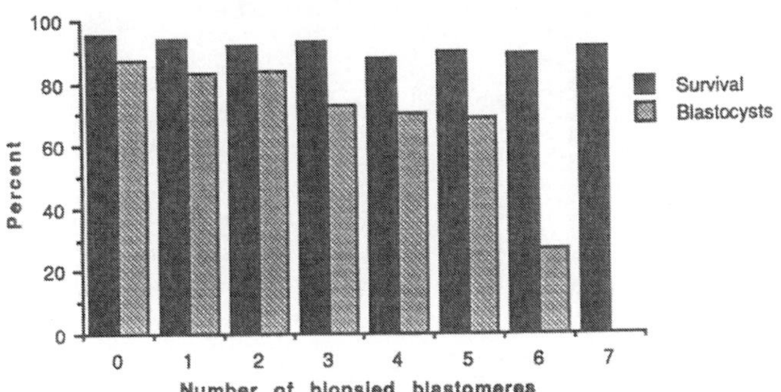

Figure 1. Survival and *in vitro* development of mouse eight-cell embryos to early blastocysts. The embryos where the zona pellucida was punctured without the removal of blastomeres is indicated as group 0 on the x-axis.

We also investigated the in vitro and in vivo developmental capacity of biopsied mouse eight-cell embryos. One, two, three, ... up to seven blastomeres were removed by micromanipulation from eight-cell mouse embryos. These biopsied embryos were then cryopreserved with the rapid freezing and slow thawing procedure using the cryoprotectant 1,2-propanediol[11]. The surviving embryos were left in culture until they reached the blastocyst stage. They were then transferred to the uterine horns of day three pseudo-pregnant mice. Appropriate controls were included at the different stages of the experiments.

The survival after freezing and thawing of biopsied embryos and the in vitro development in culture to the blastocyst stage is summarized in Figure 1. The control group consisted of embryos where the zona pellucida was punctured without removal of blastomeres. In each of the experimental conditions the number of embryos ranged between 44 and 175. Ninety-five percent of the control embryos survived the freezing and thawing. In the different experimental groups the survival was similar, ranging from 88% to 94%. After 48 h to 72 h of further in vitro culture the percentage of blastocysts was 87%. When one, two or three blastomeres had been removed prior to freezing, the in vitro developmental potential was not different from the control group. When more than three blastomeres had been removed, significantly fewer embryos could reach the blastocyst stage.

The biopsied embryos that reached the blastocyst stage were transferred into one of the uterine horns of day three pseudopregnant mice, while the controls (nonbiopsied blastocysts) were transferred in the contralateral horn. On day 17 of pregnancy we counted the number of implantation

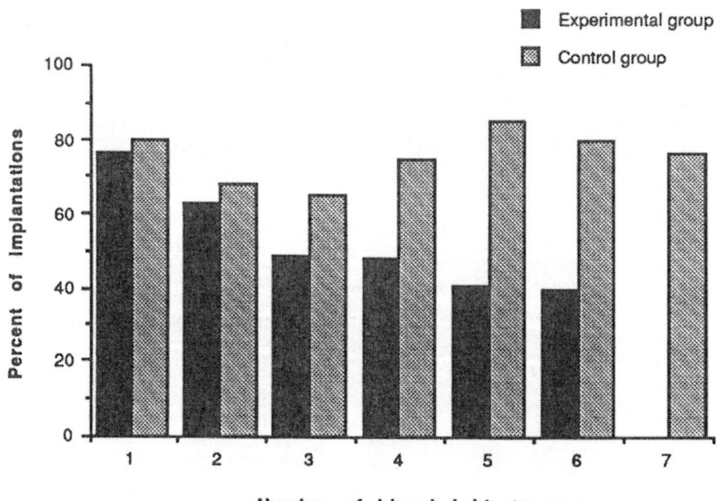

Figure 2. In vivo development of cryopreserved biopsied mouse eight-cell embryos. The percent of transferred biopsied and control embryos that had implanted is indicated for each of the experimental conditions.

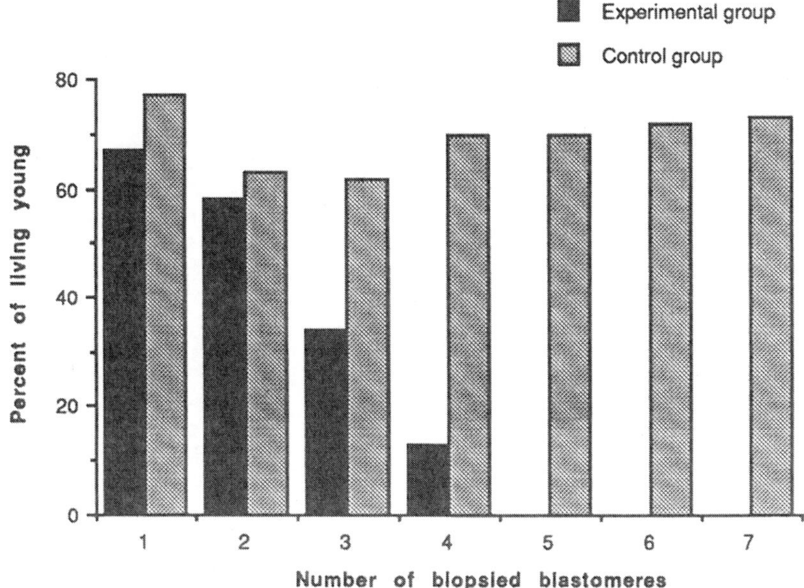

Figure 3. *In vivo* development of biopsied mouse eight-cell embryos. The percentage of living young is indicated for the experimental and the control groups.

sites and living young in each of the uterine horns (Figures 2 and 3). The experimental group had a similar number of implantation sites as the control group when one, two or three blastomeres had been removed (Figure 2). The number of living young that was generated was significantly lower than in the control group after the removal of three and more blastomeres. No living young were obtained after removal of five, six or seven blastomeres (Figure 3).

SUMMARY

The report analyzed three features that will influence the clinical application of preimplantation diagnosis: the performance of medically assisted procreation including the cryopreservation of supernumerary human embryos and the *in vitro* and *in vivo* developmental potential of biopsied and cryopreserved mouse eight-cell embryos.

Acknowledgments

The clinical, scientific and paramedical staff of the Centre for Reproductive Medicine is kindly acknowledged. This work was supported by grants 3.00356.85 and 3.0065.87 from the Belgian Fund for Medical Research.

REFERENCES

1. The Fifth Report of the Interim Licensing Authority for Human In Vitro Fertilisation and Embryology, 1990.

2. Khan, I., Camus, M., Deschacht, J., Smitz, J., Staessen, C., Van Waesberghe, L., Wisanto, A., Devroey, P. and Van Steirteghem, A.C. Success rate in gamete intra-Fallopian transfer using low and high concentrations of washed spermatozoa. Fertil Steril 50:922-927, 1988.
3. Smitz, J., Devroey, P., Camus, M., Khan, I., Staessen, C., Van Waesberghe, L., Wisanto, A. and Van Steirteghem, A.C. Addition of Buserelin to human menopausal gonadotrophins in patients with failed stimulations for IVF or GIFT. Hum Reprod 3, suppl. 2:35-38, 1988.
4. Staessen, C., Camus, M., Khan, I., Smitz, J., Van Waesberghe, L., Wisanto, A. and Van Steirteghem, A.C. An 18-month survey of infertility treatment by in vitro fertilization, gamete and zygote intrafallopian transfer, and replacement of frozen-thawed embryos. J In Vitro Fert Embryo Transf 6:22-29, 1989.
5. Palermo, G., Devroey, P., Camus, M., De Grauwe, E., Khan, I., Staessen, C., Wisanto, A. and Van Steirteghem, A.C. Zygote intra-Fallopian transfer as an alternative treatment for male infertility. Hum Reprod 4:412-415, 1989.
6. Devroey, P., Staessen, C., Camus, M., De Grauwe, E., Wisanto, A. and Van Steirteghem, A.C. Zygote intrafallopian transfer as a successful treatment for unexplained infertility. Fertil Steril 52:246-249, 1989.
7. Wisanto, A., Janssens, R., Deschacht, J., Camus, M., Devroey, P. and Van Steirteghem, A.C. Performance of different embryo transfer catheters in a human in vitro fertilization program. Fertil Steril 52:79-84, 1989.
8. Wisanto, A., Bollen, N., Camus, M., De Grauwe, E., Devroey, P. and Van Steirteghem, A.C. Effect of transuterine puncture during transvaginal oocyte retrieval on the results of human in-vitro fertilization. Hum Reprod 4:790-793, 1989.
9. Staessen, C., Van Den Abbeel, E., Carle, M., Khan, I., Devroey, P. and Van Steirteghem, A.C. Comparison between human serum and Albuminar-20 (TM) supplement for in-vitro fertilization. Hum Reprod 5:336-341, 1990.
10. Trounson, A., Mohr, L. Human pregnancy following cryopreservation, thawing and transfer of an eight-cell embryo. Nature 305:707-709, 1983.
11. Lassalle, B., Testart, J., Renard., J.P. Human embryo features that influence the success of cryopreservation with the use of 1,2 propanediol. Fertil Steril 44(5):645-651, 1985.
12. Van Steirteghem, A.C., Van den Abbeel, E., Camus, M., Van Waesberghe, L., Braeckmans, P., Khan, I., Nijs, M., Smitz, J., Staessen, C., Wisanto, A. and Devroey, P. Cryopreservation of human embryos obtained after gamete intra-Fallopian transfer and/or in vitro fertilization. Hum Reprod 2:593-598, 1987.
13. Van Steirteghem, A.C., Van den Abbeel, E., Braeckmans, P., Camus, M., Khan, I., Smitz, J., Staessen, C., Van Waesberghe, L., Wisanto, A. and Devroey, P. Pregnancy with a frozen-thawed embryo in a woman with primary ovarian failure. N Engl J Med 317:113, 1987.

14. Devroey, P., Smitz, J., Camus, M., Wisanto, A., Deschacht, J., Van Waesberghe, L. and Van Steirteghem, A.C. Synchronization of donor's and recipient's cycles with GnRH analogues in an oocyte donation program. Hum Reprod 4:270-274,1989.
15. Devroey, P., Camus, M., Van den Abbeel, E., Van Waesberghe, L., Wisanto, A. and Van Steirteghem, A.C. Establishment of 22 pregnancies after oocyte and embryo donation. Br J Obstet Gynaecol 96:900-906, 1989.
16. Van Steirteghem, A.C. and Van den Abbeel, E. Survey on cryopreservation. Ann N Y Acad Sci 541:571-574, 1988.
17. Van Steirteghem, A.C. and Van den Abbeel, E. World results of human embryo cryopreservation. In Proceedings of the VI World Congress of In Vitro Fertilization and Alternate Assisted Procreation. In Press.
18. Cohen, J., Simons, R.S., Fehilly, C.B., Edwards, R.G. Factors Affecting Survival and Implantation of Cryopreserved Human Embryos. J In Vitro Fert Embryo Transf 3:46-52, 1986.
19. Testart, J., Lassalle, B., Forman, R., Gazengel, A., Belaïsch-Allart, J., Hazout, A., Rainhorn, J.D., Frydman, R. Factors influencing the success rate of human embryo freezing in an in vitro fertilization and embryo transfer program. Fertil Steril 48:107-112, 1987.
20. Mandelbaum, J., Junca, A.M., Plachot, M., Alnot, M.O., Alvarez, S., Debache, C., Salat-Baroux, J. and Cohen, J. Human embryo cryopreservation, extrinsic and intrinsic parameters of success. Hum Reprod 2:709-715, 1987.
21. Van den Abbeel, E., Van der Elst, J., Van Waesberghe, L., Camus, M., Devroey, P., Khan, I., Smitz, J., Staessen, C., Wisanto, A. and Van Steirteghem, A.C. Hyperstimulation: the need for cryopreservation of embryos. Hum Reprod 3, suppl. 2:53-57, 1988.
22. Cohen, J., DeVane, G., Elsner, C.W., Fe, Hilly, C.B., Kort, H.I., Massey, J.B., Turner, T.G. Cryopreservation of zygotes and early cleaved human embryos. Fertil Steril 49:283-289, 1988.
23. Fugger, F., Bustillo, M., Katz, L., Dorfmann, A., Bender, S., Schulman, J. Embryonic development and pregnancy from fresh and cryopreserved sibling pronucleate human zygotes. Fertil Steril 50:273-278, 1988.
24. Mandelbaum, J., Junca, A.M., Plachot, M., Alnot, M.O., Salat-Baroux, J., Alvarez, S., Tibi, C., Cohen, J., Debache, C., Tesquier. Cryopreservation of human embryos and oocytes. Hum Reprod 3:117-119, 1988.
25. Todorow, S.J., Siebzehnruebl, E.R., Spitzer, M., Koch, R., Wildt, L., Lang, N. Comparative results on survival of human and animal eggs using different cryoprotectants and freeze-thawing regimens. II Human. Hum Reprod 4:812-816, 1989.
26. Cohen, J., Wiemer, K., Wright, G. Prognostic value of morphologic characteristics of cryopreserved embryos: a study using videocinematography. Fertil Steril 49:827-834, 1988.

27. Hartshorne, G.M., Wick, K., Dyson, H. Effect of cell number at freezing upon survival and viability of cleaving embryos generated from stimulated IVF cycles. Hum Reprod 5:857-861, 1990.
28. Lornage, J., Boulieu, D., Mathieu, C., Guerin, J.F., Pinatel, M.C., James, R., Alvarado, C. Transfers of frozen-thawed human embryos in cycles stimulated by HMG. Hum Reprod 5:60-65, 1990.
29. Siebzehnrübl, E.R., Todorow, S., Van Uem, J., Koch, R., Wildt, L., Lang, N. Cryopreservation of human and rabbit oocytes and one-cell embryos: a comparison of DMSO and propanediol. Hum Reprod 4:312-317, 1989.
30. Wright, G., Wilker, S., Elsner, C., Kort, H., Massey, J., Mitchell, D., Toledo, A., Cohen, J. Observations on the morphology of pronuclei and nucleoli in human zygotes and implications for cryopreservation. Hum Reprod 5;109-115, 1990.
31. Gordts, S., Roziers, P., Campo, R., Noto, V. Survival and pregnancy outcome after ultrarapid freezing of human embryos. Fertil Steril 53:469-472, 1990.
32. Wilton, L., Shaw, J., Trounson, A. Successful single-cell biopsy and cryopreservation of preimplantation mouse embryos. Fertil Steril 51:513-517, 1989.

UTERINE LAVAGE FOR PREIMPLANTATION GENETIC DIAGNOSIS

Bruno Brambati and Leonardo Formigli

Center for Reproductive Medicine
Milano, Italy

INTRODUCTION

There is a growing interest in early prenatal diagnosis to limit the physical and psychological involvement of the patient when facing selective abortion. By early transabdominal chorionic villus sampling and rapid analytical techniques, genetic diagnosis is now available before the 6th post-conceptional week, and selective abortion might be obtained by medical rather than surgical methods.[1,2] Nevertheless, there are couples at risk for hereditary diseases who cannot accept pregnancy termination for emotional, moral, or religious reasons. As an alternative, a number of these couples have shown a great interest in preimplantation genetic diagnosis (Table 1).

Preimplantation genetic diagnosis can be attained through either in vitro fertilization (IVF), or through uterine cavity lavage after natural conception[3-5] as 4 days after fertilization the embryo floats freely in the uterus and is readily accessible by flushing the uterine cavity (Table 2).

As a result of natural fertilization the zygote undergoes a rapid succession of mitotic divisions as it moves down the tube and into the uterine lumen, from where it may be recovered at the blastocyst stage for trophectoderm biopsy.

Table 1. Reasons of non-acceptance of post-implantation genetic tests.

- History of repeated genetic abortion
- Long term psychological side-effects following previous early or late genetic abortions
- Moral or religious objection to abortion
- Prenatal attachment (maternal-fetal bonding) has already occurred: abortion is equivalent to loss of a wanted child, therefore emotional reluctance towards interruption may follow

Preimplantation Genetics, Edited by Y. Verlinsky and A. Kuliev, Plenum Press, New York, 1991

Table 2. Methods that are being considered for preimplantation genetic diagnosis

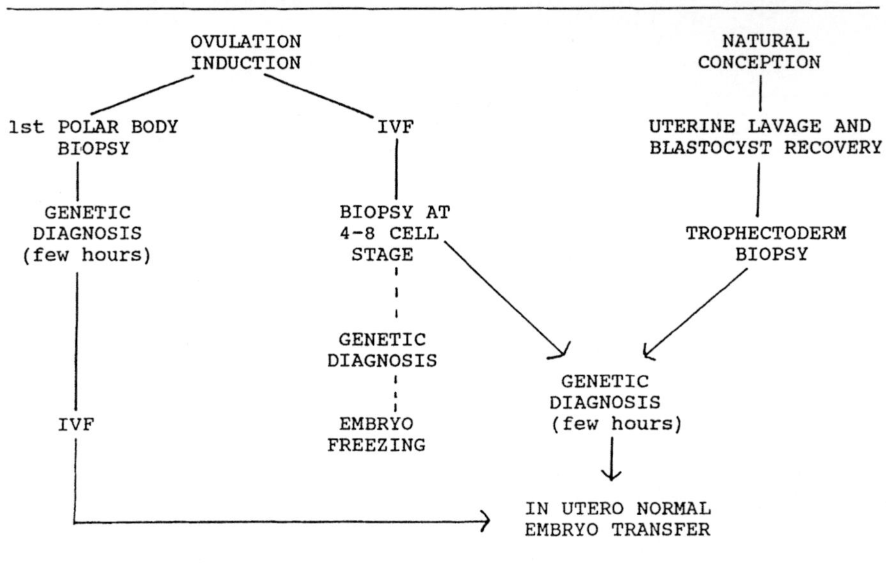

There are both advantages and inconveniences in uterine lavage compared to other methods (Table 3). Because of the incomplete knowledge regarding the efficiency and the risks of existing approaches to preimplantation diagnosis the choice of the method depends merely on the individual feeling and a social and cultural standing of a specific patient. IVF, either preceded by first polar body biopsy, or followed by blastomere biopsy seems to be mandatory only in case of a pre-existing condition of infertility. Considering that our high genetic risk couples are generally young and fertile, we find it suitable to offer them pre-implantation genetic diagnosis without significantly interfering with the quality of their reproductive and sexual life. Therefore, we set up a program including blastocyst recovery and biopsy after spontaneous oocyte maturation, normal and free intercourses, and natural conception as an alternative to cycle manipulation and invasive IVF procedures.

UTERINE FLUSHING FOR OVUM DONATION

A clinical experience in uterine cavity flushing was gained during an ovum donation program. In this program embryos recovered from the uterus of fertile female donors were transferred to infertile synchronized recipients,[6] the donor being preferably recruited from the recipient's family. The donor had to exhibit certain characteristics such as: proven fertility, good health, lack of detectable genetic or infectious diseases, regular cycles, and to be under 38 years of age. Although initially only natural cycles were used, monitoring ovulation by daily ultrasound scannings from 3-4 days before presumed ovulation made it necessary to increase the number of recovered embryos by super-ovulation. This was accomplished either by oral clomiphene citrate (CC) 100 mg daily from cycle day 5 to 9, or by human menopausal gonadotropins (HMG) 150 IU i.m. daily from cycle day 3 until the dominant follicle reached a diameter of 16-18 mm. Thereafter, human chorionic gonadotropins (hCG) 10,000 IU, i.m. was administered and ultrasound monitoring was continued to ascertain the number of ovulated follicles.

Most of the donors were artificially inseminated with 1 ml of fresh semen from the recipient's husband. In only a few cases semen from the ovum donor's husband was used, and therefore, natural intercourse took place at the time of ovulation. In natural cycles insemination started when the mean follicular diameter reached 18 mm and was repeated every other day until sonographic documentation of ovulation. On the contrary, insemination was performed only once in stimulated cycles, 36 hours after hCG injection.

The device used for uterine flushing was a 4.7 mm external diameter plastic balloon catheter with a rounded extremity, housing a stainless steel cannula threaded through an existing lateral arm.[7] The fluid was delivered into the uterine cavity by the cannula from the very tip of the catheter, and was recovered in the collecting system through lateral openings between the balloon and the tip of the catheter. The fluid was collected into a funnel shaped container and left undisturbed for 15 minutes; thereafter, a few drops of the most bottom part were poured in a petri dish for embryo detection.

Table 3. Advantages and disadvantages of different pre-implantation genetic diagnosis approaches

METHOD		ADVANTAGES	DISADVANTAGES
IVF	FIRST POLAR BODY	-DIAGNOSIS BEFORE FERTILIZATION -BIOPSY SAFE FOR FETAL DEVELOPMENT -HIGH N. OOCYTES/CYCLE	-CRITICAL DIAGNOSTIC ACCURACY -LIMITED DIAGNOSTIC FIELD
	FIRST POLAR BODY and 4 - 8 CELL STAGE BIOPSY		-MEDICATION REQUIRED -ULTRASOUND AND/OR HORMONAL MONITORING -ARTIFICIAL CONCEPTION -LOW P. RATE/EMBRYO -LIMITED TO "HIGH RESPONDERS" -OVARIAN HYPERST. SYNDROME -MINIMUM AMOUNT TISSUE -HIGH COSTS -HIGH DROP OUT RATE EXPECTED
	4 - 8 CELL STAGE BIOPSY	-SEVERAL EMBRYOS/CYCLE -EMBRYO CRYOPRESERVATION (UNLIMITED TIME FOR DIAGNOSTIC ANALYSIS)	-(?) SAFE FOR EMBRYO DEVELOPMENT (?)
UTERINE LAVAGE		-MINIMUM INVASION OF PRIVACY (NATURAL CONCEPT.) -NON INVASIVE PROCEDURE -HIGH PREGNANCY RATE/EMBRYO -HIGHER AMOUNT OF TISSUE -LOWER COSTS -LOW DROP OUT RATE EXPECTED	-LOW EMBRYO RECOVERY RATE -ONLY 1 EMBRYO/CYCLE -(?) SAFE FOR EMBRYO DEVELOPMENT (?) -ENDOMETRIUM DAMAGE

Only one uterine flushing (UF) was performed per cycle. Flushing succeeded in 37 out of 88 natural cycles. In 10 cases only unfertilized or fragmented ova were recovered. Moreover, sixteen 2 cell to morula stage embryos and 11 blastocysts were observed. With CC stimulation 39 follicles developed in 17 cycles (2.3 follicles/cycle), but only 32 follicles ruptured. Flushing was successful in 9 out of 17 cases and 13 ova were recovered. Of these, 5 were unfertilized, 3 were 2 cell to morula stage embryos and 5 were blastocysts. Using HMG stimulation 121 follicles developed in 22 cycles (5.5 follicles/cycle) and only 71 ruptured. Flushing was successful 14 times and 22 ova were recovered, 11 of which were unfertilized or fragmented, 5 were 2 cell to morula stage embryos and 6 were blastocysts.

Eighteen clinical pregnancies were obtained out of 127 flushings and 46 transfers to recipients (14.1% of flushings and 40.9% of transfers). Contrary to what was expected, the harvest of ova was minutely increased by ovulation stimulation. Moreover, this advantage was almost lost by the lower rate of cleavage in stimulated cycles and no significant difference in pregnancy rate was found (Table 4).

The mean fluid recovery rate was 97.8% (range 30-100%). In 7 out of 127 cases a conspicuous amount of fluid was probably lost in the peritoneal cavity. As the human uterus has rigid and thick walls, a very limited capacity, and a low muscular tone at the utero-tubal junction, it is quite easy to send the flushing medium into the abdominal cavity by forcing the utero-tubal junction, especially if the recovery way is narrow. Retrograde flushed embryos may more easily fall back into the abdominal cavity rather than stop in the tube. Whether this involves the danger of abdominal pregnancies is not clear since in the limited number of cases performed so far,[5,6,8] no such occurrence has been reported. On the other hand, three retained uterine pregnancies occurred in our series[6]: in the first the flushing was omitted since the single follicle did not show any sonographic evidence of rupture, the second was a biochemical pregnancy, and the third required a suction termination at 6 weeks gestation. It is possible, however, that there were other unrecovered ova which did not give rise to retained pregnancies, either because they were non-viable ova or because flushing of the uterus may have produced a detrimental environment for embryo development. An improvement of the ovum recovery device may possibly lead to a higher recovery of uterine ova.

Contamination of the medium with blood is a serious obstacle to the finding of ova under the microscope. This occurred mostly when stenosis of the internal os was present. It is possible that by reducing the washing system's external diameter the occurrence of bleeding could be considerably reduced. In such cases it is useful to dilute the first few drops of settled medium (where the embryo is usually located) with fresh medium. It might be also useful to routinely add heparin to the flushing medium in order to avoid fibrin formation, which is sometimes an obstacle to the location of ova.

Table 4. Recovery rate in donor stimulated and non-stimulated cycles and pregnancy rate in recipients

CYCLES	FLUSHINGS	FOLLICLES TOTAL	FOLLICLES RUPTURED	OVA TOTAL	OVA CLEAVED	TRANSFERS	CLINICAL PREGNANCIES
NATURAL	88	93	91	37	27(73%)	27	11 (12.5%)[a] (40.7%)[b]
CC	17	39	32	13	8(61%)	7	3 (17.6%)[a] (42.8%)[b]
HMG	22	121	71	22	11(50%)	10	4 (18.1%)[a] (40.0%)[b]
ALL CYCLES	127			72	46(64%)	44	18 (14.1%)[a] (40.9%)[b]

[a]% of flushings
[b]% of transfers

Failure to recover ova may also be the consequence of an untimely washing performed before the ovum enters the uterus, due to a certain variability in the tubal transport time. However, failure may also occur for reasons other than defects in the recovery procedure. These natural obstacles include: (1) the luteinized unruptured follicle phenomenon; (2) the empty follicle syndrome; (3) the failure of the oocyte released from the dispersed follicle; (4) loss of oocytes in the peritoneal cavity; (5) tubal defects at the fimbrial level or in other segments of the tube. This may explain, in part, a rather low monthly fecundity of the normal population, which is probably the success limit of the flushing procedure with natural cycles[9].

A NEW UTERINE WASHING SYSTEM

On the basis of previous experience and for clinical use in the preimplantation genetics program a new washing system was devised in order to optimize fluid recovery rate, minimize endometrial damage, avoid retrograde tubal and peritoneal fluid leakage, and obtain the highest patient compliance[10].

The system was made from a collecting balloon silicon catheter, 4.0 mm outer diameter, and a coaxial malleable stainless steel cannula, 0.9 mm outer diameter. The proximal portion of the silicon catheter has a separate connection for collecting and inflating tubes, while the distal end does not pass over the balloon. The distal extremity of the metallic cannula has recently been modified by applying a 3 mm diameter oval ball acting as an obturator of the collecting catheter and allowing the safest conditions when pushing forward and moving the cannula in the uterine cavity (Figure 1).

Insertion and manipulation of the system is monitored by continuous ultrasound visualization. The assembled system is gently inserted into the endocervix and stopped just after passing the internal cervical os. The balloon is inflated with 2 ml of sterile water and the flushing cannula pushed under ultrasound guidance and stopped in the proximity of the fundus.

The efficiency of the system was evaluated by injecting 50 ml of sterile, warmed saline solution from a 50 ml syringe with different pressures. The study group consisted of 14 fertile women at risk for autosomal or X-linked diseases, who accepted to undergo preliminary uterine lavage while waiting for preimplantation diagnosis. A single uterine flushing per cycle was performed a few days after the basal body temperature increase for a total of 29 attempts. As embryo collection was not the purpose of the uterine lavage the patients were invited to avoid unprotected intercourse.

The system was successfully introduced into the uterine cavity in all cases, and passed through the internal cervical os without any resistance in 61% of the cases; and with some difficulty in 39%. The recovery of the lavage fluid was always 99% or more of the injected volume. The minimal fluid wastage reported may be explained by the wetting of the internal surface of both the uterine wall and the collecting tubes. The balloon and the flushing catheter were easily visualized by ultrasound and the damage of the endometrial surface was likely minimal: light blood contamination of the recovered fluid was present in only 15% of the cases. Spotting was limited to the first hours after the procedure in all cases except one, in which it lasted until the next menstrual period. No clinical infection or menstrual irregularities were reported.

In conclusion, due to its efficiency and safety, the new uterine washing system may be recommended for use in preimplantation genetic diagnosis by blastocyst recovery and biopsy.

PREIMPLANTATION GENETIC DIAGNOSIS PROGRAM

One of the aims of the program has been to evaluate the degree of acceptability of preimplantation diagnosis in a well characterized population at high genetic risk. The Milan and Cagliari Prenatal Diagnosis Centers invited a number of their patients with previous very stressful

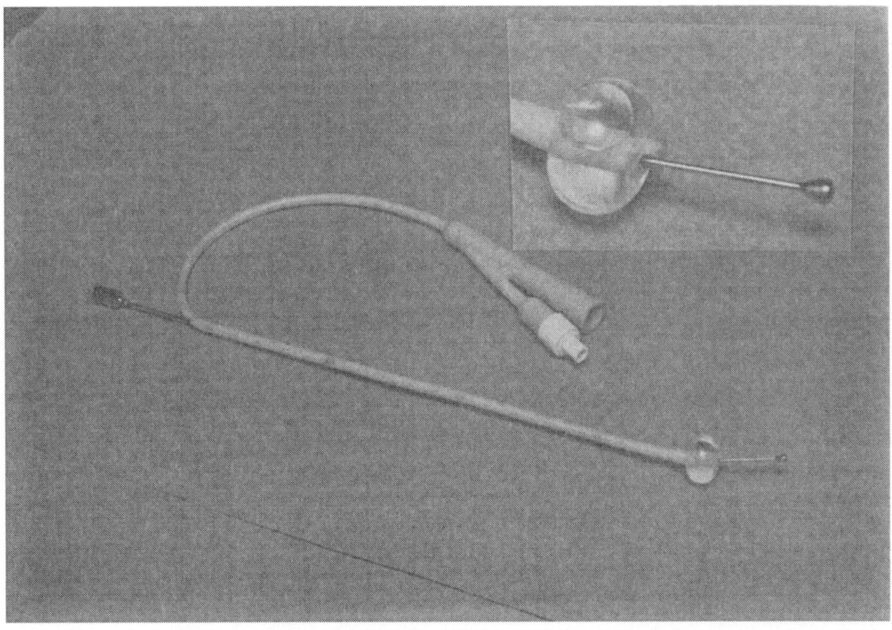

Figure 1. The ovum recovery device.

experiences of selective, first or second trimester abortion, and with unhealthy babies to attend a meeting where information about a new abortion-free genetic diagnosis method would be provided. The rate of participation was surprisingly high (80%). After a complete information on the new experimental approach was provided each couple underwent a counselling session to further debate the technical aspects, fully evaluate the acceptability of the procedure, and understand behavioral and emotional issues. A dozen couples, i.e. 75%, signed the waiting list for preimplantation diagnosis, while the remaining cases (3) preferred to postpone any decision because of a too-recent, unfavorable diagnostic experience. A very early (6-7 gestation weeks) genetic diagnosis was offered to check the reliability of the diagnostic result. Moreover, a sham uterine lavage was proposed to produce familiarity with the different technical aspects and to exclude evident anatomical inconveniences. It was also designed to train operators and assay the uterine lavage system.

The final point of the actual program would be to verify the potentials of the new method, not only to be used in selected and relatively small groups of patients, but also to be extended to the general population as an alternative to the post-implantation methods. For all these reasons no medication would be required to stimulate ovaries or to recover embryos. The washing procedure is very simple to perform, well tolerated by patients, very cheap and efficient. In addition, monitoring of ovulation is carried out by patients, themselves, and uterine lavage is made a simple monthly outpatient visit (Table 5).

ADVANTAGES AND POTENTIAL PITFALLS IN THE APPLICATION OF NON-SURGICAL EMBRYO RECOVERY PROCEDURE TO PREIMPLANTATION DIAGNOSIS

There is an obvious interest in comparing uterine flushing experience for ovum donation[5,6,8] to the expected results and problems when used for genetic diagnosis. A number of similarities and differences between methods should be stressed:

(1) The women of both groups-ovum donors and genetic patients- are probably endowed with the same fertility, on average, since donors are of proven fertility and genetic patients have often previously undergone therapeutic abortion, and/or succeeded in pregnancy.

(2) Ovum donation requires a synchronization of donors' and recipients' cycles, which may be less than optimal. The genetic application is characterized by a single patient, who is at the same time a donor and a recipient, with no particular problem of synchronization.

(3) Washing of the uterus involves a certain opening of the cervical canal with induced uterine contractions, potential

bacterial colonization of the uterine cavity, possible bleeding from the cervix and/or uterine cavity, and, lastly, washing out the endometrial secretions possibly required for embryo development and implantation. All these considerations are of no importance in ovum donation, whereas in the genetic application they may be of a great importance; particularly if one wishes to replace the embryo immediately after the diagnosis during the same flushing cycle.

(4) Microbiopsy of the embryo has no place in ovum donation whereas in genetic application this step is of primary importance and may lead to a lowering of the pregnancy success rate.

(5) Freezing of the embryo may be necessary in case of uterine bleeding after uterine flushing for genetic diagnosis. This step could further decrease the pregnancy success rate.

In conclusion, by comparing the positive and the negative aspects of the two applications of uterine lavage one may anticipate that the genetic one has lower chances of success than the ovum donation application. Therefore, we should be very cautious when counselling about the efficiency of uterine lavage for preimplantation genetic diagnosis.

Table 5. Protocol of the Clinical Preimplantation Diagnosis Study

(1) Monitoring of the ovulation is achieved by instructing patients to perform daily urine LH test using commercially available home kits.

The daily test starts 3 days before presumed ovulation; monitoring should continue for 2 days after the first detection of LH increase.

Ultrasound monitoring of ovulation is offered to the candidates who live in the neighborhood of the clinical center.

(2) Intercourses should occur in the 24 hours after the detection of LH 'Peak'; abstinence is recommended during the late follicular phase.

(3) Uterine lavage for embryo recovery is scheduled approximately 6 days after LH 'peak': the exact timing is determined by studying LH curve; no more than 1 lavage per cycle should be done.

(4) Embryo recovery and biopsy, genetic analysis, and embryo replacement are performed in the same cycle.

REFERENCES

1. B. Brambati, L. Tului, G. Simoni, and M. Travi, Prenatal diagnosis at 6 weeks, Lancet, ii:397 (1988).
2. UK Multicentre Trial, The efficacy and tolerance of mifepristone and prostaglandin in first trimester termination of pregnancy, Br. J. Obstet. Gynaecol., 97:480 (1990).
3. A. H. Handyside, E.H. Kontogianni, K. Hardy, and R.M.L. Winston, Pregnancies from biopsied human preimplantation embryos sexed by Y-specific DNA amplification, Nature, 344:768 (1990).
4. C. M. Strom, Y. Verlinsky, S. Milayeva, S. Evsikov, J. Cieslak, A. Lifchez, J. Valle, J. Moise, N. Ginsberg, and M. Applebaum, Preconception genetic diagnosis of cystic fibrosis, Lancet, ii:306 (1990).
5. J. E. Buster, M. Bustillo, I.A. Rodi, S.W. Cohen, M. Hamilton, J.A. Simon, I.H. Thorneycroft, and J.R. Marshall, Biologic and morphologic development of donated ova recovered by nonsurgical uterine lavage, Am. J. Obstet. Gynecol., 153:211 (1985).
6. L. Formigli, C. Roccio, G. Belotti, A. Stangalini, M.T. Coglitore, and G. Formigli, Non-surgical flushing of the uterus for pre-embryo recovery: possible clinical applications, Hum. Reprod., 5:329 (1990).
7. L. Formigli, G. Formigli, and C. Roccio, Donation of fertilized uterine ova to infertile women, Fertil. Steril., 47:162 (1987).
8. M. V. Sauer, R.E. Anderson, and R.J. Paulson, A trial of superovulation in ovum donors undergoing uterine lavage, Fertil. Steril., 51:131 (1989).
9. D. W. Cramer, A.M. Walker, and I. Schiff, Statistical methods in evaluating the outcome of infertility therapy, Fertil. Steril., 32:80 (1979).
10. B. Brambati, and L. Tului, Preimplantation genetic diagnosis: a new simple uterine washing system, Hum. Repro., 5:448 (1990).

ASSESSING LOSS RATES IN THE PREIMPLANTATION STAGES OF GESTATION

Joe Leigh Simpson

University of Tennessee, Memphis
Department of Obstetrics and Gynecology
853 Jefferson Avenue, Memphis, TN 38163

INTRODUCTION

Assessing safety of preimplantation genetics procedures requires a clear understanding of the natural history of pregnancy losses. Two aspects in particular will be addressed here - 1) What is the natural history of losses, both *in vivo* as well as *in vitro*? 2) What potential confounding variables need to be taken into account in order to judge loss rates in the four - to eight - cell embryo or in the blastocyst subjected to manipulation?

LOSS RATES DERIVED FROM *IN VIVO* COHORTS

For several decades Hertig and Rock[1-4] recovered embryos either prior to implantation or immediately thereafter. Of 8 preimplantation embryos recovered over many years, 4 were morphologically abnormal. Of 26 implanted embryos (prior to approximately day 23 to 25 of the menstrual cycle), 8 were morphologically abnormal.

To what extent can the data from Hertig and Rock[1-3] be used to derive a priori loss rates which in turn might be useful in assessing safety of embryo biopsy procedures. The key question is the extent to which we should assume that abnormal preimplantation embryos implant. If abnormal preimplantation embryos do not implant, losses from the preimplantation stage onward are very high, much greater than 50% (Figure 1). If abnormal preimplantation embryos implant, loss rates prior to 25 days are lower, perhaps only 30%. That is, in the latter abnormal preimplantation embryos would have become abnormal implantation embryos; thus, the total loss rate approximates that of abnormal implantation embryos alone.

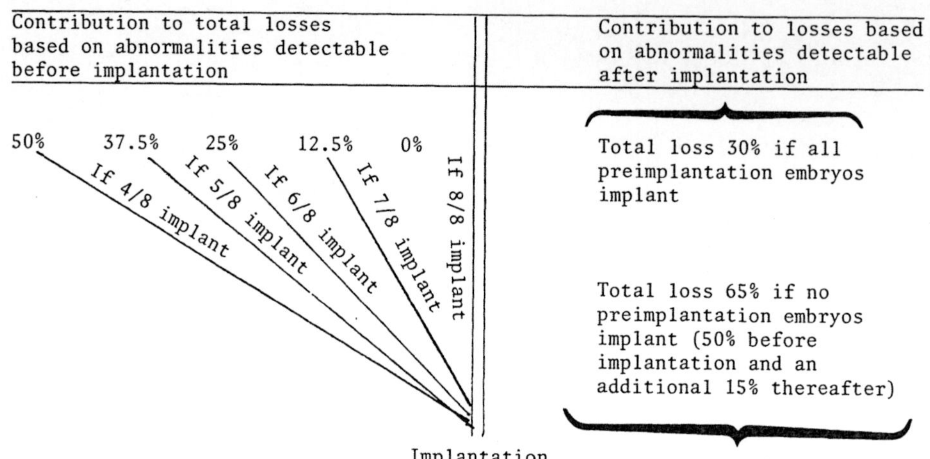

Figure 1. A schematic graph indicating possible extent of embryonic losses, based upon data derived from human embryos recovered from normally fertile women by Hertig and Rock.[1-4] Four (4) of 8 preimplantation embryos were morphologically abnormal, as were 8 of 26 implanted embryos. The overall loss rate depends upon likelihood that abnormal preimplantation embryos implant, the total loss rate is only 30%. If none implant, the total loss rate is 65% (50% of all embryos before implantation plus 30% of the remaining 50% for an overall total of 65%). If half the abnormal preimplantation embryos implant, an intermediate loss rate would be derived.

What other data can help us to decide between the above possibilities? If morphologically abnormal preimplantation embryos of the type recovered by Hertig and Rock[1-4] produce hCG, one would assume that they can implant. Relevant data can be derived from cohort studies of patients attempting to become pregnant underwent ß-hCG assays during the presumed cycle of conception.[5-8]

Early data were inconsistent (Table 1). In the study of Edmonds et al,[6] 118 cycles showed ß-hCG, but only 51 clinical pregnancies. Whitaker et al[7] found far fewer losses. Differences between these studies presumably reflect vicissitudes in ß-hCG assays. Thus, one might prefer more recent studies. Wilcox et al[7] studied 707 cycles, 28% of which showed elevated ß-hCG. Of the 198, 43 (22% of the original sample) experienced a loss prior to clinical recognition of pregnancy. Eighteen more (18/198 or 9% of the original sample) subsequently experienced clinical loss. The loss rate was 31%. The figures are strikingly similar to those "optimal" cases studied by Hertig et al[4]. This subgroup consisted of women in whom there was no evidence of uterine pathology, and in which coitus occurred on the day of ovulation or shortly thereafter. Among 107 cycles there were 24 normal preimplantation and early implantation embryos; 10 were abnormal, giving an abnormality total rate of about 33%. Overall, one might conclude that many if not almost all morphologically abnormal preimplantation embryos probably will implant. If so, loss rates from preimplantation stage onward are probably not dramatically higher than 30%.

What happens in subsequent weeks of pregnancy? In one cohort study patients recruited prior to pregnancy using basal body temperature and underwent ß-hCG testing 14 days after ovulation if menstruation failed to occur.[9] Loss rates were 16% in this sample, which was ascertained at approximately 28 to 30 days of pregnancy. Given that these patients were detected a few days later than those of Wilcox et al[8] (21 to 25 days), the difference in loss rates between the two studies (32% vs. 16%) would lead one to suggest that the fourth gestational week is an interval of very high mortality.

Utilizing the same cohort alluded to above, our group also routinely performed ultrasound to confirm viability of pregnancy at 8 gestational weeks (6 embryonic weeks). Of 220 viable pregnancies, there were only 8 clinical losses (3.2%).[10] That is, from 8 weeks until the end of gestation the loss rate was only 3%. Extrapolating backwards, many of the 16% clinical losses occurred after 4 weeks but before 8 weeks. Pregnancies surviving to 8 weeks gestation had a very high likelihood of surviving until the end of gestation.

Overall, the loss rates after ß-hCG detection is approximately 30%. About half of the total loss rates are likely to be recognized clinically.

CHROMOSOMAL ABNORMALITIES IN EARLY EMBRYOS

What is the etiology of these early pregnancy losses? A. Dyban (this volume) has reviewed data indicating that specific murine chromosomal abnormalities are characteristically lost at specific stages of gestation. Utilizing mice doubly heterozygous for Robertsonian translocations, monosomies and trisomies for all murine chromosomes can be recovered. Early in the preimplantation period, nullisomies and haploids were recovered by Dyban (this volume) and Gropp.[11] Monosomies were recovered around implantation, but only briefly thereafter. Trisomies next begin to be recovered, with all except a few restricted to the time before parturition. It follows that in humans a similar timetable probably exists.

Table 1. Pregnancy Losses

Investigators	ß-hCG	Clinical Pregnancies	Clinical Pregnancy Losses
Miller et al, 1980[5]	152	102	14(14%)
Edmonds et al, 1982[6]	118	51	6(12%)
Whitaker et al, 1983[7]	92	85	11(13%)
Wilcox et al, 1988[8]	198	155	18(9%)

LOSS RATES DERIVED FROM *IN VITRO* DATA

Another source of data is that derived from *in vitro* fertilization (IVF). Recall that the IVF "take home baby" rate is relatively low even in the best of hands. Data from the United States IVF and Embryo Transfer (American Fertility Society) show a 12% delivery rate (1988),[12] however, this rate is not success per embryo but rather success per patient or per cycle. The success rate of delivery in terms of cryopreserved embryos was only 7%.

Data more relevant to preimplantation genetic diagnosis have been published from Bourn Hall,[13] in which the clinical pregnancy rate per transfer in 1985 was about 15%. Walters et al[14] sought to partition losses into those due to (a) inability to sustain a pregnancy (uterine factors) (b) inability of the embryo to develop or to implant (genetic or embryonic factors). The maximum likelihood calculation showed that the proportion lost due to inability to sustain the pregnancy (a, above) was 0.43. Taken in aggregate, we arrive at the conclusions that the IVF clinical pregnancy rate per embryo is lower than the background rate in normal couples. However, if corrected for those pregnancies lost because of the inability to sustain the pregnancy, clinical pregnancy rates *in vitro* and *in vivo* are more nearly equal. This is especially relevant to couples undergoing IVF for preimplantation genetics, for lack of endometrial pathology and optimal maternal age would produce optimal circumstances for IVF success. It follows that ostensible difference in pregnancy success between the *in vivo* studies of Hertig and Rock,[1-4] ß-hCG cohort studies[5-8] and IVF cohorts[13,14] may narrow in the cohort of patients undergoing preimplantation genetics.

CONFOUNDING VARIABLES

The second major topic to which we must allude are potential confounding variables that need to be taken into account in deriving data on safety. Applicable for any invasive procedure, these variables are reviewed elsewhere in detail.[14,15]

Let us first consider gestational age. Suppose we wish to compare procedures in the first trimester, perhaps weeks 9 through 12. A traditional but, unfortunately, simplistic method is to pool all patients who undergo a procedure at a given interval and to compare two procedures. But suppose that the distribution of patients per week undergoing procedure A differs from the distribution of those undergoing procedure B (Table 2). The sample undergoing procedure B has a higher a priori risk than the sample undergoing procedure A because the former has more patients of earlier gestational age. One needs to perform a life table correction to take into account such differences. This potential pitfall may not be terribly relevant for preimplantation genetics because we are dealing with a restricted window of gestation. However, one could not simply compare losses after 8 cell embryo biopsy to losses after trophectoderm biopsy.

Table 2. Gestational Age Bias

Week	Procedure A	Procedure B
9	10	40
10	20	30
11	30	20
12	40	10
	100	100

Another pitfall occurs if one fails to take into account that in a procedure group (but not in the control) abnormal embryos are being removed. A priori loss rate in the two groups would not be comparable; thus, one should not compare the background loss rate with the loss rate in a cohort in embryos possibly destined to abort spontaneously were selected against therapeutically. One might well find spuriously conclude safety after a procedure.

Another pitfall is failure to take into account the yearly increase in loss rate with increasing maternal age. One should eschew the frequent practice of grouping patients within 5 year intervals (e.g., aged 35-39 years). In fact, such a situation arose in the U.S. collaborative trial comparing transcervical CVS versus amniocentesis.[16] The amniocentesis cohort had more patients aged 33 to 34 than the CVS cohort, who were much more likely to be nearer age 40.

Other variables may or may not need to be taken into account in assessing preimplantation losses. Prior pregnancy history and exposure to toxins like cigarette smoking are examples, clearly affecting first trimester losses but do not necessarily affect preimplantation stage. However, it should be assumed that they could be relevant.

CONCLUSION

Given the above, what recommendations are appropriate for assessing preimplantation embryo biopsy? Information exists concerning loss rates, but we still lack conclusive information on background preimplantation rates. I believe that we should proceed with clinical investigations, assuming that clinical pregnancy rates after biopsy should approximate those of IVF in the same institution. Comparisons should be corrected for maternal age, gestational age, status of transferred conceptions and potential confounding variables known to affect clinical loss rates.

REFERENCES

1. A. T. Hertig, J. Rock: A series of potentially abortive ova recovered from fertile women prior to the first missed menstrual period. Am. J. Obstet. Gynecol.,58:968 (1949).

2. A. T. Hertig, J. Rock, E.C. Adams, W.J. Mulligan: On the preimplantation stages of the human ovum. A description of four normal and four abnormal specimens ranging from the second to the fifth pathologic ova. Contrib. Embryol., 35:199 (1954).
3. A. T. Hertig, J. Rock, E.C. Adams, M.C. Menkin: Thirty-four fertilized human ova, good, bad and indifferent, recovered from 210 women of known fertility. A study of biologic wastage in early human pregnancy. Pediatrics, 25:202 (1959).
4. A. T. Hertig: The overall problem in man. In Benirshcke K (ed). Comparative Aspects of Reproductive Failure. New York, Springer-Verlag, pp 12 (1967).
5. J. F. Miller, E. Williamson, J. Glue: Fetal loss after implantation: A prospective study. Lancet, 2:554 (1980).
6. D. K. Edmonds, K.S. Lindsey, J.R. Miller, et al: Early embryonic mortality in women. Fertil. Steril., 38:447 (1982).
7. P. G. Whitaker, A. Taylor, T. Lind: Unsuspected pregnancy loss in healthy women. Lancet, 1:1126 (1983).
8. A. J. Wilcox, C.R. Weinberg, J.F. O'Connor, et al: Incidence of early pregnancy loss. N. Engl. J. Med., 319:189 (1988).
9. J. L. Mills, J.L. Simpson, S.G. Driscoll, et al: Incidence of spontaneous abortion among normal women with insulin-dependent diabetic women whose pregnancies were identified within 21 days of conception. N. Engl. J. Med., 319:1617 (1988).
10. J. L. Simpson, J.L. Mills, L.B. Holmes, et al: Low fetal loss rates after ultrasound-proved viability in early pregnancy. JAMA, 258:2555 (1987).
11. A. Gropp: Chromosomal animal model of human disease. Fetal trisomy and development failure. In Berry L, Poswillo DE (eds). Teratology, Berlin, Springer-Verlag, pp 17 (1975).
12. Medical Research International and Society for Assisted Reproductive Technology: In vitro fertilization-embryo transfer in the United States: 1988 results from the IVF-ET Registry. Fertil. Steril. 53:13 (1990).
13. D. E. Walters, R.G. Edwards, M.L. Meistrich: A statistical evaluation of implantation after replacing one or more human embryos. J. Reprod. Fertil., 74:557 (1985).
14. J. L. Simpson: Incidence and timing of pregnancy losses: Relevance to evaluating safety of early prenatal diagnosis. Am. J. Med. Genet., 35:165 (1990).
15. J. L. Simpson, S.A. Carson: Genetic and nongenetic causes of spontaneous abortion. In Sciarra JW (ed), Gynecology and Obstetrics, Vol. 5. Chapter 91. Philadelphia, Lippincott, In Press.
16. G. G. Rhoads, L.G. Jackson, S.E. Schelesselman, et al: The safety and efficacy of chorionic villus sampling for early prenatal diagnosis of cytogenetic abnormalities. N. Engl. J. Med., 320:609 (1989).

PRACTICAL CONSIDERATIONS IN PREIMPLANTATION DIAGNOSIS

Laird Jackson

Thomas Jefferson University, Jefferson Medical College, Division of Medical Genetics, 1100 Walnut Street, Philadelphia, PA 19107

INTRODUCTION

Anne McLaren, a pioneer in the study of embryo manipulation, wrote the following summary of the prospects for preimplantation genetic diagnosis: "Although possibilities exist for preimplantation diagnosis by means of embryonic biopsy, much research on human material would need to be done to find out whether they were feasible, or whether they reduced unduly the chances of a successful pregnancy. Since they necessarily involve in vitro manipulation and subsequent embryo transfer, it seems unlikely that they would ever become the method of choice except for couples who were undertaking IVF for other reasons, or who had absolute objections to post-implantation termination of pregnancy, combined with considerable determination and financial resources. For most couples, chorionic villus biopsy as early in the post-implantation period as is safe and practicable would seem to offer better prospects."(1)

Two years later, in 1987 the comments on these prospects, were as follows: "The prospects for preimplantation diagnosis have advanced substantially since the topic was last reviewed, and it is therefore likely that the relevant animal experiments will soon be supplemented by studies on human material within the guidelines laid down by the Voluntary Licensing Authority: (a UK agency responsible for monitoring this scientific area)(2).

As relevant studies have demonstrated, human preimplantation genetic diagnosis has been accomplished with at least two pregnancies having come to term (3-6). This experience has largely concentrated on the scientific and technological aspects of the subject and the progress documented is truly impressive. The impetus for such intensive investigation derives, from the considerable distress of some patients in their experience with amniocentesis for managing high genetic risks in the process

Preimplantation Genetics, Edited by Y. Verlinsky and A. Kuliev, Plenum Press, New York, 1991

of reproduction(2). Although CVS and altered performance of amniocentesis to move the gestational timing of prenatal diagnosis into earlier weeks have made for improvements, preimplantation diagnosis offers to move the decision point to the beginning of gestation.

The technology, therefore, has clearly been demonstrated to be up to the task and the indication is there for at least a select group of patients. Questions now are: what are the goals of the new technology other than proving it doable, and what is the balance of expected return versus cost? As with other invasive prenatal diagnostic technologies, assessment of the long term safety and efficacy of the approach needs to be done. Importantly, in assessing efficacy, one should question whether or not what can be done should be done just because it is possible.

Assessment of the safety of a perinatal technology or procedure usually includes examination of both maternal and fetal risks. Maternal risks are related to the invasive aspects of oocyte recovery and embryo transfer as well as any secondary procedure such as later CVS for cytogenetic confirmation. These risks can probably be considered to be minimal as considerable experience with both of these aspects prior to the incorporation of preimplantation diagnosis is already documented to be acceptably safe. The additional manipulation of the embryo in preimplantation diagnosis does not alter the maternal involvement of physical risk.

Fetal risks have always dwelled on the risk of fetal loss post procedure and this could be translated into the risk of failure of embryonic implantation and growth. Further fetal risks include any risk for induction of developmental defect due to embryo manipulation and the often mentioned risks of ultrasound exposure.

Evaluation of the fetal aspect of the safety of invasive prenatal diagnostic procedures during pregnancy usually concentrates on the risk of postprocedure fetal loss or compromise. Loss is usually considered to occur through the most common path of 'spontaneous' abortion although a smaller risk of later loss through stillbirth or neonatal death can also occur (considered together as 'perinatal loss'). Fetal compromise generally focuses on the potential for induction of teratologic or other mechanism leading to the production of a birth defect in the liveborn child. As in the case with maternal risks, much of the risk to the fetus is already documented as the physically invasive procedures have already been utilized for some time in either IVF procedures or in standard prenatal diagnostic procedures. Again, in both cases the risks are known and considered acceptable, except the addition of the preimplantation diagnostic manipulations of the embryo which does not seem to alter considerably the physically invasive process nor its attendant risk.

The embryo manipulation approach to preimplantation diagnosis does pose two new areas for 'risk' consideration, however. The first is the simplest and asks the question, does the manipulation alter the risk for embryo survival and development in any way compared to the nonmanipulated embryo? In the limited experience to date the answer appears to be 'no'. Reports from Handyside, for example, seem to suggest that the rate of successful embryo transfer and subsequent development is excellent as compared to the usual IVF figures (6). However, the second question, the risk of losing a pregnancy that is viable at the time of the diagnostic procedure, is more difficult. Here, one is dealing with the dilemma of different starting points for the consideration of fetal 'safety'. Preimplantation diagnosis starts from the point of the viable oocyte or embryo. Ordinarily, one starts with an established pregnancy and considers the effect of the procedure on survival. Problems with this somewhat simplistic concept first arose in comparing CVS to amniocentesis when it was realized that 7 week fetuses do not have the same survival potential as 16 week fetuses and that we do not have accurate statistics on that difference. Now, we may have to deal with the questions of relative success in healthy fetuses per oocyte or embryo as opposed to success per 'pregnancy'. A 65-75% loss of fetuses would be considered abysmal when compared to even the gloomiest of early CVS attempts while a 25-35% success rate in pregnancies per transferred embryo reported by Handyside (this volume) is quite remarkable compared to early IVF success. So it would seem to be an open question as to what criteria will be appropriate for evaluating the success or the risk of preimplantation diagnosis. But even with this open question, some aspects of risk evaluation will have to be faced when considering any expansion of the application of preimplantation diagnosis. So far, the risks attendant to the IVF procedures now utilized in preimplantation diagnosis have been entirely acceptable against the background of the patient's inability to reproduce by any other means. Accepting the rates of IVF success (or 'risk') in the face of no other reproductive option is quite different from choosing between the current vagaries of preimplantation diagnosis with its attendant embryo manipulation or alternatively utilizing CVS, amniocentesis or other prenatal diagnostic techniques etc. with reasonably well understood limitations.

Risk of developmental effect on surviving infants is another concern in evaluation of safety. IVF, to my knowledge, has not directly caused or contributed to problems in this regard. Monitoring of the development of infants following preimplantation diagnosis is again a straightforward process. However, it is expensive and traditionally difficult to accomplish and will require moderate dedication to purpose if it is to be done well. Experience with IVF seems to suggest that there is not likely to be a serious question of teratogenic effect. It seems logical that interference at this early stage of embryologic development is likely to produce an all or none effect, preventing implantation or not.

Evaluation of efficacy means taking a look at whether you achieved the desired outcomes, including not only accuracy but also economic feasibility and impact on quality of life. This opens an elastic box of questions. Accuracy of diagnosis is fairly straightforward to consider but involves things such as developmental timing of gene expression for enzymatic assays in diagnosis. Questions of sensitivity and susceptibility to 'contamination' must be examined for DNA methods(7,8). Methods to work on single or small population cell samples exist and are in current use by many workers but the application to a blind diagnosis without the usual laboratory duplicate controls will alter the rules somewhat. All current experience suggests that these concerns will be dealt with successfully and that most significant single gene genetic disorders will probably be as amenable to accurate preimplantation diagnosis as they are to current methods. Significant problems may still arise in translating this optimism to a more routine procedure, however. For example, in the case of inborn errors of metabolism, one must establish the timing of gene expression if enzyme assay is to be established. According to Braude (9), HPRT is not expressed until after the 4 cell or perhaps the 8 cell stages. Further investigation of the potential for enzymology is clearly desired because of the long-standing usefulness of diagnosis by assessment of the functional protein. Both Tay-Sachs and Lesch-Nyhan disease may be diagnosed by direct gene analysis but prior knowledge of the individual mutations is required in that approach. Carrier screening currently does not always include mutation analysis so this consideration must be included in future plans.

Cytogenetic diagnosis in the preimplantation period will have its own set of problems. As reported by Handyside, this assessment is still required for fetuses whose primary preimplantation genetic diagnosis has already been established (Handyside, this volume). Karyotyping single cells is feasible but technically difficult and unsatisfactory at present(10-12). Secondarily, there is always the question of whether mosaicism would be a greater or lesser problem under those conditions. Ledbetter(13) presented a very elegant approach to cytogenetics of small cell samples and perhaps this type of methodology will become practical soon. However intriguing, these questions remain for the future to answer.

The area of effect on the quality of life is probably the most difficult to evaluate. It certainly produced spirited debate and feelings in recent CVS trials(14,15). Whether this becomes an issue in preimplantation genetics is a matter of wait and see. For current patients whose motivation is high, this is not so likely to be relevant. They are being chosen to participate in preimplantation diagnosis by prior experience and circumstance. Will that change with wider appreciation of the technique and wider experience and success in an expanding diagnostic arena? To what extent will patients be recruited for this approach and what expectations will this secondary group have? Hence cost will likely emerge as a factor in 'quality of life' measurement. Are you achieving the desired endpoint? The cost at the moment is frightening in comparison to other approaches to achieve this same endpoint. However, technology has a way of driving down costs and there is no doubt that this will occur with preimplantation diagnosis, given enough time. How soon and how effective that will be is something to watch. Resources are not unlimited and this aspect of consideration will demand attention.

One of the things that has driven the present technology is the relevance to individual patients at very high risk and those patients' desires to avoid the difficulties that are inherent in other forms of prenatal diagnosis (mainly 'time'). Many of such patients would agree with Bernadette Modell (see reference 2), who said that her thalassemia patients suggest that what would improve their situations most would be 'To be able to start a pregnancy feeling committed to it, in the knowledge that it would not be affected'. In a sense, this would be true if one had a really effective, clinically applicable method of preimplantation diagnosis. But so far, for example, even in Handyside's successful cases, those patients subsequently underwent an invasive prenatal diagnostic procedure, CVS, to look at the chromosomes of the fetus when some other (sexing) assay had been performed by preimplantation diagnosis(6). So we are a long way from being able to tell the patient that they can be committed to the pregnancy from the start. I suspect that what happens to some of these patients is that if one were able to make this clinically applicable, they would not be such a direct participant in the decision to cast off a pregnancy. It would be psychologically removed in some ways from their power since their decision after assay would be to implant or not and this can be differentiated psychologically from aborting. However, even here I would urge caution as there will be others who assess this psychology. The observers, the remainder of society, will contain those who object to meddling with life and decisions not to implant will still be meddling. For example, our own experience with fetal 'reductions' after multiembryo implantation and survival in IVF suggests that the 'accessory' genetic counselor has a less tolerant attitude than the physician or patient in these cases after repeated exposure to deliberate 'over-transfer' as opposed to the accidental occurrence of a multifetal pregnancy. It may be a case of who is manipulating who. Therefore, I would suggest that we keep an ear tuned to the voice of contemporary society while we seek to support the most important person in all of this, our patient.

REFERENCES

1. A. McLaren, Prenatal diagnosis before implantation: Opportunities and problems, <u>Prenat. Diag.</u> 5:85-90 (1985).
2. R. Penketh, A. McLaren, Prospects for prenatal diagnosis during preimplantation human development, <u>Ball Clin. Obstet. Gynecol.</u> 1:747-64 (1987).
3. M. Monk, K. Hardy, A. Handyside, D. Whittingham, Preimplantation diagnosis of deficiency of hypoxanthine phosphoribosyl transferase in a mouse model for Lesch-Nyhan syndrome, <u>Lancet</u> 11:423-5 (1987).
4. M. Monk, C. Holding, Amplification of ß-haemoglobin sequence in individual human oocytes and polar bodies, <u>Lancet</u> 335:985-8 (1990).

5. L. J. Wilton, J.M. Shaw, A.O. Trounson, Successful single-cell biopsy and cryopreservation of preimplantation mouse embryos, Fert. Ster. 51:513-7 (1989).
6. A. H. Handyside, E.H. Kontogianni, K. Hardy, R.M.L. Winston, Pregnancies from biopsied human preimplantation embryos sexed by Y-specific DNA amplification, Nat. 344:768-70 (1990).
7. R. J.A. Penketh, J.D.A. Delhanty, J.A. Van den Berghe, E.M. Finkleston, A.H. Handyside, S. Malcolm, R.M.L. Winston, Rapid sexing of human embryos by non-radioactive in situ hybridization: Potential for preimplantation diagnosis of X-linked disorders. Prenat. Diag. 9:489-500 (1989).
8. K. Porter-Jordan, C.T. Garrett, Source of contamination in polymerase chain assay, Lancet 335:1220 (1990).
9. P. Braude, V. Bolton, S. Moore, Human gene expression first occurs between the four- and eight-cell stages of preimplantation development, Nat. 332:459-61 (1988).
10. R. R. Angell, A.A. Templeton, F.J. Aitken, Chromosome studies in human in vitro fertilization, Hum. Genet. 72:333-9 (1986).
11. J. L. Watt, A.A. Templeton, I. Messinis, L. Bell, P. Cunningham, R.O. Duncan, Trisomy 1 in an eight cell human pre-embryo, J. Med. Genet. 4:60-4 (1987).
12. H. Wramsby, K. Fredga, P. Liedholm, Chromosome analysis of human oocytes recovered from preovulatory follicles in stimulated cycles, N.E.J.M. 316;121-4 (1987).
13. D. H. Ledbetter, Molecular cytogenetic techniques, J. IVF and ET, 7;195 (1990).
14. Canadian CVS Group, Multicentre randomized clinical trial of chorion villus sampling and amniocentesis, Lancet 1:1-6 (1989).
15. G. G. Rhoads, L.G. Jackson, S.E. Schlesselman et al, The safety and efficacy of chorionic villus sampling for early prenatal diagnosis of cytogenetic abnormalities, N. Engl. J. Med. 320:609-17 (1989).

PART V

GENE EXPRESSION IN PREIMPLANTATION DEVELOPMENT AND GENE THERAPY

EVOLUTIONARY ANALYSIS OF MAMMALIAN HOMEOBOX GENES

Claudia Kappen[1] and Frank H. Ruddle[1,2]

Departments of Biology[1] and Human Genetics[2]
Yale University, New Haven, CT 06511

INTRODUCTION

The homeobox was first discovered as a common DNA sequence element in homeotic genes of the fruit fly Drosophila melanogaster (McGinnis et al., 1984, Scott and Weiner, 1984). These genes regulate the development and morphogenesis of the fly, and mutations in such genes lead to characteristic phenotypes, so called homeotic transformations as for example the transformation of antenna into legs. The resulting malformations change the identity of a given body segment without affecting the total number of segments. The identity of a segment with regard to its position along the anterior-posterior axis appears changed. In concordance with their effects in certain regions of the body, homeotic genes are expressed in respective regions of the embryo that later give rise to the affected structures (for review, Gehring, 1987). Homeotic genes are organized in two gene clusters in the Drosophila genome, the Ant-C and BX-C clusters. It is intriguing to note that the linear order of the genes on the chromosome in these complexes is collinear with the gradient of expression of homeotic gene products along the anterior-posterior axis in the embryo.

Homeotic gene products are transcription factors that regulate the activity of their target genes during developmental and differentiation processes. They also regulate the activities of other homeotic genes and control their own activities in an autoregulatory fashion. Thus, homeotic genes of the fly encode for a network of regulatory interactions that functions to control gene activity during development. In concordance with the genetic evidence, it has recently been shown that homeodomain proteins bind DNA. The homeodomain consists of 61 amino acids and forms a helix-turn-helix structure similar to the DNA-binding domain of procaryotic repressor molecules (reviewed in Scott et al., 1989). Thus, on the molecular level, the action of homeotic genes is mediated by their role as DNA-binding proteins controlling the transcriptional activity within the homeotic gene network as well as the expression of target genes in determining segment identity during fly development.

Preimplantation Genetics, Edited by Y. Verlinsky and
A. Kuliev, Plenum Press, New York, 1991

MAMMALIAN HOMEOBOX-CONTAINING GENES

Homeobox genes have subsequently been isolated from quite a variety of organisms including frogs, chicken, mice and humans. According to their sequences as well as their biological features, they can be grouped into different categories. Homeodomains have been identified in ubiquitous transcription-factors such as Oct-1 as well as in tissue-specific transcription factors. Examples for the latter are Pit-1, a pituitary-specific transcriptional activator of the gene for growth hormone, and Oct-2 which regulates the transcriptional activity of immunoglobulin genes. Genes of the so called Antennapedia-class are the most similar to the Drosophila homeotic genes and have generally been isolated based on their sequence similarities rather than their biological activities. The similarity to the Drosophila genes, however, suggests that they may also play important roles in development in vertebrates (Fienberg et al., 1987).

Structure and Expression of Mammalian Homeobox Genes

A homeobox gene in vertebrates typically is about 10 kb long and consists of at least two exons. The spliced product encodes a protein of the average size of 240 amino acids. The homeodomain resides in the carboxyterminal part of the protein. Additionally, the aminoterminus is conserved between several proteins, as well as a hexapeptide sequence preceding the homeodomain. This sequence has been implicated in mediating dimerization of homeodomain proteins.

The expression patterns of homeobox genes are quite complex and cannot be adequately covered in the context of this article (for review, see Shashikant et al., 1990). The most striking feature of the expression patterns of homeobox genes in developing embryos is the region-specific accumulation of their mRNAs. Each homeobox gene has a characteristic anterior boundary of expression in the spinal cord, and while the posterior boundary is less well defined, a gradient can be observed in the succession of anterior expression levels. This graded expression is also found in the mesoderm. These patterns are reminiscent of those observed in Drosophila in that homeobox gene expression is detected in structures originating from different germ layers, such as neural tissue, which is of ectodermal origin, and somitic mesoderm. Moreover, as in the fly, the region of expression is reflected by the chromosomal position of the respective gene (Graham et al., 1989), Doboule and Dollé, 1989).

Genomic Organization of Mammalian Homeobox Genes

The genomic organization of mammalian homeobox genes is described in Fig. 1. Homeobox-containing genes reside in four clusters on different chromosomes, **HOX-1** on chromosome 7 in human (6 in the mouse), **HOX-2** on chromosome 11 (17), **HOX-3** on chromosome 12 (15), and **HOX-4** on chromosome 2 (2) (Ruddle, 1989). The transcriptional orientation is the same for all genes on one cluster. They are spaced approximately every 10 kb and no pseudogenes have been identified.

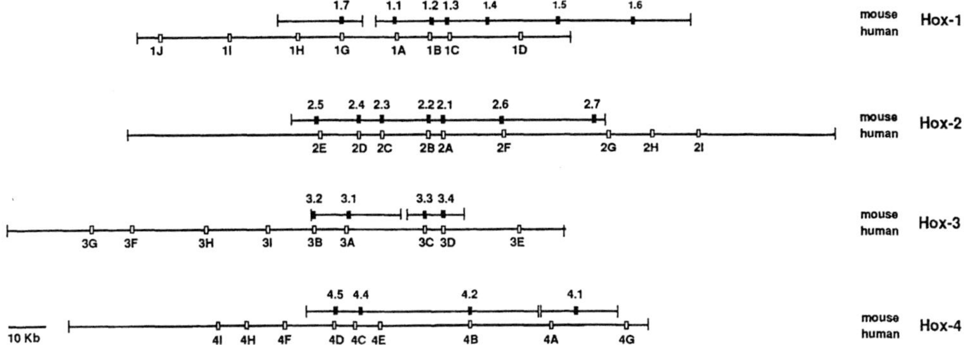

Figure 1. Organization of mammalian homeobox-containing genes. Closed symbols: murine homeobox genes, open symbols: human homeobox genes. Information taken from Kappen et al., 1990, Acampora et al., 1990 and Boncinelli et al., personal communication.

Each of the genes in a given cluster has closely related cognate genes located on at least one of the other clusters. Thus, the highest similarity between genes is found across clusters rather than along one cluster. These similar genes form subfamilies which are termed cognate groups. Moreover, the physical order of the corresponding genes is identical along the four clusters. This parallel arrangement has served as evidence that the four clusters of homeobox genes have arisen through duplication events that involved entire clusters. A virtually identical organization is found in mouse and humans (Kappen et al., 1989a). The arrangement as well as the overall spacing of the genes are extremely conserved (see Fig.1). This indicates that the duplication event leading to four clusters must have taken place at least before the divergence of these two species. Analysis of all available sequence information of vertebrate homeoboxes supports this hypothesis. Furthermore, homeobox sequences representing genes from at least three clusters have been isolated from non-mammalian vertebrate species such as frog, chicken, and zebrafish. Information on the genomic organization of those sequences is limited but in all cases compatible to the mammalian situation. These data strongly suggest that lower vertebrates may also have four clusters of homeobox genes similar to those in mammals.

DUPLICATION OF HOMEOBOX GENE CLUSTERS DURING VERTEBRATE EVOLUTION

A more detailed calculation of the timing of the duplication events with respect to vertebrate evolution has been derived from analysis of nucleotide sequences of mammalian homeoboxes (Kappen et al., 1989b). When the human and mouse sequences for each of the homeoboxes are compared in pairs, they differ in an average of 11.5 nucleotides. More than 90% of these differences represent silent substitutions, they do not result in amino acid differences. These data show that human and mouse homeobox sequences have been very strongly conserved since the divergence of the two species. Comparisons of cognate homeoboxes located on different clusters in the mouse show that each pair of sequences differs by an average of 32.4 nucleotides of which 84.6% account for silent substitutions. Such a comparison for human sequences yields an average of 37 differences (87.3% silent) between cognate pairs (Kappen, unpublished).

These figures represent the distance between homeobox clusters as measured by the number of nucleotide differences (Table 1). The four clusters in the mouse appear equally distant. If one of the clusters arose from a very recent duplication it should differ by a smaller number of nucleotides from its sister cluster than from the remaining clusters. However, such a difference is not observed in mouse (Table 1) or human (Kappen, unpublished). Therefore, it is likely that the four clusters arose in a rapid succession of duplications rather than during a long time period.

Table 1. Comparison of cognate mouse homeobox sequences. The figures represent the numbers of nucleotide differences between two cognate sequences from the respective clusters. Cognate groups are designated according to a prominent member sequence of the Hox-2 cluster.

Cognate group	"2.5"	"2.4"	"2.3"	"2.2"	"2.1"	"2.6"	"2.7"	"2.9"	mean (+s.d.)
Cluster comparison									
Hox-1 vs. Hox-2	34		27	27	30	32	19	38	29.6 (±5.6)
Hox-1 vs. Hox-3	32				29	43			34.7 (±6.0)
Hox-1 vs. Hox-4	32					32	19	44	31.8 (±8.8)
Hox-2 vs. Hox-3	30	37		33	45				36.3 (±5.6)
Hox-2 vs. Hox-4	30					37	14	48	32.3 (±12.3)
Hox-3 vs. Hox 4	33								

The interpretation of these data is based on two assumptions. The first argument assumes that nucleotide substitutions which do not cause amino acid changes are not subject to evolutionary selection. Then, the accumulation of silent nucleotide substitutions increases over time and the number of differences can be correlated to evolutionary time. As a second condition, the rate of nucleotide substitution has to be relatively constant over long periods of time. It has been argued that the rate of mutation may be different for different organisms (Britten, 1987). However, in the calculations presented here, only sequences from the same organism (only mouse or only human) are compared to each other. Therefore, rate differences during evolution should apply equally to all the sequences in a given comparison. Additionally, it appears that the overall number of differences between cognate sequences within either mouse or human is similar arguing against dramatically different substitution rates in the homeobox sequences of these two species after their divergence. All potential rate fluctuations before the divergence timepoint apply equally to mouse and human sequences since these two species arose from the same ancestor organisms.

TIMING OF THE CLUSTER DUPLICATIONS

Following these assumptions we have attempted to calculate the evolutionary timescale of the cluster duplications by taking the mouse-human divergence time as a reference point. Mouse and human species are thought to have diverged at least 70 Million years ago, a figure which recently has been corrected up to 100 Mio. years (Li et al., 1990). During this period an average of 10 silent nucleotide differences has accumulated between mouse and human homeobox sequences. Based on the distance between individual mouse clusters of 27.4 silent nucleotide differences (and 32.3 silent nucleotide differences for the distances between human clusters) the minimum time progressed since the cluster duplications can be calculated to be around 300 Million years (Table 2). This figure represents a minimum estimate since it is based on silent nucleotide differences only. However, if the occurrence of multiple mutations at the same site and the occurrence of 15.4% (in the mouse, 12.7% in human) replacement changes are taken into account the timepoint for the duplications might be assumed even earlier at around 350 - 400 Million years ago. Such an estimate coincides well with the evidence on homeobox genes from lower vertebrates suggesting that the duplication events happened early during or possibly preceding vertebrate evolution (Kappen and Ruddle, 1990).

COMPARISON OF MAMMALIAN AND DROSOPHILA HOMEOBOX GENES

In this context it is interesting to compare mammalian homeobox clusters to the homeotic gene complexes of Drosophila in which homeoboxes were first detected. As described above, both in the fly as well as in mammals the domains of expression in the embryo along the anterior-posterior axis are reflected in the relative chromosomal position of the genes. The more anterior a gene is expressed, the more 3' it is located in the mammalian Hox-cluster, and as depicted in Fig. 2, this situation is similar to that in the fly. However, in Drosophila the homeotic genes are located in two physically separated complexes on the same chromosome. Furthermore, the Ant-C complex contains a number of additional genes that are not homeotic, and in some instances do not contain homeoboxes. For example, bicoid is a homeobox gene that is non-homeotic, and amalgam does not contain a homeobox and is not a homeotic gene. Such "extra" genes are absent from the mammalian Hox-clusters. Additionally, along each cluster, the transcriptional orientation of mammalian Hox-genes is the same. In Drosophila, however, several genes are transcribed in directions opposite to each other, for example Antp and ftz as well as Scr and Dfd.

When the amino acid and nucleotide sequences of the fly homeobox-containing genes of these complexes are compared to vertebrate homeobox sequences, it becomes apparent that not all sequences have direct counterparts. Good matches are found for the following groups: Abd-B and the **Hox-2.5** cognate group, Dfd and the **Hox-2.6** cognate group, labial and the **Hox-2.9** cognate group, and pb has **Hox-2.8** as a

Table 2. Timing of Hox-cluster duplications.
The number of silent nucleotide differences obtained from sequence comparisons (Kappen, unpublished) was correlated with evolutionary time. The divergence time for the rodent-human divergence was taken from Li et al., 1990.

Sequence comparisons	silent nucleotide differences	species divergence time
Human - mouse 23 cognate pairs	**10.9**	**70 - 100 Mio years**
Human - human 34 cognate pairs	**32.3**	estimated time after duplications
Mouse - mouse 23 cognate pairs	**27.4**	**300 Mio years**

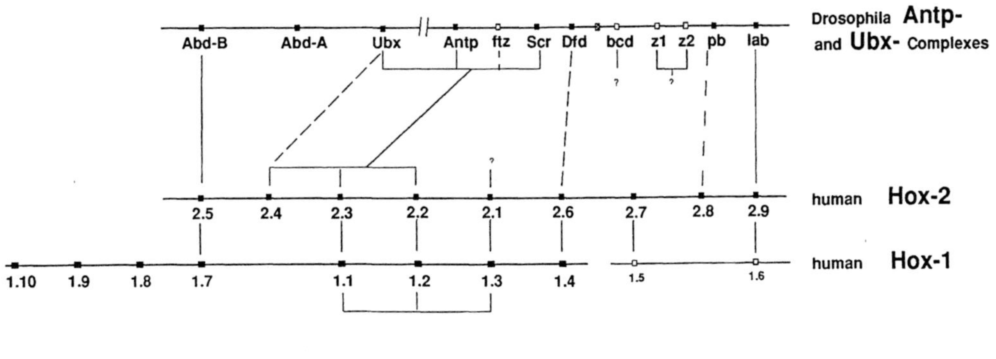

Figure 2. Comparison of the genomic arrangement of human homeobox genes to the Drosophila homeotic gene clusters. In the fly clusters closed symbols represent homeotic genes; open symbols represent homeobox-containing genes; the filled box represents the non-homeotic gene amalgam. In the mammalian clusters, closed boxes represent human genes, open boxes represent information additionally available from the mouse. The bottom line shows the minimal number of ancestor genes contained in the proposed primordial cluster.

counterpart. The strong conservation of these homeobox sequences between the fly and mammals has been taken as an indication that they represent homologous genes (Lobe and Gruss, 1990). However, the mammalian homeobox genes in the middle of the clusters cannot easily be correlated to the Drosophila sequences, and for more 5' genes such as human **HOX-1.10**, **HOX-1.9**, **HOX-1.8**, and **HOX-3.7** and **HOX-3.6** (Acampora et al., 1990) or murine **Hox-4.6** and **Hox-4.5** (Duboule and Dolle, 1989) no direct counterparts have been described in the fly. These differences can be taken as an indication for separate lines of evolution in the arthropod and vertebrate lineages starting out from an ancestral cluster of a few genes. In arthropods this cluster acquired more genes by gene duplications, followed by inversions or transpositions of genes into the Ant-C complex. The Ant-C and BX-C are split into separate complexes in Drosophila, whereas genetic evidence suggests that they are still closely linked in the red flour beetle (Beeman et al., 1989). In the lineage leading to vertebrates, an independent increase in the number of homeobox genes was achieved by gene duplications along a primordial cluster. These local duplications (Kappen et al., manuscript in preparation) involved genes for the more posteriorly expressed group (**Hox-2.5** and further 5' cognate groups), genes that are located in the middle of the clusters, and only to a limited degree the more anteriorly expressed genes (creation of the **Hox-2.7** cognate group). Alternatively, it may be possible that ancestor genes for the more posterior genes were present before the arthropod-vertebrate divergence but were subsequently lost in the insect lineage (Acampora et al., 1990). Finally, in the vertebrate lineage a further increase in the number of homeobox genes was brought about by duplications of the primordial complex to form at least four clusters of genes.

DUPLICATIONS OF LARGE GENOMIC REGIONS

The chromosomal extension of the duplicated regions is unclear at present, however, it appears that a sizable number of genes in linkage with the homeobox clusters have been duplicated as well and are found on the corresponding chromosomes in human and mouse (Ruddle et al., 1987). This has been taken as evidence that large genomic regions if not entire chromosomes or genomes have been duplicated during vertebrate evolution (Schughart et al., 1989). The early timepoint of such duplications for the homeobox clusters within the vertebrate lineage lends support to the speculation that these cluster duplications were crucial in providing new regulatory circuits to set up the more complicated vertebrate body plans. It is interesting in this respect to note that no homeobox genes of the Antennapedia-class have so far been reported to be expressed in the head. It has been argued that the vertebrate head was an evolutionary invention created out of neural crest (Gans and Northcutt, 1983). It may be speculated that the homeobox genes present from before the divergence of arthropods and vertebrates may not have been suitable to perform the specific functions required. This would invoke the postulation of a separate group of regulatory genes expressed in the brain (Lonai and Orr-Urtreger, 1990), and indeed several homeobox-containing genes transcribed in the brain have recently been identified in mammals (He et al., 1989, Suzuki et al., 1990, Murtha and Ruddle, unpublished).

CLUSTER DUPLICATIONS AS AN EXAMPLE OF "REGULATORY EVOLUTION"

Generally, the acquisition of new functions by homeobox genes after duplication events can be conceptualized in different ways:

First, the amino acid sequence of the protein itself can change. If the homeodomain is affected, the DNA-binding specificity of the protein will change. If regions outside of the homeodomain are altered, the activity of the protein in the context of transcriptional regulation (alone or in conjunction with accessory proteins) may be affected. Thus, altered DNA-binding may result in altered regulation of a target gene or the direction of regulatory activity towards a new target gene involved in different developmental processes.

A second way to achieve a new function for a regulatory gene would be a change in its temporal or spatial expression pattern. In this case, the product of the homeobox gene would still bind to the same target sequences but in different cell types or tissues or regions of the developing organism or at an altered time during development. Such changes may occur either in enhancers or promoters of the homeobox gene itself or in the proteins that regulate its expression. The new or ectopic expression of a homeodomain protein would then result in ectopic regulation of its target gene(s), and that may potentially lead to development of new structures.

These two alternatives presume that the original function of the given duplication ancestor is preserved in one of the duplicated copies while the other diverges or that both copies diverge to develop new functions.

A third way of generating new functions for homeoproteins would not require a gene duplication. It concerns the homeodomain binding sites in target genes. Such sites could appear in conjunction with different or new types of genes thus involving an essentially unchanged (regulatorily or structurally) homeobox gene in completely new differentiation pathways. A similar result would be achieved if a change would occur in only the accessory proteins influencing homeobox gene function.

When the situation of Drosophila homeotic genes is considered, it is difficult to disseminate these alternatives since each gene has a distinct sequence and a distinct expression pattern. Moreover, although in vitro, quite diverged homeodomains appear to be able to bind the same DNA-sequence, in vivo however, their specificity is tightly controlled. This evidence suggests that the gene duplications as established in the fly homeotic gene system involved both changes in the primary structure and the regulation of expression of homeobox genes. Compatible with this notion, homeotic genes of the fly are considerably larger than mammalian homeobox genes, mostly accounted for by cis-flanking regulatory regions.

The situation in mammals is more revealing in that the protein sequences encoded by homeobox genes belonging to the same cognate group are very similar to each other (Kessel et al., 1988, and Kappen, unpublished data) and virtually identical in their homeodomain sequences, especially in Helix 3, the DNA-binding portion. Thus, it seems likely that products of cognate genes bind to the same target sequence and thus regulate the same target genes. Diversity between homeodomain sequences is found along each cluster suggesting that the target gene(s) for each cognate group may be different. It cannot be ruled out that the products of cognate genes actually compete with or augment each other in regulatory activity since during embryonic development they appear to be expressed at least at some stages in overlapping domains.

The expression patterns of cognate genes belonging either to the **Hox-2.1** or the **Hox-2.6** cognate group have been studied in detail (Gaunt et al., 1989, 1990). **Hox-1.3**, **Hox-2.1**, and **Hox-3.4** all belong to the same cognate group but show varying anterior expression boundaries in the central nervous system and differences in tissue distribution. These data are consistent with observations on the expression patterns of **Hox-1.4**, **Hox-2.6** and **Hox-4.2** (belonging to the **Hox-2.6** cognate group) which show similar anterior expression domains in the spinal cord but striking stage- and tissue-dependent differences.

It can be concluded from these data that cognate homeobox genes in vertebrates differ in their expression patterns.

Thus, it appears that acquisition of new functions by duplicated vertebrate homeobox genes was mediated through changes in expression patterns rather than changes in binding specificity. These changes in expression patterns of very similar proteins created new sites for regulation of the same target gene as before. It is conceivable that the acquisition of new expression domains contributed to the formation of new or altered tissues or body parts.

Alternatively, one could hypothesize that the target binding sites for the duplicated homeobox genes reside in different loci. This would imply that at the same time when a duplicated homeobox gene copy was created, new binding sites should have been created. This is a difficult process to envision in the case of the cluster duplications since they involved most likely more than 10 genes at the same time, thus 10 changes of target genes would have to be postulated concomitant with the duplication event.

In summary, it appears more likely that the "upstream" genes for members of a given cognate group are different and that the "downstream" genes may be the same. Then, one duplicated copy would appear as the ectopically expressed version of the original ancestor, or both copies would acquire new expression patterns. In conclusion, the evolution of the vertebrate homeobox gene system, after the duplication events, appears to have mainly been a "regulatory evolution". It has been argued (Davidson, 1982) that such regulatory changes would be much more effective in creating variability during evolution than the structural modification only of the existing parts.

CONSERVATION IN A NETWORK REGULATORY SYSTEM

Several features of the homeobox regulatory network are conserved in organisms as diverse as fruit fly and mammals. Sequences of the homeodomains of homologous genes are very similar. The genomic organization in clusters has been well preserved in the arthropod as well as vertebrate lineages. The collinearity of genomic organization and the succession of expression domains along the anterior-posterior axis have also been strongly conserved. Following the sequence comparison results it appears that at least five homeobox genes were present in an ancestral cluster prior to the divergence of arthropods and vertebrates. It is conceivable that these genes were involved in a small network of genes regulating the transcription of target genes. The strong sequence conservation during evolution suggests that at least some of the target genes and that those "upstream" genes which regulate the anterior-posterior gradient of expression of homeobox genes may also turn out to be conserved. Such a rudimentary regulatory network was then independently further diversified in the arthropod as well as vertebrate lineages. The conserved function of homeotic fly genes and mammalian homeobox genes in specifying positional identity along the anterior-posterior axis suggests that this particular process of pattern formation during embryonic development was established very early in the evolution of metazoans and has been exploited subsequently to allow the development of diverse body plans.

CONCLUSION

Homeotic genes in the fruit fly Drosophila are involved in pattern formation during embryonic development. Baséd on sequence similarities more than 30 homeobox-containing genes of the Antennapedia-class have been identified in mouse and man. The genomic organization of this multigene family in gene clusters located on different chromosomes supports the hypothesis that the homeobox gene complexes evolved by duplications. We present evidence that these cluster duplications happened early during vertebrate evolution. Based on a comparison of the arthropod homeotic and vertebrate homeobox genes we argue that the evolution of the vertebrate homeobox gene network proceeded as a "regulatory" evolution and may have been crucially involved in the development of new body plans.

Acknowledgements - We are grateful to E. Boncinelli (Naples) for making data available to us before publication. We thank Miri Einat for critically reading the manuscript and J. Michael Salbaum for discussion. This work was supported in part by a postdoctoral fellowship of the Deutsche Forschungsgemeinschaft to C.K. and NIH-grant GM09966 to F.H.R.

REFERENCES

Acampora, D., D'Esposito, M., Faiella, A., Pannese, M., Migliaccio, E., Morelli, F., Stornaiuolo, A., Nigro, V., Simeone, A. and Boncinelli, E. (1990): The Human Hox Gene Family. Nucl. Acids Res. 17:10385-10402.

Beeman, R.W., Stuart, J.J., Haas, M.S. and Denell, R.E. (1989): Genetic analysis of the homeotic gene complex (HOM-C) in the beetle Tribolium castaneum. Dev. Biol. 133:196-209.

Britten, R.J. (1986): Rates of DNA Sequence Evolution Differ Between Taxonomic Groups. Science, 231:1391-1398.

Davidson, E.H. (1982): Evolutionary Change in Genomic Regulatory Organization: Speculations on the Origins of Novel Biological Structure. in:J.T. Bonner (Ed.) Evolution and Development, Dahlem Konferenzen, Springer Verlag, Berlin, Heidelberg, New York, pp. 65-85.

Duboule, D. and Dollé, P. (1989): The structural and functional organization of the murine HOX gene family resembles that of Drosophila homeotic genes. Embo. J., 8:1497-505.

Fienberg, A.A., Utset, M.F., Bogarad, L.D., Hart, C.P., Awgulewitsch, A., Ferguson-Smith, A., Fainsod, A., Rabin, M. and Ruddle, F.H. (1987): Homeo box genes in murine development. Curr. Top. Dev. Biol., 23:233-256.

Gans, C. and Northcutt, R.G. (1983): Neural Crest and the Origin of Vertebrates: A New Head. Science, 220:268-274.

Gaunt, S.J., Krumlauf, R. and Duboule, D. (1989): Mose homeo-genes within a subfamily, **Hox-1.4**, **-2.6** and **-5.1**, display similar anteroposterior domains of expression in the embryo, but show stage- and tissue-dependent differences in their regulation. Development, 107:131-141.

Gaunt, S.J., Coletta, P.L., Pravtcheva, D., Sharpe, P.T. (1990): Mouse Hox-3.4: homeobox sequence and embryonic expression patterns compared with other members of the Hox gene network. Development, 109:329:339.

Gehring, W.J. (1987): Homeo boxes in the study of development. Science, 236:1245-52.

Graham, A., Papalopulu, N. and Krumlauf, R. (1989): The murine and Drosophila homeobox gene complexes have common features of organization and expression. CELL, 57:367-78.

He, X., Treach, M.N., Simmons, D.M., Ingraham, H.A., Swanson, L.W. and Rosenfeld, M.G. (1989): Expression of a large family of POU-domain regulatory gene in mammalian brain development. Nature, 340:35-42.

Kappen, C., Schughart, K. and Ruddle, F.H. (1989): Mammalian Antennapedia Class Homeobox Genes: Organization, Expression, and Evolution. in: M.R. Capecchi (Ed.) Molecular Genetics of Early Drosophila and Mouse Development, Current Communications in Molecular Biology, Cold Spring Harbor Laboratory Press, pp.67.

Kappen, C., Schughart, K. and Ruddle, F.H. (1989b): Two steps in the evolution of Antennapedia-class vertebrate homeobox genes. Proc. Natl. Acad. Sci. USA., 86:5459-63.

Kappen, C. and Ruddle, F.H. (1990): Duplication of homeobox gene clusters in early vertebrate evolution. abstract and poster presented at "Evolution of molecules and developmental systems", Indiana University, Bloomington.

Kessel, M., Fibi, M. and Gruss, P. (1988): Organization of homeodomain proteins. in: Cellular Factors in Development and Differentiation: Embryos, Teratocarcinomas, and Differentiated Tissues. Alan R. Liss, Inc. pp. 93-104.

Li, W.-H., Gouy, M., Sharp, P.M., O'hUigin, C. and Yang,Y.-W. (1990): Molecular phylogeny of Rodentia, Lagomorpha, Primates, Artiodactyla, and Carnivora and molecular clocks. Porc. Natl. Acad. Sci. USA, 87:6703-6707.

Lobe, C.G. and Gruss, P. (1989): Mouse versions of fly developmental control genes: Legitimate or illegitimate relatives? The new biologist, 1:9-18.

Lonai, P. and Orr-Urtreger, A. (1990): Homeogenes in mammalian development and the evolution of the cranium and central nervous system. The FASEB Journal, 4:436-1443.

McGinnis, W., Garber, R.L., Wirz, J., Kuroiwa, A. and Gehring, W.J. (1984): A homologous protein-coding sequence in Drosophila homeotic genes and its conservation in other metazoans. CELL, 37:403-8.

Ruddle, F.H. (1989): Genomics and evolution of murine homeobox genes. in: The Physiology of Growth, Tanner, J.M. and Priest, M.A. (Eds.), Cambridge University Press, pp. 47-66.

Ruddle, F.H., Hart, C.P., Rabin, M., Ferguson-Smith, A.C. and Pravtscheva, D. (1987): Comparative genetic analysis of human homeo-box genes in mouse and man. in: New Frontiers in the Study of Gene Functions, Poste, G. and Crooke, S.T. (Eds.), Plenum Publishing Corporation pp. 73-86.

Scott, M.P., Tamkun, J.W. and Hartzell, F.W. 3d.(1989): The structure and function of the homeodomain. Biochim. Biophys. Acta, 989:25-48.

Scott, M.P. and Weiner, A.J. (1984): Structural relationships among genes that control development: Sequence homology between the Antennapedia, Ultrabithorax and fushi tarazu loci of Drosophila. Proc. Natl. Acad. Sci. USA, 81:4115.

Schughart, K., Kappen, C. and Ruddle, F.H. (1989): Duplication of large genomic regions during the evolution of vertebrate homeobox genes. Proc. Natl. Acad. Sci. USA, 86:7067-71.

Shashikant, C.S., Utset, M.F., Violette, S.M., Wise, T.L., Einat, P., Einat, M., Pendleton, J.W., Schughart, K. and Ruddle, F.H. (1990): Homoeobox Genes in Mouse Development. CRC Critical Reviews, in press.

Suzuki, N., Rohdewohld, H., Neuman, T., Gruss, P. and Schöler, H. (1990):Oct-6: a POU transcription factor expressed in embryonal stem cells and in the developing brain. The EMBO J., 9:3723-3732.

GENE EXPRESSION IN PREIMPLANTATION EMBRYOS

Robert P. Erickson

Department of Pediatrics
University of Arizona Health Sciences Center
Tucson, Arizona 85724

Gene expression in preimplantation embryos has been excellently reviewed several times in recent years. Particularly useful are the reviews by Magnuson and Epstein (1), Schultz (2), and Fleming and Johnson (3). Although preimplantation genetic diagnosis is likely to be based on polymerase chain reaction (PCR) methodologies which are highly useful because the DNA is the same in each cell, there may be some use for diagnosis based on gene products. Thus, this review will try to outline some of the general conclusions which come from a vast amount of work by a large number of researchers in the last several decades. One general conclusion is that maternal RNA has a brief role in the mammalian embryo. In this regard, Dyban (this volume) and I share some beliefs about the role of maternal RNA and have some differences of opinion. I think that maternal mRNA, at least in the mouse, cannot do very much related to differentiation except, perhaps, to turn on many of the transacting factors that are going to lead to the panorama of gene expression that occurs in a short time. A second general conclusion is that there is a complex array of gene expression in the preimplantation embryo. If we take the mouse by the 8-cell stage, all of the vagaries of gene expression that we have to worry about with the adult phenotype, such as pleiotropism or genomic imprinting, can already be seen. The preimplantation embryo is not a simple system and one cannot hope to avoid words like "penetrance" and "expressivity" when we're talking about gene expression in the preimplantation embryo.

There have been hundreds of studies looking at protein products or particular enzymes, and/or their messages, in preimplantation embryos. A general finding has been that if maternal messages are present, they are expressed in the very early embryo, and that these enzyme activities in general, and their messages in particular, decrease markedly in amount early in embryogenesis (4). In mice these events occur at the 2-cell stage. Also, in this most frequently studied species, it has been abundantly demonstrated that many of the maternal mRNAs that will persist briefly in the

Preimplantation Genetics, Edited by Y. Verlinsky and A. Kuliev, Plenum Press, New York, 1991

embryo are polyadenylated with resumption of meiosis. Many of these mRNAs are also translated with meiotic maturation and then degraded at the 2-cell stage. It has been shown that the 3' portion of these mRNA's is important in control of this turnover. In particular, Strickland et al, have shown that an antisense RNA directed to the 3' region of the tissue plasminogen activator (TPA) gene blocked translation when the antisense RNA was injected into single oocytes (5). Antisense RNA directed to sequences 5' of the coding sequence was without effect on the expression of TPA (which can be measured on single cells). In addition, in separate experiments, these authors showed that the injection of 3' antisense RNA to TPA prevented mRNA degradation. Thus, in this experimental system, turnover of RNA has been shown to be controlled by the 3' region.

Studies of changes in gene expression have shown that after this degradation of maternal mRNA, while levels of many enzymes decrease, large increases in several other enzymes occur and, particularly, that there are major changes in multiple unidentified proteins. Our studies by 2-dimensional gel protein electrophoresis provide examples of this. In this work, we focused on a chromosomal 17 protein expressed in testes, Tcp-1 (6). As can be seen in the figure 1, the ^{35}S-labelled proteins from a single blastocyst can be studied. When newly synthesized proteins were studied at the 2-cell stage, neither the paternal nor maternal allele of this gene could be detected. However, by the morula stage, both alleles could be detected. Such analyses by 2-D gel electrophoresis have shown major changes in gene expression between the 2-cell and 8-cell stage (7). Some particular proteins have been shown to have the paternal allele specifically expressed starting at the 8-cell stage and, sometimes, as early as the 2-cell stage. For instance, in the case of β_2-microglobulin, the work of Epstein's lab has shown that the paternal allele can be detected at the 2-cell stage (8) and this was suggested, although not definitive, for ß-glucuronidase (9). As Braude has presented at this symposia, the data suggest that in humans there is a slightly later expression with paternal genome expression starting at the 4-cell stage (10).

The clear evidence of early gene expression for particular genes was predicted because of the lethality of monosomies and trisomies and because of the early defects seen with a number of developmental mutations. Thus, it is clear that multiple gene products are needed at these early stages for the embryo to develop properly. Dyban (this volume) has presented evidence that there are effects of "background genes" on expression of the aneuploid phenotypes at these early stages. I believe that this shows how complete a range of gene expression occurs at these very early stages. Another example of the expression of modifying genes early in development, to join those given by Dyban, concerns the lethal albino deletions. This comes from Lewis' work on the $\underline{c}^{25H}$ homozygous deletion (11). When it was present on its own inbred background, it was an eight-cell lethal. But by simply making an outcross to the A/HeJ inbred strain so as to have it on this F-1 background, and then crossing these mice to create homozygotes on the 50% A/HeJ background, it became a two-cell lethal. Thus, as was shown in the

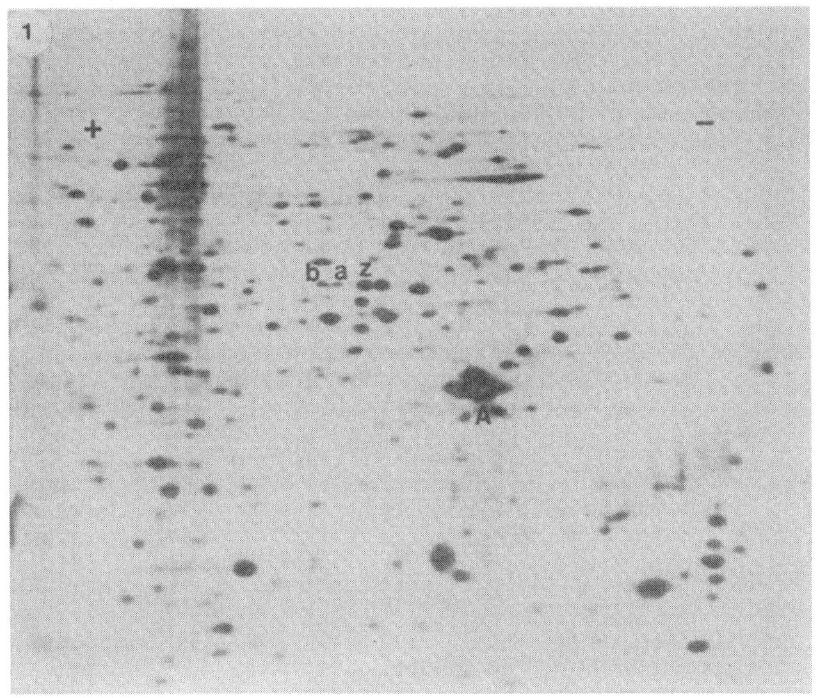

Figure 1. Fluorograph of NP-40 soluble proteins from a single "hatched" blastocyst resolved by 2-D gel electrophoresis. Basic end (+) on left, acidic end (-) on right. Relative molecular masses increase from bottom to top. p 63/6.9 and related proteins are designated a, b, and z, while actin is designated A. From ref (6) with permission.

aneuploid situations, there are marked effects of other genes on gene expression even at these very early stages.

One example of complex gene expression that we didn't know how to correctly interpret at the time when we performed the experiments occurred when we were looking at the expression of the F9 antigen (12). We were testing the claim that $\underline{t}^{12}$, one of the mutations in the $\underline{T}$-complex, in homozygous condition, prevented expression of this antigen. Lewis and I, in studying over a thousand embryos found something that we could only believe because all these embryos had been read "blind". When we looked at the time of F9 antigen expression by day 3-1/2, all embryos were expressing the antigen including crosses in the first two rows of Table 1, in which up to 50% of the embryos were $\underline{t}^{12}$ homozygotes. However, both crosses generating the experimental embryos and the crosses generating controls could be distinguished by whether or not $\underline{T}$ was transmitted from the male. What we were surprised to find was that when $\underline{T}$ came from the male, there was delayed expression of the antigen so that the percentage of the embryos expressing F9 was low at day 2-1/2 while 100% expressed the antigen by day 3-1/2. The effect of $\underline{T}$ coming from the father in delaying the expression of this antigen was seen in the controls as well, although here the effect was perhaps not quite as large (by chi-square, it was equally significant). Although we weren't "sharp" enough to think of germ line imprinting back when we were doing this, the result is easily explained in terms of germ line imprinting which also explains T-hairpin (T^{hp}). T^{hp} is a dominant mutation which, when transmitted from the mother, is lethal (13). The viable transmission of T^{hp} by the male, even in the same litter with lethal T^{hp} from the female, allows maintenance of the mutation. The simplest explanation is that this region of chromosome 17 is germ line imprinted and that it not only affects the phenotypic expression of the T-hairpin mutation, but can also affect F9 antigen expression.

Table 1. F9 Antigen Expression on Preimplantation Embryos from Crosses Involving t^{12} (from ref. 12)

Cross	Day 2 1/2		Day 3 1/2	
	# Pos./Total (# Litters)	%Pos.	# Pos./Total (# Litters)	%Pos.
Homozygous t^{12}/t^{12} present (excluding T from male) $+/t^{12}$ X $+/t^{12}$	27/71 (6)	38	97/99 (10)	98
	}x^2 = 9.2, P<0.01			
Homozygous t^{12}/t^{12} present (including T from male)[a]	1/29 (3)	3	10/10 (2)	100
Controls (excluding T from male)[b]	34/41 (4)	83	36/36 (3)	100
	}x^2 = 12.4, P<0.01			
Controls - T from male[c]	20/52 (6)	38	11/11 (2)	100

[a] T/t^{12} X T/t^{12} and $+/t^{12}$ X T/t^{12}; [b] $T/+$ X $+/t^{12}$ and $+/+$ X $+/t^{12}$
[c] $+/t^{12}$ X $T/+$, $+/+$ X T/t^{12}, and $T/+$ X T/t^{12}

One other vagary of gene expression in the preimplantation embryo is one that relates to superovulated versus naturally ovulated eggs. There are data that embryos resulting from superovulation are different (14). The most dramatic effect I know of is the work of Spielmann on an antigen on the surface of embryos, Fig. 2. The 6 very pale embryos are the superovulated ones, the much darker embryos were naturally ovulated (15). This was a difference that would first appear to be the result of the anti-LDH-X. However, when Spielmann's group very carefully removed all the antisera to LDH-X with an LDH-X affinity column, they still saw these differences. We are probably looking at a difference in a Forssman antigen which the superovulated eggs have not developed whereas the naturally ovulated ones express. It is likely to be a Forssman antigen because it could be absorbed out with splenocytes; these were rabbit antisera and rabbits frequently make anti-Forssman components. Thus, this is a reminder that there is a potential to find biochemical differences as well as morphological and other differences in superovulated compared to naturally ovulated eggs and embryos.

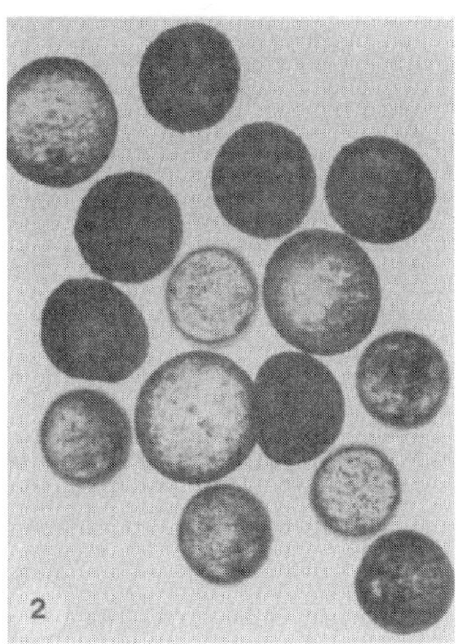

Figure 2. Nonspecific immunohistochemical staining of normally ovulated 1-cell ova (dark staining; 3 fertilized and 3 unfertilized) compared to lack of reaction of superovulated 1-cell ova (3 fertilized and 3 unfertilized) and blastocysts (3 large embryos). Embryos were incubated with rabbit anti-LDH-X-IgG followed by goat anti-rabbit-IgG coupled to horseradish peroxidase followed by the peroxidase reaction. From ref (15) with permission.

Our own current efforts have been directed at the ablation of gene expression in preimplantation embryos by the antisense approach. This has been a very powerful approach in studying the Drosophila embryo where one can make phenocopies by this method. We chose ß-glucuronidase, as a model system because we could do single embryo assays and thus answer different kinds of questions. In work published some time ago we showed that when the antisense RNA was directed to 1.4 kb of coding sequence of ß-glucuronidase, we could get at most a 75% inhibition of expression (16). This was found after trying many different variations and was the result of injecting each blastomere at the four-cell stage with as much in vitro synthesized antisense RNA as there is total poly-A RNA in each cell of the embryo, i.e. a massive excess. We then identified the DNA fragment containing the start codon using the antisense approach (17). When we started this work, the 5' of the ß-glucuronidase gene had not yet been found in a genomic cosmid known to contain the complete gene. The work led to the conclusion that we could get a 90% inhibition with an antisense to just 350 base pairs when it covered the start codon. This was not surprising since these are cytoplasmic injections. There is good evidence that if you "cover" the ribosome binding site, then you can readily inhibit translation. However, if you just "cover" sequences down stream from the ribosome binding site, the ribosome can "peel off" the antisense RNA. Recently, we have been working with transgenic mice to see what sort of effects these same constructs will exert in the nucleus. In this case, instead of having a huge excess of antisense RNA in the cytoplasm, we are talking about 500 to 1,000 copies of a transgene in the nucleus from which one or a few of them may then integrate and be expressed while others may be transiently expressed. When we injected a transgene driven by the Met promoter (metallothionine-1 is somewhat of a constitutive promoter in preimplantation embryos), there was little effect with the sense construct. The antisense construct with the 1,400 base pair coding region led to results in which almost all of the embryos were somewhat inhibited in their expression of ß-glucuronidase, presumably because of transient expression, and some were inhibited a great deal. On the other hand, when the transgene construct with the Met promoter contained only the 350 base pair fragment which covered the start codon, it did not have much effect (18). We think this difference has to do with the different mechanisms by which antisense RNA can effect gene expression in the nucleus compared to the cytoplasm. In the nucleus, binding of the antisense RNA to the sense RNA presumably blocks processing and transport out of the nucleus. Thus, it makes sense that the more antisense you have in the construct, the more it can inhibit these processes. Given the 1,400 base pairs versus the 350 base pairs, we think that the antisense transgene does better with the longer coding sequence. We have also looked at the use of synthetic oligodeoxynucleotides and we have not been successful in inhibiting embryonic gene expression using them (18). We have used 20-mers directed across the starting codon and directed to the coding sequence of exon 7. We have injected them in the cytoplasm or cultured the embryos with them. They have been used singly or together and we have not been able to inhibit ß-glucuronidase expression in this system. We found that radiolabelled oligodeoxynucleotides added to the culture media were poorly taken up.

We have also shown some effects on early development in one experimental system - gap junction protein (19). Basically, in this work we were able to replicate all the findings that previously had been obtained with antibodies to gap junctional protein by using antisense RNA (19). We used a complete cDNA to the liver, connexin 32 (also known as the 27 kilodalton gap junction protein because of the discrepancy between its apparent mobility on SDS-PAGE gels and its predicted size). When we injected two-cell embryos with antisense RNA to connexin 32, with ß-glucuronidase as control, there was an inhibition of compaction. There also was significantly more degeneration in experiments in which embryos received antisense to connexin 32. Since the experimental RNA is made in the same kind of test tube and with the same enzyme as the control, we think there are significant toxic effects from inhibiting the expression of connexin 32. If injections were into most of the blastomeres of the eight cell stage, blastulation (as well as compaction), was also inhibited markedly. Another phenomenon that we saw, which had also been seen with the injection of antibodies, is that if you injected one blastomere, an excluded blastomere resulted. This did not happen with control antisense RNAs. It is very rare, in my experience, to see an embryo with one blastomere excluded from a compacted embryo. To study the effects of the antisense RNA on gap junctional communication, we used

Lucifer yellow, a small molecular weight dye, that can go through the gap junctions and distribute throughout the embryo. If one blastomere is injected with antisense to connexin 32 and followed by injection of the Lucifer yellow, you see one of two things; you see embryos in which you "hit" the same cell which had received the antisense RNA and the Lucifer yellow is held in that cell and does not diffuse throughout the embryo. Sometimes you "hit" a different cell and all the cells "light" up, except one cell. This approach allowed us to answer the question that had not been possible to answer with the antibody and that was to study turnover times of connexin proteins in the early embryo. We injected the antisense RNA and then, at time intervals thereafter, injected Lucifer yellow. By one hour, there was an inhibition of the transfer of Lucifer yellow, which did not occur with the control injections. From this we concluded that the connexin 32 mRNA was turning over at a quite rapid rate - on the order of an hour. In liver, connexin 32 has turnover of about four hours. Given the active metabolism of preimplantation embryo, the one hour turnover is not too surprising. Real aficinados of gap junctions might be confused by this result with an antisense to connexin 32 since Kidder found connexin 43 to be newly synthesized in embryos (20). The 5' regions of the connexins are so closely related sequence-wise that our antisense might be inhibiting connexin 43. Alternatively, in *Xenopus* an embryonic connexin has been cloned (21) and there may well be an embryonic connexin in mice and our antisense RNA to connexin 32 might cross-hybridize to it.

In conclusion, I would like to agree with Dyban (this volume) that maternal mRNA is important and that it perhaps includes genes for transacting factors involved in the full panorama of gene activity that occurs even during preimplantation development. But I think we also have to emphasize the very large difference from *Drosophila* in which maternal mRNAs have laid down positional information which is very important for the whole sequence of developmental gene expression in that organism. All the clinical results reported at the meeting, or the experimental results on mouse models, show that mammalian development is regulative. We can take out half the blastomeres at an eight-cell stage and still get a normal embryo which is not possible in *Drosophila*. Thus, mammalian maternal mRNA cannot provide the same kind of positional information that it does in some invertebrates.

Acknowledgements: My thanks to Ms. Debbie Bohlender for secretarial assistance and NIH grant HD 26454 for support.

References

1. T. Magnuson and C.J. Epstein, Genetic Control of very early mammalian development, *Biol. Rev.* 56:369 (1981).
2. G. A. Schultz, Utilization of genetic information in the preimplantation mouse embryo, *in*: "Experimental approaches to mammalian embryonic development", J. Rossant and R.A. Pedersen, eds., Cambridge University Press, Cambridge.

3. T. P. Fleming and M.H. Johnson, From egg to epithelium, Ann. Rev. Cell Biol. 4:459 (1988).
4. M. B. Dworkin and E. Dworkin, Functions of Maternal mRNA in Early Development, Molec. Reprod. and Develop. 26:261 (1990).
5. S. Strickland, J. Huarte, D. Belin, A. Vassalli, R.J. Rickles, and J.D. Vassalli, Antisense RNA directed against the 3' noncoding region prevents dormant mRNA activation in mouse oocytes, Science 241:680 (1988).
6. E. R. Sanchez and R.P. Erickson, Expression of the Tcp-1 locus of the mouse during early embryogenesis, J. Embryol. Exp. Morph. 89:113 (1985).
7. J. Levinson, P. Goodfellow, M. Vadeboncoeur, and H. McDevitt, Identification of stage-specific polypeptides synthesized during murine pre-implantation development, Proc. Nat'l Acad. Sci., U.S.A., 75:3332 (1978).
8. J. A. Sawicki, T. Magnuson, and C.J. Epstein, Evidence for expression of the paternal genome in the two-cell mouse embryo, Nature. 294:450 (1982).
9. L. Wudl and V. Chapman, The expression of ß-glucuronidase during preimplantation development of mouse embryos, Develop. Biol. 48:104 (1976).
10. P. Braude, V. Bolton, and S. Moore, Human gene expression first occurs between the four and eight-cell stages of preimplantation development, Nature. 332:459 (1988).
11. S. E. Lewis, Developmental analysis of lethal effects of homozygosity for the c^{25H} deletion in the mouse, Develop. Biol. 65:553 (1978).
12. R. P. Erickson and S.E. Lewis, Cell surfaces and embryos: expression of the F9 teratocarcinoma antigen in T-region lethal, other lethal, and normal pre-implantation mouse embryos, J. Reprod. Immunol. 2:293 (1980).
13. D. R. Johnson, Hairpin Tail: A case of post-reductional gene action in the mouse egg: Genetics. 76:795 (1974).
14. P. Hyttel, T. Greve, and H. Cullesen, Ultra-structural aspects of oocyte maturation and fertilization in cattle, J. Reprod. Fert. Suppl. 38:35 (1989).
15. H. Spielmann, H.G. Eibs, C. Mentzel and D. Nagel, Studies on the binding of antibody against mouse lactate dehydrogenase (isoenzyme X) by preimplantation mouse embryos, J. Reprod. Fert. 50:47 (1977).
16. A. Bevilacqua, R.P. Erickson, and V. Hieber, Antisense RNA inhibits endogenous gene expression in mouse preimplantation embryos: lack of double-stranded RNA "melting" activity, Proc. Nat'l Acad. Sci., U.S.A. 85:831 (1988).
17. A. Bevilacqua and R.P. Erickson, Use of antisense RNA to help identify a genomic clone for the 5' region of mouse ß-glucuronidase, Biochem. Biophys. Res. Comm. 160:937 (1989).
18. A. Ao, R.P. Erickson, J. Karolyi, and A. Bevilacqua, Antisense inhibition of ß-glucuronidase expression in preimplantation mouse embryos: a comparison of transgenes and oligodeoxynucleotides, Antisense Research and Development, in press.

19. A. Bevilacqua, R. Loch-Caruso, and R.P. Erickson, Abnormal development and dye coupling produced by antisense RNA to gap junction protein in mouse preimplantation embryos, Proc. Nat'l. Acad. Sci., U.S.A. 86:5444.
20. D. J. Barron, G. Valdimarsson, D.L. Paul, and G.M. Kidder, Connexin 32, a gap junction protein, is a persistent oogenetic product through preimplantation development of the mouse, Develop. Genetics. 10:318 (1989).
21. L. Ebichara, E.C. Beyer, K.I. Swenson, D.L. Paul, and D.A. Goodenough, Cloning and expression of a *Xenopus* embryonic gap junction protein, Science. 243:1121 (1989).

SOMATIC GENE THERAPY

Arthur Bank

Columbia University
College of Physicians and Surgeons
Department of Medicine, and
Department of Genetics and Development
New York, New York

INTRODUCTION

The initial social and political barriers preventing the use of gene therapy in humans have essentially been overcome by a variety of committees, and by both the NIH and the FDA. During the past year, at least two experiments have been approved for somatic gene therapy. In principle, gene therapy does not differ from any other form of therapy; e.g., giving a drug to a cancer patient or an antibiotic to a patient with infection. The requirements for ethical and scientifically valid gene therapy experiments, as in all other clinical research, involve an estimation of the risk:benefit ratio. If the risk is small and the benefit potentially great, then the experiment should be approved; if the risk is great and potential benefits small, the experiment should not be approved. In addition, if there is no appropriate way to evaluate the effects of an experiment, it should not be performed. It should be noted that in the currently approved experiment involving a child with adenosyldeaminase (ADA) deficiency, this latter criterion is not met since the child is on a form of therapy involving modified ADA enzyme with polyethylene glycol (PEG), and the effect of adding gene therapy to the treatment of this patient will be at best difficult if not impossible to evaluate. However, in summary, the climate for performing somatic gene therapy has improved dramatically over the past several years and there is every hope that if successful it will be an acceptable form of treatment.

METHODOLOGY AND REQUIREMENTS FOR SOMATIC GENE THERAPY

There are two forms of somatic gene therapy that are currently conceivable. One involves the correction of a gene defect by the addition of a normal gene and homologous recombination correcting the defective gene in situ. This is a most exciting potential therapy since all the elements surrounding the defective gene would remain intact and the simple correction of the gene defect in its original position in the genome has a great chance of curing a given disease. However, it is estimated currently that the

Preimplantation Genetics, Edited by Y. Verlinsky and A. Kuliev, Plenum Press, New York, 1991

efficiency of homologous recombination is that only one in 10^6-10^7 involved cells are corrected. This is too low a figure to pursue somatic gene therapy at the present time.

The other methodology and the one that I will focus on is the addition of a normal gene while the defective gene remains. The goal of this treatment is to provide a normal gene expressed at a high level with the assumption that the normal gene will provide the function of the defective gene.

Almost weekly, genes are cloned that are responsible for disease. The recent findings of the cystic fibrosis and the neurofibromatosis genes are examples of these advances. In addition, there are a large number of diseases in which the genetic defect has been identified. These include the hemophilias, immunodeficiency disorders such as ADA, urea cycle disorders, the anemias especially ß thalassemia and sickle cell anemia, emphysema due to α-1-antitrypsin deficiencies, lysosomal storage diseases such as glucocerebroside deficiency, and other metabolic disorders such as phenylalanine hydroxylase deficiency and related disorders (1).

Thus, there are a large number of diseases that one can approach with gene therapy. One important aspect of gene therapy is whether the target cells which must be corrected to cure the disease are available. Table 1 lists a variety of target cells for gene therapy. These include bone marrow stem cells, skin fibroblasts, hepatocytes, endothelial and muscle cells, and lymphocytes (1). All these target cells, except for bone marrow stem cells and lymphocytes, present difficulties in transferring genes to a significant number of cells, or in the limited life span and death of cells containing the gene. The only exception to these limitations currently is the use of bone marrow stem cells for gene therapy.

Stem cells in the bone marrow, if properly infected with a new gene, will give rise to all hematopoietic elements (red cells, white blood cells, and platelets) and will contain the added gene. These stem cells will continue to divide, proliferate, and provide the infected gene to the progeny of the animal throughout its lifetime. Therefore, if one can transfer a gene into most of the stem cells and have that gene appropriately expressed in those cells, gene therapy should be achieved. In the case of the ß thalassemias and sickle cell anemia, the usual method of allogeneic bone marrow transplantation leads to cure, therefore, it is almost a certainty that autologous transplantation would cure these patients. It would also alleviate the great problem of finding compatible donors for patients with these disorders.

Table 1. Target Cells for Gene Therapy

Bone Marrow Stem Cells
Skin Fibroblasts
Hepatocytes
Endothelial Cells
Muscle Cells
Lymphocytes

To summarize, the requirements for somatic gene therapy are: 1) One has to have normal genes cloned and available in large amounts; 2) It is important to have a method for the efficient transfer of the gene into most of if not all of the patient's bone marrow cells since the number of stem cells which are self-renewing is small; 3) One needs a method to select for those cells which have acquired the normal gene; if one does not infect a majority of or all of the stem cells, defective stem cells will be reimplanted into the animal or the patient; there is no evidence that the cells with the normal gene will outgrow those with defective genes; thus, there has to be a selection for infected stem cells; and 4) After proper infection of the gene into a large number of stem cells, normal expression of the newly transferred gene must be achieved. With the gene integrated into a variety of positions in the human genome, this may be extremely difficult. In the case of the disorders of hemoglobin, 30-50% of the normal level of hemoglobin production is probably required to convert a patient homozygous for an illness into a relatively normal heterozygote.

The above requirements for gene therapy can be summarized into two major areas: (1) Efficient and safe gene transfer; and (2) High-level and appropriate gene expression.

Efficient and Safe Gene Transfer

Retroviruses are the most efficient vectors for transferring genes into cells. They are also extremely efficient in providing a means of obtaining efficient integration of the infected genes into the genome. Usually, in retroviral infection, one gene enters each cell. This makes the procedure relatively safe in terms of the number of retroviral components that can harm any individual stem cell. The use of retroviruses is limited in two ways: first, the amount of DNA (converted into RNA) that can be inserted into retroviral vectors is limited; second, it is possible that if experiments are not done properly intact retroviruses may be produced that could lead to a dangerous proliferative retroviral infection. This latter possibility must be avoided in both animal and human studies since if significant numbers of retroviruses are allowed to proliferate they will eventually integrate near undesirable genes such as oncogenes which would then be activated and could give rise to tumors (as demonstrated in animals).

The retrovirology involved in these experiments is relatively simple. A so-called "packaging cell" is produced by transfecting mouse 3T3 cells (fibroblasts) with a plasmid containing the retroviral genes _gag_, _pol_, and _env_. _Gag_ genes are required for the production of viral core proteins; _pol_ genes for reverse transcriptase; and _env_ genes for viral envelope proteins. Viral long terminal repeats (LTRs) are also required.

Richard Mulligan and his collaborators at Massachusetts Institute of Technology produced a retroviral packaging line in which the psi sequence, required for the packaging of RNA into a virus, was mutated so that it was inactive (2).

While the psi2 packaging cell line produced viral proteins, the viral RNA cannot be packaged into an intact virus. To provide the gene of interest, an intact retroviral vector containing the gene of interest, long terminal repeats (LTRs), and an intact psi sequence are added. Most of these retroviral vectors also utilize a selectable marker, generally the neomycin resistance (neo^R) gene. When neo^R is expressed in recipient cells, the cells become resistant to neomycin and its analogues; a major useful analogue is a compound called G418. The retroviral vector contains both the neo^R gene and the gene of interest, e.g., the human ß globin gene. After retroviral infection into bone marrow target cells, G418 is added and the cells are grown. Cells resistant to G418 are most likely to contain both the human ß globin and the neo^R genes since both genes are on the same piece of DNA. All cells that do not contain the transferred genes are killed; this provides excellent selection for the cells containing the gene of interest for use in gene therapy.

The problem with using the psi2 cell line and an incoming retroviral genome is that a single recombination between the defective psi sequence and the intact psi sequence of the incoming retrovirus containing the genes of interest can lead to the production of wild-type virus. This wild-type virus, as described previously, is extremely dangerous.

To avoid this problem, D. Markowitz (in my laboratory) devised modified packaging lines (3,4). First, the _gag_, _pol_, and _env_ genes were separated on different plasmids. Then a _gag_-_pol_-containing plasmid was transfected into 3T3 cells; subsequently an _env_-containing plasmid was transfected into these cells. The 3' LTRs were removed. Two different types of _env_ DNA were used: one containing an ecotropic sequence which permits infection of only mouse cells, and another containing an amphotropic _env_ which allows infection of all cells including monkey and human. In this system, three recombinations instead of one are necessary to make an intact viral genome: _Gag_ and _pol_ must recombine with _env_, the psi mutation must be corrected, and the 3' LTR must recombine with the incoming retroviral genome containing this sequence. This cell line has been shown to be extraordinarily safe as well as providing high viral titers similar to cell lines in which the three viral genes are on the same plasmid. We have sent this packaging line to more than 170 laboratories world-wide over the past two years. None of them has yet reported the generation of intact virus. Thus, we now have available a safe and efficient packaging line which can be used in experiments to transfer genes into recipient cells. The original psi2 line and, indeed, the packaging line used in patients at NIH, PA317, which requires two mutations to generate intact virus, produce intact virus when these populations are expanded.

High Level and Appropriate Gene Expression

The results of experiments using retroviral gene transfer to express ß globin and ADA genes in live animals have been disappointing thus far. There is no question from a large variety of experiments that when retroviral vectors containing either the ß globin or the ADA gene are transferred into erythroid or lymphocytic cells in culture, these cells are easily infected and do produce high levels of the appropriate proteins - ß globin and ADA. The difference between the ability to achieve high-level expression in cells in living animals as compared to the results in cell lines is poorly understood but is presently being investigated by several laboratories.

The major model system for studying gene therapy is to use irradiated mice and to transfer stem cells from congenic mice which have been infected with retroviruses containing the gene of interest (1,5). When the human ß globin gene has been transferred into mouse stem cells, and an irradiated mouse repopulated with these stem cells, the human ß globin gene is expressed at less than one percent of the mouse ß globin gene; this is in contrast to 40 to 50% as much human ß globin gene expression compared to that of mouse ß globin in mouse erythroleukemia cells (MELC) in culture. The most positive result in live mice is that the human ß globin gene is only expressed in appropriate tissues, i.e., blood, spleen, and bone marrow.

In our laboratory, as in several others, we have used the neo^R gene in a virus as a model system for gene transfer into intact animals (5). We have been able to transfer the neo^R gene into irradiated mice after infection of the bone marrow stem cells with retrovirus containing the neo^R gene. Thirteen posttransplantation individual spleen colonies containing hematopoietic elements were isolated from mice and most shown to contain the neo^R gene if G418 selection had been performed prior to transplantation of the infected stem cells. These individual spleen colonies are assumed to represent single stem cells which have integrated into the spleen and then proliferated to form a colony. The neo^R probe is used on a DNA blot to determine the presence of the neo^R gene. When a restriction enzyme cuts outside the retroviral vector containing the neo^R gene, the total number of integration sites indicates the total number of stem cells which have been infected. In this analysis, we have shown that one to six infected stem cells are all that are required to completely repopulate the entire population of the mouse with infected cells (5). Indeed, this result is quite impressive and suggests that it may be possible to obtain a comparable outcome if high-level infection can be attained using retroviral vectors containing other genes of interest.

There appear to be at least two major problems in the high-level transfer and expression of globin genes and indeed of other genes. First, the size of the retrovirus must be small since large retroviruses will not give rise to viral titers which are high enough to infect most of or all of the small number of stem cells in the bone marrow population. Second, as indicated earlier, the expression of the retroviral vectors used in live mice and containing genes such as globin and ADA has been low to date, and this problem is currently under intense investigation. In the globin system, the discovery of a specific DNA sequence upstream from the most embryonic of the ß -like globin genes, the ϵ gene, has greatly improved the possibility of successful high-level gene transfer (6). This particular sequence, known as the locus control region (LCR), allows tissue-specific, i.e., erythroid-specific, high-level expression of the gene to which it is linked regardless of the chromosomal position into which the construct was integrated. This result should allow experiments to be done in which this region is linked to a ß globin gene, inserted into a retroviral vector, and optimistically have it function at this high level. To date, this has not been possible because of presumed idiosyncracies in the ability of retroviruses containing this sequence to synthesize intact viral RNA for packaging.

In summary, gene transfer through retroviruses can reconstitute the entire bone marrow of a mouse with cells containing a gene of interest, to date, the neo^R gene. High-level expression of cellular genes such as globin and ADA has been difficult to achieve in live animals but studies are underway to solve these problems.

Last, I would like to discuss a gene that provides great potential not only in cancer but also as a selectable marker in _in vivo_ as well as in _in vitro_ experiments. This gene, called the multiple drug resistance (mdr) gene, provides resistance to certain classes of chemotherapeutic agents utilized in cancer. These include the anthracyclines (adriamycin and daunomycin), the vinca alkaloids including velban and vincristine, and the epotosides including VP16. The cDNA for mdr codes for a glycoprotein and has been cloned from both human and mouse genomes. Retroviral gene transfer and expression of the mdr gene in bone marrow cells could theoretically provide increased resistance of these bone marrow cells to suppression following chemotherapy. Human bone marrow cells normally contain low amounts of mdr. Autologous bone marrow transplantation is routinely performed in patients requiring high-dose chemotherapy for resistant cancers. The marrow is removed so that the patient's bone marrow can be reconstituted after high-dose chemotherapy. In a human gene therapy scenario, the marrow when removed would be infected with a retrovirus containing the mdr gene. After treatment with high-dose chemotherapy, the bone marrow cells infected with this gene would be replaced in the patient as usual. The hope is that when and if the cancer recurs, a significantly higher level of chemotherapy can be given to the patient without the need for subsequent bone marrow transplantation.

To date, transgenic mice containing the mdr gene in bone marrow cells have been produced by injection of the mdr gene into fertilized eggs (7). The bone marrow cells of one of these mice has a greater resistance to white blood cell count suppression after a dose of adriamycin than a normal mouse. Thus, the mdr gene appears to function in live mice and provides great potential for this system's use in cancer chemotherapy. In addition, using this system, it is conceivable that a second gene can be introduced on the same retroviral vector as the mdr gene in somatic gene therapy using bone marrow. The selection, by adriamycin or other mdr-sensitive compounds _in vivo_ and _in vitro_, then allows the preferential survival of bone marrow stem cells not only containing the mdr gene but the co-infected gene of interest as well. Of great potential significance is that repeated chemotherapy _in vivo_ could favor the establishment of a population of bone marrow cells in which most of or all of the cells contain the mdr gene and any other gene infected.

CONCLUSION

The globin, ADA, and mdr genes have been discussed to provide examples of the manner in which gene therapy might be utilized to help treat a variety of human disorders. Progress has been great in all areas of somatic gene therapy experimentation over the past several years. Therefore, it should be expected that our ultimate goal of providing effective gene therapy for a variety of both genetic and acquired disorders is increasing at a significant rate and may be useful in the treatment of human disease in the not too distant future.

REFERENCES

1. A. D. Miller, Progress toward human gene therapy, _Blood_ 76:271 (1990).
2. R. Mann, R. C. Mulligan, and D. Baltimore, Construction of a retrovirus packaging mutant and its use to produce helper-free defective retrovirus, _Cell_ 33:153 (1983).
3. D. Markowitz, S. Goff, and A. Bank, A safe packaging line for gene transfer: Separating viral genes on two different plasmids, _J. Virol._ 62:1120 (1988).
4. D. Markowitz, S. Goff, and A. Bank, Construction and use of a safe and efficient amphotropic packaging cell line, _Virology_ 167:400 (1988).
5. C. Hesdorffer, M. Ward, D. Markowitz, and A. Bank, Efficient gene transfer in live mice using a unique retroviral packaging line, _DNA & Cell. Biol._ in press (1990).
6. F. Grosveld, G. B. V. Assendelft, D. R. Greaves, and G. Kollias, Position-independent, high-level expression of the human ß globin gene in transgenic mice, _Cell_ 51:975 (1987).
7. H. Galski, M. Sullivan, M.C. Willingham, K-V Chin, M.M. Gottesman, I. Pastan, and G. T. Merlino, Expression of a human multidrug resistant cDNA (mdr1) in the bone arrow of transgenic mice: Resistance to daunomycin-induced leukopenia, _Mol. Cell. Biol._ 9:4357 (1989).

STRATEGIES FOR HUMAN GERM LINE GENE THERAPY

Jon Gordon

The Mount Sinai Medical Center
Mount Sinai School of Medicine
1 Gustave L. Levy Place, Box 1175
New York, NY 10029

INTRODUCTION

Human germ line gene therapy, described in general terms as the heritable genetic modification of the human organism, is already technically feasible. However, such intervention, which presumably would be undertaken for medical therapeutic purposes, cannot be responsibly implemented until a variety of technical and ethical problems are properly addressed. From a technical standpoint, while it is feasible with current technology to insert a new genetic material into the human germ line, it is not possible to perform such manipulations without incurring significant risk of adverse health effects relating to the gene transfer event. Moreover, the predicted efficiency of gene transfer, extrapolated from the results of animal experiments, is not high enough to be acceptable as a form of medical therapy. From an ethical perspective, the problems are complex and multi-faceted, but may be divided into three broad categories: those relating to the issues of informed consent and other aspects of the patient-doctor relationship; societal concerns that germ line gene therapy could endow an individual and his/her descendants with an unfair genetic advantage; and, finally, philosophical questions relating to whether gene insertion violates "divine intent", or, put in more secular parlance fundamental laws of Nature.

While some would argue that no situation will ever exist to justify transfer of new genetic material into the germ line of the human species, the present paper will dismiss this contention forthwith as both reflective of a lack of open-mindedness and entirely in conflict with the purpose of the paper. Were this position considered valid, there would of course be no purpose to a discussion of germ line gene therapy strategy.

With these ground rules established, the paper will consider how to approach the issue of human germ line gene insertion. As the technical aspects of the problem must

Preimplantation Genetics, Edited by Y. Verlinsky and A. Kuliev, Plenum Press, New York, 1991

first be confronted, the paper will review the current approaches to gene transfer into the mammalian germ line and their various advantages and deficiencies as potential technology for health. Next, genetic changes to be theoretically appropriate to justify alteration of a human genome must be discussed. The paper will attempt to develop the argument that correction of "single gene" defects (barring major unforseen advances in gene transfer technology) is not a rational use of gene insertion. Rather, it will offer a personal bias that genes might be inserted in order to render an individual's inherent genetic disadvantage more easily managed by standard medical approaches. Finally, areas of social and medical ethics that impact upon the continuing research in gene transfer will be addressed.

TECHNICAL APPROACHES TO GERM LINE GENE INSERTION

Microinjection

Microinjection is by far the most widely used of gene transfer approaches. First developed in 1980 (Gordon et al., 1980), this technique almost always entails microinjection of purified DNA into the pronucleus of the fertilized egg. We showed that this procedure results in retention of the donor genetic material throughout development in mice (Gordon et al., 1980). Soon after, an integration and transmission of the new material to subsequent generations of progeny was demonstrated (Gordon and Ruddle, 1981; Costantini and Lacy, 1981, Brinster et al., 1981). Animals genetically transformed in this manner were called "transgenic" (Gordon and Ruddle 1981), a term that is now applied to similarly altered species of all classes and kingdoms.

A salient feature of transgenic animals, as first convincingly shown by Brinster et al. (1981), is that if appropriately constructed, foreign genes are efficiently expressed in the adult animal. Moreover, the tissue distribution and developmental regulation of gene expression are determined by discrete regions of DNA in and around the gene sequence that are often (though by no means always) situated close to the promoter. Importantly, these promoter/enhancer elements can be grafted to heterologous coding sequences by recombinant DNA technology to target expression of the coding element (Brinster et al., 1981). These features have made it possible to express a variety of microinjected genes in patterns typical of their expression in the organism from which they were cloned, and to direct expression of a variety of genes so as to reveal the presence of promoter/enhancer elements, or engineer phenotypic alterations into transgenic animals (for reviews see Palmiter and Brinster, 1986; Gordon, 1989). These findings indicate that genes could be inserted and expressed in human beings with resultant alterations of phenotype. In fact, experiments with mice have already demonstrated that it is possible to correct disorders such as ß-thalassemia (Costantini et al., 1986) or gonadotrophin-releasing hormone deficiency (Mason et al., 1986) by microinjection. Thus, it is theoretically possible to microinject genes into pronuclei of human zygotes in _in vitro_ fertilization programs and obtain physiologically significant expression of the donor gene in adult transgenic humans.

However, microinjection currently suffers from several technical deficiencies that render it unsuitable as a therapeutic tool. Current problems include the relatively low efficiency of gene integration (around 15%), embryo mortality secondary to mechanical disruption of zygote cytoarchitecture, the fact that inserted genes are not always located in a position within the recipient genome that is permissive for expression, and the fact that integration is apparently random. This latter problem poses the hazard of insertional mutagenesis, a phenomenon characterized by disruption of host genes by integration of the new DNA (e.g., Woychik et al., 1985; McNeish et al., 1988; Krulewski et al., 1989; see Gordon, 1989, for review). While insertional mutagenesis occurs in only about 5% of cases, this frequency is excessive when viewed as a side effect of medical intervention.

Because only 15% of embryos retain injected genes, an important advance in this area would be an attainment of the ability to identify those zygotes destined to become transgenic. It is here that recent advances in preimplantation genetic diagnosis could play an important role. The polymerase chain reaction (PCR, Saiki et al., 1985) might be utilized to identify transgenes in embryos prior to implantation. This technique has already been used successfully to predict the sex of mouse (Bradbury et al., 1990) and human (Handyside et al., 1990) embryos. The current problem with such strategy, in addition to the common problems attendant upon attempts to identify genes in single cells, is that even embryos that will not become transgenic will have been microinjected with hundreds or thousands of copies of the foreign gene one desires to introduce. The problem, therefore, is to distinguish transgenic embryos from those harboring residual microinjected DNA that is ultimately to be lost.

One possible solution to this problem lies in the fact that foreign DNA fragments are frequently integrated as head-to-tail concatemers (Costantini and Lacy, 1981). It seems that formation of concatemers is a certain prelude to integration, a hypothesis based upon the early finding of Burki and Ullrich (1982) that in one of only two mouse fetuses of a series of about 60 that became transgenic, concatemers were apparently integrated at different sites in the placenta and somatic components of the animal. The fact that two integration events could occur in one animal and not at all in so many others indicates that once the concatemer was formed integration was inevitable (Gordon, 1989). If concatemer formation leads to integration, it may be possible to detect impending integration by designing PCR primers on either side of the head-to-tail junction. In this way, PCR would only be successful if the link was formed between the head of one DNA fragment and the tail of the next. Our laboratory is currently performing experiments to determine the feasibility of this strategy.

Therefore, microinjection is attractive as a method of

gene therapy because of the high rates of expression of integrated genes and the absence of an identifiable limit to the size of DNA fragment that can be inserted (Palmiter et al., 1982). However, several technical drawbacks, some of which might be overcome by advances in preimplantation genetic diagnosis, still exist.

Retrovirus-mediated gene transfer

This strategy is based upon the finding of Jaenisch (1976) that infection of mouse preimplantation embryos with Moloney murine leukemia virus leads to integration of the proviral DNA into the mouse genome. Several years later, methods were developed for generating infectious retroviral particles from replication defective recombinant genomes that contained heterologous genes (Mann et al., 1983). Such recombinant retroviruses can be used to infect embryos (van der Putten et al., 1985; Jahner et al., 1985); however, expression of the substituted gene(s) is often low (Jahner et al., 1985), and the size of the DNA fragment that can be inserted is limited by the packaging constraints of the retrovirus to about 9 kilobases (kb) of DNA. These drawbacks render unlikely the use of retroviruses for transformation of human embryos.

Embryonic stem (ES) cell mediated gene transfer

The approach to gene insertion relies upon the fact that ES cell lines established from blastocysts placed in culture (Evans and Kaufman, 1981; Martin, 1981) can be genetically altered and then inserted into the cavity of a normal blastocyst, whereupon they can colonize the inner cell mass of the recipient, participate in normal development and give rise to all cell lineages of the resultant animal including germ cells (Robertson et al., 1986). The most attractive feature of ES cell mediated gene transfer is that homologous recombination, which occurs rarely but detectably between the donor and corresponding recipient DNA after DNA transfection (Smithies et al., 1985), can be exploited to target genes in ES cells (Thimpson et al., 1989; Ziljstra et al., 1990; DeChiara et al., 1990). Thus, the formidable problem of random integration and insertional mutagenesis can be overcome with this system, where gene replacement can be successfully accomplished.

Several intriguing difficulties are associated with the use of such system for genetic engineering of the human germ line. First and foremost is the fact that the founder organism harboring the ES cell derivatives is a genetic mosaic composed of descendants of both the ES line and the blastocyst into which it was inserted. Such mosaics in a human being could pose a variety of problems. For example, XX<-->XY mosaicism could lead to sterility or hermaphroditism, mosaicism at the loci controlling pigmentation could lead to a mottled, piebald or even striped appearance, and mosaicism at the HLA locus could result in tolerance to more that one HLA haplotype. Some of these problems might be overcome by generating an ES line in which the B2-microglobulin gene was destroyed by homologous recombination, as has already been done in mice (Ziljstra et al., 1990). This could negate the ability of the ES line to present antigens that influence gonadal sex or the immune response. However, it is unclear whether ES cells without HLA function would be vectors for spread of infections that are normally eliminated by killer T cells that utilize HLA determinants for identification of virally infected cells. Thus, the genetically treated individual might be chronically ill, with immunologically invisible ES cell derivatives providing a reservoir of infectious viruses.

Another strategy for controlling the side effects of mosaicism would be to design ES cell lines that were not capable of colonizing all tissues of the newborn. The strategy of preventing germ line transmission of ES derivatives is not only feasible but particularly attractive, as ES-cell mediated gene transfer would then act as a form of somatic gene therapy. The drawback of this approach is that human ES lines have not yet been made at all. Other problems include the fact that if one ES line were used for several gene therapy procedures, the human blastocyst which gave rise to the line would then effectively be genetically cloned in the treated individuals. This happenstance poses monumental ethical dilemmas that are not altogether lacking in their amusing facets. For example, the use of such gene therapy might lead to competition among couples to have their blastocysts chosen to generate a donor ES cell line and thereby confer immortality upon their genotypes.

Therefore, although ES cell technology poses some interesting prospects for human germ line gene therapy, it is unlikely that such an approach will be practicable in the near future, and its use would pose some unique and profound ethical problems.

WHAT KIND OF GENES SHOULD WE INSERT?

Notwithstanding the many difficulties posed by human germ line gene insertion, the current technical feasibility of such a procedure, manifested in the production of numerous transgenic mice and other animals, compels us to contemplate the issue. If the premise is accepted that conditions could theoretically arise where such a process was desirable, the kinds of genetic alterations that would be most advantageous and acceptable need careful consideration.

Barring unforseen advances in targeted gene integration, it seems unwise to consider inserting a gene in order to correct a genetic lesion. First, as it is now becoming possible to identify genetic anomalies in preimplantation embryos, it will be possible to detect genetic diseases at endogenous loci and exclude from transfer those embryos destined to develop genetic disease. Second, if a new gene is inserted to compensate for a genetic deficiency (e.g., the wild type globin gene as a cure for sickle cell anemia), and integration does not occur at the corresponding host locus, the individual would be triploid at the locus. While this poses no certain hazard to the recipient of the new DNA (assuming the problems of insertional mutagenesis are solved), it does pose a threat to subsequent generations. For example, a transgenic individual homozygous for sickle cell anemia, but with a third globin locus donated by gene transfer, could transmit both the normal transgene and one mutant locus to his/her progeny. If the partner of this individual was not a carrier of sickle cell trait, the child would then carry two copies of the wild type globin gene and one sickle gene. After another generation of transgene transmission under similar circumstances, grandchildren with three normal ß-globin genes and two α-genes would then emerge, and these children would then have a functional α-thalassemia. Thus, in the absence of homologous recombination and gene replacement, transfer of DNA with the goal of correcting genetic disease creates problems for subsequent generations of progeny.

It is likewise not ethically feasible to insert genes that would provide an individual with a distinct competitive advantage. While single loci affecting such traits as intelligence may very well be nonexistent, it is possible that genes or small groups of genes will be identified, allelic combinations of which enhance some aspect of performance. Given that even standard medical technology is not available equally, it would be highly unethical to offer such profound genetic modification opportunities to a select group within the population.

Given these practical and ethical constraints associated with various gene insertion strategies, it is reasonable to conclude that the genes one would insert into the human germ line would not have an endogenous counterpart, and that the "patients" who would receive such genes would already be at a significant health disadvantage. The purpose of the gene transfer, then, would be to alleviate the disadvantage.

For example, genes could be inserted into embryos from parents known to be transmitting a significantly increased risk of malignant disease to their children. These genes would confer tolerance to more aggressive chemotherapy protocols. Genes targeting resistance to the toxic effects of chemotherapy may allow the affected individual to resist such toxicity when and if a malignancy developed. Transgenic mice exemplary of such a strategy have already been generated. We have produced the mice carrying dihydrofolate reductase genes that encode high resistance to methotrexate, and observed systemic resistance to the drug (Isola and Gordon, 1986). Similarly, Galski et al., (1989) have inserted the multiple drug resistance gene (mdr1) driven by an actin promoter and found that transgenic mice express the foreign gene in bone marrow and resist the toxic effects of doxorubicin. Such genes would best be inducible, so that possible untoward effects of constitutive expression could be avoided.

Another example where a disadvantage could be relieved is currently under exploration in our laboratory, and also seeks to reduce the risk of death from cancer. This strategy involves the insertion of genes encoding potentially toxic products, driven by regulators that would be expected to function only in malignant cells. In this circumstance, malignant cells would express the foreign gene and be killed. This kind of gene could be inserted into embryos from families with a known predisposition to development of specific malignancies such as breast, intestinal or retinal cancer.

SUMMARY AND CONCLUSIONS

Steady advances in germline gene transfer technology have rendered feasible the insertion of genes into the human germ line. Of the current available technologies, microinjection appears to be the most promising potential tool for human germ line gene insertion. Perhaps the most important drawback to this approach, as with retrovirus-mediated gene transfer, is the fact that insertion of a new DNA can cause insertional mutations within the host genome. This problem is not shared by ES cell mediated gene transfer, where gene replacement by homologous recombination has already been successful in mice. However, ES cell mediated gene transfer has its own unique problems including the fact that human ES cell lines have not been generated and that the treated individual would be a genetic mosaic.

A problem to be confronted with all gene transfer strategies is that transfer of new genetic material can create genetic imbalances in the recipient which might not be manifest for several generations. Because of this, it is logical that in the absence of a feasible approach to gene replacement, information to be transferred would not correspond to any existing host locus, but would contain new coding information that would have beneficial phenotypic effects.

Current social realities also demand that genes to be inserted not confer upon individuals advantages that would allow themselves and their descendants to compete more effectively in society. Because of this imperative, it might be suggested that new genes would be inserted solely for the purpose of overcoming particular genetic handicaps, such as a predisposition to disease. Examples would include insertion of chemotherapy resistance genes into families with an inordinate genetic risk for developing malignant disease.

Finally, it is not possible to leave this subject without some mention of the profound ethical issues raised by the prospect of placing new coding information into the human germ line. Some believe that even generation of transgenic animals is either inordinately cruel or a transgresses unwritten principles of appropriate use of scientific knowledge. The notion that such procedures might be extended to humans can thus engender fear and anger. It should be considered appropriate to question the advisability of human germ line gene insertion, and proper for members of a democratic society to voice their personal reactions to such a measure. However, we must, as we approach this most important scientific frontier, not allow such concerns to be exploited by those who would curtail the progress of science. Fear is that the emotion may be easily manipulated to foster the notion that knowledge itself is harmful. Quite the opposite, the acquisition of knowledge is the most reliable method of overcoming fear and of making rational decisions. In this regard, it is important to keep in mind that mastery of gene transfer technology knowledge does not by itself create the impetus for action. Thus, as we move closer to human germ line gene insertion, we must separate the capability to perform such a procedure from the need or desirability to do so. And, in order to fully comprehend all of the possible consequences of germline gene therapy, as well as to better understand the basic mechanisms of mammalian development, differentiation and disease, we must continue an active program of animal experimentation.

Acknowledgements: Work toward preparation of this manuscript was supported by National Institutes of Health grants HD20484, HD25136 and CA42103, and March of Dimes grant 1-1026. This is manuscript No. 56 of the Bookdale Mcenter for Molecular Biology, Mt. Sinai Medical Center.

REFERENCES

Bradbury, M.W., Isola, L.M., and Gordon, J.W., 1990, Enzymatic amplification of a repeated DNA sequence from the mouse Y chromosome allows the determination of sex in a single blastomere, Proc. Natl. Acad Sci. USA 87:4053-4057.

Brinster, R.L., Chen, H.-Y., Trumbauer, M.E., Senear, A.W., Warren, R., and Palmiter, R.D., 1981, Somatic expression of herpes thymidine kinase in mice following injection of a fusion gene into eggs, Cell 27:223-231.

Burki, K., and Ullrich, A., 1982, Transplantation of the human insulin gene into fertilized mouse eggs, EMBO J. 1:127-131.

Costantini, F., Chada, I., and Magram, J., 1986, Correction of murine B-thalassemia by gene transfer into the germ line, Science 233:1192-1194.

Costantini, F., and Lacy, E., 1981, Introduction of a rabbit B-globin gene into the mosue germ line, Nature 233:92-94.

DeChiara, T.M., Efstratiadis, A., and Robertson, E.J., 1990, A growth-deficiency phenotype in heterozygous mice carrying an insulin-like growth factor II gene disrupted by targeting, Nature 345:78-80.

Evans, M.J., and Kaufman, M.H., 1981, Establishiment in culture of pluripotential cells from mouse embryos, Nature 292:154-156.

Galski, H., Sullivan, M., Willingham, M.C., Chin, K.-V., Gottesman, M.M., and Pastan, I., 1989, Expression of a human multidrug resistance cDNA (MDR1) in the bone marrow of transgenic mice: resistance to daunomuycin-induced leukopenia, Mol. Cell. Biol. 9:4357-4363.

Gordon, J.W., 1989, Transgenic animals, Int. Rev. cytol. 115:171-229.

Gordon, J.W., and Ruddle, F.H., Integration and stable germ line transmission of genes injected into mouse pronuclei, Science 214:1244-1246.

Gordon, J.W., Scangos, G.A., Plotkin, D.J., Barbosa, J.A., and Ruddle, F.H., 1980, Genetic transformation of mouse embryos by microinjection of purified DNA, Proc. Natl. Acad. Sci. USA 77:7380-7384.

Handyside, A.H., Kontogianni, E.H., Hardy, K., and Winston, R.M.L., 1990, Pregnancies from biopsied human preimplantation embryos sexed by Y-specific DNA amplification, Nature 344:768-770.

Isola, L.M., and Gordon, J.W., 1986, Systemic resistance to methotrexate in transgenic mice carrying a mutant dihydrofolate reductase gene. Proc. Natl. Acad. Sci. USA 83:9621-9625.

Jahner, D., Haase, K., Mulligan, R., and Jaenisch, R., 1985, Insertion of the bacterial gpt gene into the germ line of mice by retgroviral infection, Proc. Natl. Acad. Sci. USA 82:6927-6931.

Krulewski, T.F., Neumann, P.E., and Gordon, J.W., 1989, Insertional mutation in a transgenic mouse pedigree allelic with pcd, Proc. Natl. Acad.Sci. USA 86:3709-3712.

Mann, J.R., Mulligan, R.C., and Baltimore, D., 1983, Construction of retrovirus packaging mutant and its use to produce helper-free defective retrovirus, Cell 33:153-159.

Martin, G.R., 1981, isolation of a pluripotent cell line from early mouse embryos cultured in medium conditioned by teratocarcinoma stem cells, Proc. Natl. Acad. Sci. USA 78:7634-7638.

Mason, A.J., Pitts, S.L., Nikolics, IK., Szonyi, E., Wilcox, J.N., Seeburg, P.H., and Stewart, T.A., 1986, The hypogonadal mouse: reproductive functions restored by gene therapy, Science 234:1372-1378.

Palmiter, R.D., and Brinster, R.L., 1986, Germ-line transformation of mice, Ann. Rev. Genet. 465:465-499.

Saiki, R.K., Scharf, S., Faloona, F., Mullis, K.B., Horn, G.T., Eflich, H.A., and Arnheim, H., 1985, Enzymatic amplification of B-globin genomic sequences and restriction site analysis for diagnosis of sickle cell anemia, Science 230:1350-1354.

Smithies, O., Grett, R.G., Boggs, S.S., Koralewski, M.A., and Kucherlapati, R.S., 1985, Insertion of DNA sequences into the human chromosomal B-globin locus by homologus recombination, Nature 317:230-234.

Thompson, S., Clarke, A.R., Pow, A.M., Hooper, M.L., and Melton, D.W., 1989, Germ line transmission and expression of corrected HPRT gene produced by gene targeting in embryonic stem cells, Cell 56:313-321.

Van der Putten, H., Botteri, H., Miller, F.M., Rosefeld, A.D., Fan, M.G., Evans, R.M., and Verma, I., 1985, Efficient insertion of genes into the mouse germ lie via retroviral vectors, Proc. Natl. Acad. Sci. USA 82:6148-6152.

Woychik, R.P., Stewart, T.A., Davis, L.G., D'Eustachio, P.D. and Leder, P., 1985, An inherited limb deformity created by insertional mutagenesis in a transgenic mocuse. Nature 318:36-40.

Ziljstra, M., Box, M., Simister, N.E., Loring, J.M., Raulet, D.H., and Jaenisch, R., 1990, B2-microglobulin deficient mice lack CD-8+ cytolytic T cells, Nature 344:742-746.

PART VI

ETHICAL AND LEGAL ASPECTS OF PREIMPLANTATION GENETICS

ETHICAL ISSUES IN THE CONTROL OF GENETIC DISEASES

Anver Kuliev* and Bernadette Modell**

* WHO Collaborating Centre, Medical Genetics Research Centre AMS USSR, Moscow; and
** WHO Collaborating Centre, Faculty of Clinical Sciences, Prenatal Centre, University College London

INTRODUCTION

Considerations on ethical and legal issues are of a great importance for policy-making decisions in the planning and organization of genetic services. These considerations have evolved together with the evolution of the technology for the control of genetic diseases and have now become one of the key subjects in discussing the acceptability of preconception and preimplantation genetic diagnosis. Ethical and legal issues would determine to a considerable extent whether these new approaches will be promoted to become an integral part of preventive genetic service or will be waived on ethical grounds. Clearly their future would first of all depend on the safety of the methods involved in preimplantation diagnosis and the accuracy of genetic analysis, which are being dealt with in other chapters of this volume.

Data presented in the above sections seem to produce an evidence that there is a considerable progress in developing safe methods for gametes and pre-embryo biopsy as well as in evaluating the accuracy of genetic analysis in single cells. Although many important technical problems are still remain to be solved, the present state of research and developments makes it timely to analyze major ethical and legal problems involved in possible implementation of preimplantation diagnosis for the control of genetic disorders.

* Present Address: Reproductive Genetics Institute, 836 W. Wellington, Chicago, IL 60657

Preimplantation Genetics, Edited by Y. Verlinsky and A. Kuliev, Plenum Press, New York, 1991

HEALTH BURDEN OF GENETIC DISEASES

Most of industrial countries have now reached a level of health care where congenital and hereditary disorders contribute greatly to infant mortality and chronic morbidity. In Europe, over half a million infants are born annually with different types of severe congenital disorder (Kuliev, Modell, 1990). Over half die very early or have some kind of chronic disability: many children with congenital malformations benefit from pediatric surgery, but the rest need continuous medical or social support (table 1).

The ethical and legal issues in connection with the control of congenital and hereditary diseases will depend on the way individuals (societies) cope with congenital handicap which varies according to differences in the level of medical care, socioeconomic circumstances, cultural and religious background. Clearly, the likely impact of genetic factors on morbidity and the related necessary measures for their management will vary greatly according to the general level of development (as indicated by infant mortality and birthrate), which vary tremendously in different countries: European and North American countries and Japan are at one hand of a "global developmental curve", formed by relationship of infant mortality and birthrate, whereas most African countries are on the opposite side; South American and Asian countries are scattered along the curve, differences being as much as 10-50 times (WHO, 1985b). No doubt that ethical problems regarding management of genetic disorders could not be similar in countries at different sides of the curve, so the emphasis in this paper is on the general objectives and principles of genetics services and major ethical issues involved in the planning and organization of genetic disease control programs.

OBJECTIVES AND PRINCIPLES IN THE CONTROL OF GENETIC DISEASE

The ethical issues need to be considered in the context of the general objective of genetic services, which according to the recommendations of the World Health Organization is to help genetically disadvantaged people live and reproduce as normally as responsibly as possible (WHO, 1985b). This includes both striving to provide a fulfilling existence for affected individuals, and providing people with genetic information that can help them to maintain their own health and that of their family.

The guiding principles of clinical geneticists, that are included in professional teaching and adhered to in practice, have been studied empirically and summarized in a set of guidelines for genetic practice. Core principles are the autonomy of the individual or the couple, their right to adequate and complete information, and the maintenance of the highest standards of confidentiality (WHO, 1985b; Fletcher et al., 1985). The choices are to be made by the individual or the couple and the genetic advice should be provided in a non-directive manner. It does not mean simply giving people the facts and letting them to make up their

Table 1. Cogenital anomalies: classification and estimate of total disability generated annually in Europe.

Category of anomaly	Births per 1000	Annual births	Early mortality		Chronic problems		Successful treatment		Main therapeutic needs
			%	number	%	number	%	number	
Congenital malformation	30	408000	22	89800	24	98000	54	220500	paediatric surgery
Chromosomal disorder	3	43500	34	14800	64	28000	2	900	social support
Inherited disease	7	95200	58	55200	31	29500	11	10500	medical treatment and support
Total	40	546700	29	159800	28	155500	42	232900	

Table 2. Percentage of recognized pregnancies voluntarily aborted at different maternal ages (Kuliev, Modell, 1990).

Country	Pregnancies with induced abortion (%)	Pregnancies aborted in maternal age group (%)[a]									Abortions in nulliparous women (%)
		< 15	15-19	20-24	25-29	30-34	35-39	40-44	45-49	>50	
Denmark	29	0	62	27	17	25	49	77	84	-	43
Finland	17	82	52	17	8	11	22	48	75	100	53
Iceland	12	66	25	9	9	9	20	--------37---------			41
Netherlands	11	38	11	6	9	21	42	--------46---------			-
Norway	22	94	51	20	12	17	33	62	----74-----		49
Sweden	25	97	60	24	15	20	37	64	----87-----		39
UK(England & Wales)	17	84	39	15	9	12	24	41	----40-----		55
France	18	-	39	15	12	17	29	54	----66-----		-
Italy	27[b]	91	32	21	20	29	46	62	70	79	-
FRG	13	72	28	11	8	11	23	45	62	58	45
Czechoslovakia	31	45	20	17	31	49	68	82	87	79	11
Hungary	38	48	32	22	31	54	76	88	----93------		19

[a]Maternal age in years.
[b]Illegal abortions still occur in Italy. The figure for all abortions is estimated to be about 30% higher than the official figures used here.

own mind but a special skill requiring an appropriate personality and training, and involves accompanying counsellees in decision-making by helping them to take account of their unique medical, social and moral situation. The principle of confidentiality should also be used not only to protect the privacy, but also to serve the true interests of the people involved, and other members of the family. The maintenance of confidentiality is of particular importance in the establishment of genetic registers.

THE ISSUE OF ABORTION

The ethical issues connected with medical genetics services are very broad and cannot be fully discussed in this paper. So here these issues will be considered in selective manner, with emphasis on a few outstanding topics. For practical purposes that are commonly raised may be divided into speculative and the real. For example, the question of abortion because of an acceptable risk of bearing a severely affected child is often considered in the context of genetics services. However the abortion issue is in fact extraneous to debate about genetics practice. The acceptability of abortion under specific circumstances is decided on general social and medical, rather than on genetic grounds, and genetics service, particularly prenatal diagnosis, can operate only within the framework of such general decisions. As shown in table 2 about 40% of recognized pregnancies are aborted in Eastern European countries, these being predominantly among older women, whereas in Western Europe abortions are less common varying from 11% in Netherlands to 29% in Denmark, being predominantly in young women wishing to postpone childbearing. The examples of the distribution of pregnancies aborted in comparison with childbearing in the different age groups in Denmark and Hungary are shown in fig 1. In fact, in most countries abortion because of a severely affected fetus is accepted as a legitimate medical procedure; and clearly in those countries where abortion is legal when the pregnancy is unwanted, it must also be legal if the pregnancy is unwanted because the child is severely ill or malformed. In this context preimplantation genetic diagnosis makes the measures for controlling genetic diseases more ethically acceptable as it avoids the abortion as a necessary measure in the cases when the gametes or the pre-embryos are affected.

NONFINANCIAL COSTS AND BENEFITS IN CONTROL OF GENETIC DISEASE

In discussing ethical issues analysis of costs and benefits becomes of a particular importance as the attempts are being made to demonstrate mainly financial savings by preventing a handicap, which is actually a price paid by a couple for having a healthy child. Having a healthy child or even a desired family size is one of the major nonfinancial benefits of preventive genetic program. Nonfinancial benefits include also relief of anxiety, replacement of affected child with a healthy one and provision of a possibility for a couple to have a normal family life. These services are highly acceptable to couples, are ethically indicated, can be organized and monitored efficiently, and are highly cost-effective at the same time. In the WHO ongoing genetic diseases control programs, taking thalassemia as an example, analysis of some of the financial costs and benefits of screening for genetic reproductive risk shows that the birth of an accepted affected child to informed parents is a benefit (RCP, 1989), in contrast to some speculative considerations on this issue.

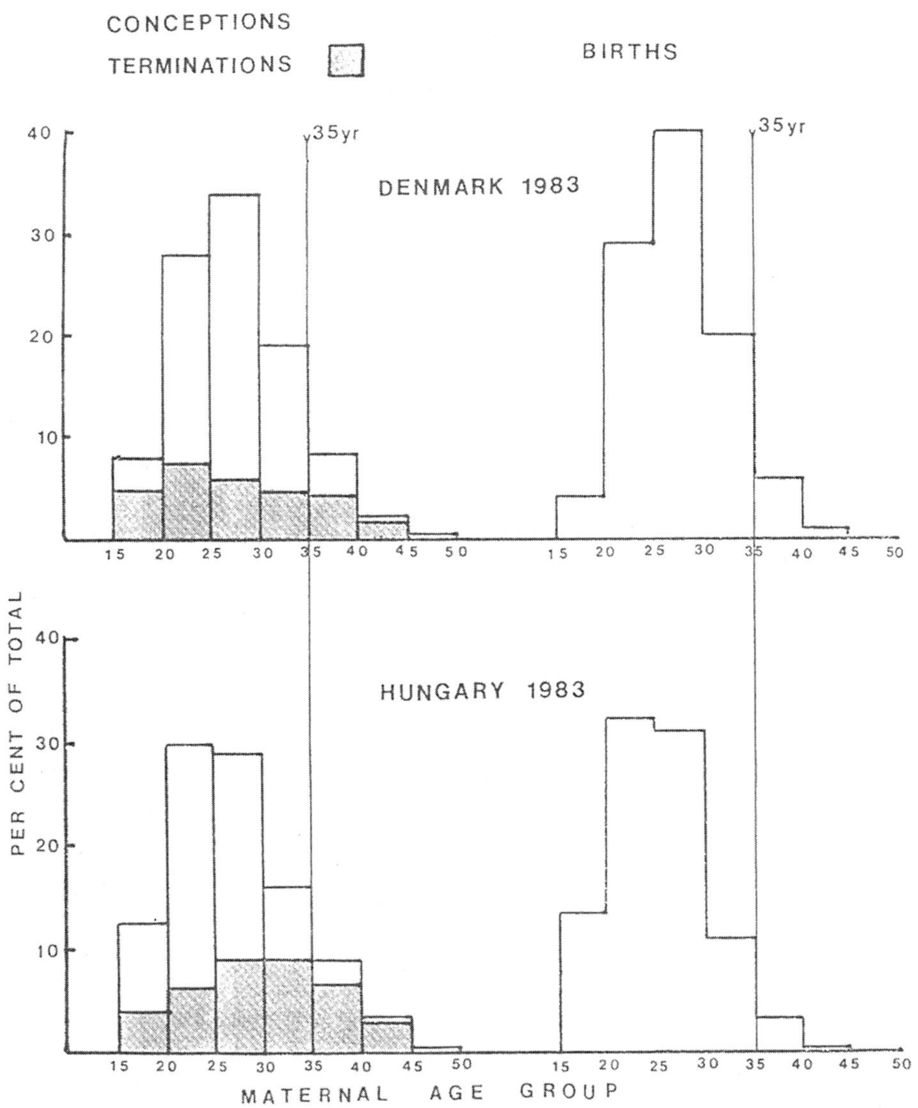

Fig. 1. Proportion of conceptions terminated and distribution of births in different maternal age groups in Denmark and Hungary.

At the same time the major nonfinancial cost for having an unaffected child is elimination of affected fetus which as mentioned could be avoided by preconception or pre-embryo diagnosis.

OPTIONS AVAILABLE TO AVOID HAVING AFFECTED CHILDREN WITH GENETIC DISEASE

A concern has been also raised, that selective abortion for severe genetic disorders may alter attitudes towards the rights of handicapped people, though there is no evidence to support this: no country where prenatal diagnosis has been established has reduced its efforts to care for handicapped. There are intrinsic reasons why the existence of prevention makes it more important, and more feasible, to provide optimal care for the handicapped, and in many countries there are now firmer recommendations than ever that individuals with congenital disorders should be given optimal treatment. In reality, as shown by the best thalassemia control programs in Cyprus, Sardinia and Greece, heterozygote diagnosis, prenatal diagnosis and patient care are complementary, not competing, approaches to helping people with a genetic disadvantage (WHO, 1987). A WHO working model of the delivery of genetics services, followed in the above control programs mentioned (Modell et al., 1990) demonstrated that provided that individuals and couples are informed of the risk of bearing an abnormal fetus they may choose from a number of different options shown in table 3 (RCP, 1989).

With the introduction of preimplantation diagnosis these options will become wider and those who cannot accept prenatal diagnosis and abortion may choose the gamete or the pre-embryo diagnosis so that no pregnancy has to be selectively interrupted.

The evolution of different policies used in prevention of genetic disease (using hemoglobinopathies as an example) may be followed from a baseline in which there is neither treatment, no genetic counselling or prenatal diagnosis, as in most developing countries today (fig. 2 policy 1). In the next stage with the introduction of treatment and with the provision of the information on recurrence risk when many couples either refrain from reproduction or terminate their pregnancies if family planning failed, genetic fitness falling to less than half of the population norm (fig. 2, policy 2).

With the introduction of prenatal diagnosis provided initially only retrospectively (after identification of affected child in the family) genetic fitness improved, returning to the original level of about 75% the population norm, but at the cost of abortion of affected children (fig. 2: policy 3 and 4; policy 3 includes prenatal diagnosis in the second trimester, while policy 4, in the first trimester of pregnancy).

By contrast further introduction of the community-based screening and first trimester fetal diagnosis offered the possibility of near-normal genetic fitness for couples at

Table 3. Possibilities open to carriers of an inherited disease to avoid having affected children (taken from RCP 1989).

Time of discovering risk	Possible action
Before marriage (uncommon)	1 Remain single (uncommon) 2 Avoid selecting another carrier as partner (very uncommon) 3 Select a partner in the usual way (the commonest choice)
After marriage (more common)	4 Remain childless (common only for severe disease when PND impossible) 5 Take the chance (common for less severe diseases) 6 Use PND (very common) 7 Use AID or other form of assisted reproduction (very uncommon) 8 Separate and find another partner (very uncommon indeed)
After birth of an affected child (commonest)	Options 4-8 above for further reproduction, plus: 9 Accept infant and treatment (usual) 10 Accept infant, but reject treatment (sometimes) 11 Reject infant (can happen)

PND, Prenatal diagnosis: AID, Artificial insemination by donor.

risk, but again for the price of terminating almost a quarter of their wanted pregnancies (fig. 2: policy 5 and 6; policy 5 refers to prenatal diagnosis available prospectively in the second trimester, while policy 6 - to that in the first trimester of pregnancy).

The possible impact of preimplantation diagnosis on the existing policies for the control of genetic disease is shown in fig. 2 (policy 7): in addition to achieving a normal genetic fitness it eliminates the price of terminating 25% of wanted pregnancies, such as in the cases of autosomal recessive diseases.

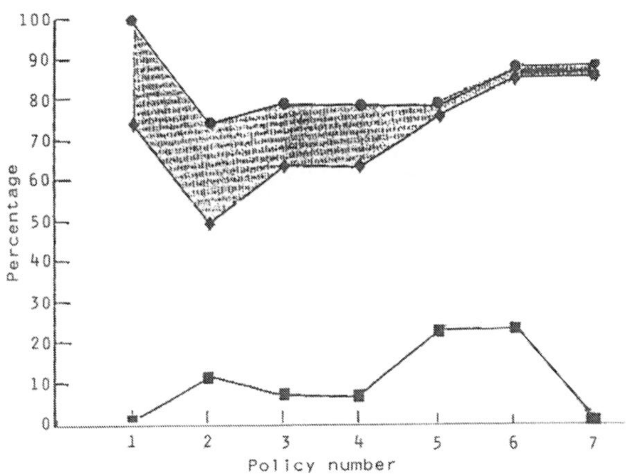

Fig. 2. The successive stages in the evolution of genetic services in relation to the genetic fitness of at-risk couples (explanation in the text).

SEX SELECTION

An ethical issue in connection with fetal sexing, which provides an important example of non-pathological genetic diagnosis has been thoughtfully discussed elsewhere (Fletcher, 1983; Kuliev et al., 1984; WHO, 1985b). Termination of pregnancy in the mid-trimester on grounds of fetal sex is generally not accepted, but the possibility of sex selection in first trimester, at a stage when termination of pregnancy on social grounds is widely available, is a new issue. There seems no contra-indication to providing information on fetal sex within the context of genetic diagnosis. On the other hand, sex selection as a primary indication for prenatal diagnosis is generally viewed as an undesirable expression of inequality of the sexes, and there can be little doubt that its practice by geneticists would produce a negative social reaction and endanger the entire genetics service.

This will become even more sensitive after introduction of preimplantation diagnosis as it makes possible to achieve the desired sex of a child even without interruption of pregnancy.

AVAILABILITY OF GENETICS SERVICES - REAL ETHICAL ISSUE

The broadest of the real ethical issues is the limited availability of genetics services. The care of the handicapped and support for their families are universally deficient, and at the same time the services, that can now be provided for responsible family planning are not effectively delivered, and are unequally distributed even in industrialized countries. The level to which present possibilities are realized varies considerably even in different sections of populations. The full implementation of such services in the European countries would reduce the annual number of births of children with serious congenital disorders by tens of thousands. However serious problems arise because in order to be effective genetics services are needed to be realized at the community level through the general health care and obstetric services, while the necessary organization for these purposes has not been developed. In most countries there is no national policy on genetic disorders, no recognized body is responsible for service delivery, and adequate resources are not available. In addition there is a serious lack of the genetic education both of the health professionals and the population.

The poor delivery of prenatal diagnosis to older mothers is one of the examples of uneven delivery of the services. For example, an uptake of prenatal diagnosis about 80% is observed among such women at specialist genetic centers in Copenhagen, Rotterdam and Paris. A low uptake rate in many other populations suggests an inadequate provision of the information about the potential risk by the medical profession. In reality, in many countries obstetricians and practicing doctors are not trained for this kind of counselling. One of the means of overcoming this obstacle is to develop professional codes of practice, or other authoritative guidelines to improve service delivery.

Experience shows clearly that genetic services are delivered most effectively in areas where definite recommendations have been made concerning the standards of practice, such as routine obligatory ultrasound scanning during pregnancy in the FRG (RCOG, 1984), and the officially-stated obligation of obstetricians in Denmark to inform pregnant women over the age of 35 about the risk of chromosomal abnormalities and the fact that fetal diagnosis is available. The exceptional effectiveness of thalassemia control program in Cyprus, where almost no new affected infants have been born since 1986, is also partly due to ecclesiastical ruling which stipulates that all couples be tested for thalassemia prior to marriage, and counselled appropriately when indicated (Angastiniotis et al., 1986).

An important ethical issue is the right of the individual to know every reliable new possibility for controlling severe disease which should be made available for the communities to decide themselves whether to use such information or not. Yet the responsibility for organizing the control programs rests with policy makers who decide the need for the program on the grounds of health priorities and costs. As an example, if, in a certain area, screening of maternal alpha-fetoprotein levels is not practiced, and, therefore, couples are not able to avoid the birth of a child with Down's syndrome, who should take the responsibility for such a tragedy on the population level? In many countries with the change of the maternal age distribution the proportion of preventable births with Down's syndrome related to maternal age is less than 30%, particularly in Eastern European countries; however combining this approach with maternal serum screening in younger age groups increase this proportion up to 60% (Kuliev, Modell, 1990). If such services are not provided, many couple at risk will have children with Down's syndrome those that are preventable. These individuals have the right to ask why they were not given the genetic information and who is responsible: in fact, the most common lawsuits in this area are brought by parents who were not advised of detectable genetic risks and so unnecessarily delivered a severely affected child. Such suits have occurred in many industrialized countries, and the settlements awarded have promoted an increased readiness to offer genetic services, as it is the responsibility of the medical profession to inform the population as soon as preventive measures become available.

ETHNIC MINORITIES

One of important ethical problems is the provision of genetic services to ethnic minorities. For example, the hemoglobinopathies are very prevalent in former malarial areas (WHO, 1983). However because of migration and emigration, hemoglobinopathies have now also become prevalent in most European countries, reflecting the proportion of those ethnic groups with a high frequency of hemoglobinopathy genes (WHO, 1987). Nevertheless, appropriate facilities for treatment and prevention of the disease are not provided in most of these countries. The same is true regarding a provision of screening services for Tay-Sachs disease among the groups at risk, the program which has been highly effective in the USA, leading to over 90% fall in the birth incidence of this disease. Still, this has not been organized in many European countries, where estimated numbers of Jews at risk for Tay-Sachs disease are considerable (table 4).

Another ethical issue related to the ethnic minorities is the problem of consanguineous marriages, also as a result of migration and immigration processes of ethnic groups in whom conventional consanguinity is very high. In fact, at least 14% of the world population and 19% of all births are located in the areas where consanguineous marriages are very high (WHO, 1985b). Any attempt to discourage consanguineous marriages on genetic grounds in these ethnic minorities would do certainly more harm than good. On the contrary, ethically more appropriate is to use such information for planning genetic services, in order to offer genetic advice to the extended families of any individual in whom genetic disease is diagnosed, and to promote counselling for the young related couples prior to their marriage or reproduction.

EFFECTIVENESS OF GENETIC DISEASE CONTROL MEASURES

Much could be achieved by monitoring the effectiveness of the genetic services, which is regularly being carried out for ongoing thalassemia control programs according to the WHO evaluation criteria (WHO, 1985a). The analysis of this experience shows that the major reason for the residual births of affected infants is the ignorance of medical personnel (26%) and individuals (population) (55%) about the

Table 4. Estimated numbers of Jews at risk for Tay-Sachs disease in European countries (taken from Japhet 1989).

Country	Number (thousands)[a]	Country	Number (thousands)[a]
Albania	0.2	Austria	12.0
Belgium	12.0	Bulgaria	5.1
Czechoslovakia	12.0	Denmark	7.0
Finland	1.0	FRG	28.0
GDR	0.55	France	700.0
Greece	5.0	Hungary	85.0
Ireland	1.6	Italy	35.0[b]
Luxembourg	1.0	Malta	0.06
Netherlands	30.0	Norway	1.05
Poland	8.0	Portugal	0.3
Romania	23.0	Spain	12.0[b]
Sweden	16.0	Switzerland	20.0
UK	330.0	Yugoslavia	6.0
Europe	>1375.9		
Turkey	24.0		
Israel	3653.1		
USSR	>2200.0		
Total	>7253.0		

[a] The figures are an absolute minimum. They do not take account of people of Jewish extraction unknown to Jewish organizations, nor of people of mixed parentage.
[b] Jewish populations predominantly of Sephardic origin, and not at increased risk of Tay-Sachs disease.

availability of screening test and the possibility of prevention, while only 9% of such births were due to laboratory error (mainly in heterozygote detection) and 10% - either due to refusal of fetal diagnosis or due to choice not to terminate an affected pregnancy. This raises one of the major problems in the organization and delivery of genetics services which could be overcome through the development of educational aids and programs for medical profession and the community. For example, in most of neonatal screening programs, as in the case of sickle cell disease where both carriers and affected infants are usually detected, information for parents prior to as well as after testing has been neglected. The lack of appropriate information for participants of screening programs will certainly lead to serious ethical and social problems in conditions such as cystic fibrosis, where the diagnostic tests are not yet reliable and particularly in even less predictable conditions such as alpha-1-antitrypsin deficiency (McNeil et al 1988).

CONCLUSION

As demonstrated above the ethical issues follow in sequence all new developments in the control of congenital and genetic disorders entering now a new stage of considerations with the development of preconception and preimplantation genetic diagnosis. While some of the longstanding ethical problems would probably be solved by preimplantation diagnosis, such as the abortion issue which thank to this new approach will be avoided, others could become a serious obstacle, particularly those related to antiscience movements due to fears in the populations of a "brave new world". These considerations are highly relevant to the subject of preimplantation genetic diagnosis and probably will be with us as long as we proceed with further development of appropriate technology for controlling genetic disability.

REFERENCES

Angastiniotis MA, Kyiakidou S, Hadjiminas M. 1986. How Thalassemia was controlled in Cyprus. World health Forum 7: 291-297.

Fletcher JC. 1983. Is sex selection ethical? Research Ethics. Alan R. Liss Inc. N.Y. 333-338.

Fletcher JC, Berg C, Tranoy KE. 1985. Ethical aspects of Medical Genetics. Clinical Genetics 27: 199-205.

Japhet R (Ed), 1989 The Jewish Yearbook 1989 (London: Jewish Chronicle Publications).

Kuliev AM, Modell B. 1990. Problems in the control of genetic disorders. Biomedical Sciences 1: 3-17.

Kuliev AM, Modell B, Galjaard H. 1984. Perspectives in Fetal diagnosis of congenital disorders. Arns Serono Symposia.

McNeil TF, Sveger T, Thelin T. 1988. "Psychosocial aspects of screening for somatic risk: the Swedish alpha-1-

antitrypsin experience" Thorax 43: 505-507.
Modell B, Kuliev AM, Wagner M. 1991. Community Genetics Services in Europe. WHO Regional Office of Europe, Public Health in Europe Series. In press.
RCOG, 1984 "Report of the RCOG Working Party on routine ultrasound examination in pregnancy" Royal College of Obstetricians and Gynecologists. London, UK.
RCP 1989. Prenatal diagnosis and genetic screening. Community and Service implications. Report of the Royal College of Physicians. London, UK.
WHO 1985. Community Control of hereditary Anaemias: Memorandum from WHO meeting. Bulletin of the World Health Organization 61:63-80.
WHO 1985a. "Update of Hemoglobinopathies Control". Report of the Third and Fourth Annual Meetings of the WHO Working Group for the community control of Hereditary Aneamias. Unpublished WHO document HMG/WG/85.8.
WHO 1985b. "Community Approaches to the control of hereditary diseases". Report of a WHO Advisory Group; unpublished WHO document HDP/WG/85.10.
WHO 1987. "The Hemoglobinopathies in Europe" report of two meetings of the WHO/Mediterranean Working Group on Hemoglobinopathies (March 1986 in Brussels, and March 1987 in Paris) offset publication IPC/MCH 110, WHO Regional Office for Europe, Copenhagen, Denmark.

ETHICAL AND SELECTED MEDICAL ASPECTS OF PREIMPLANTATION GENETIC DIAGNOSIS

Aubrey Milunsky

Center for Human Genetics
Boston University School of Medicine
Boston, MA 02118

INTRODUCTION

In the United States, two government commissions, the National Commission for the Protection of Human Subjects[1] and the President's Commission for the Study of Ethical Problems in Medicine and Biomedical and Behavioral Research,[2] recognized three ethical principles that basically influence the formulation of public policy. The first is the respect owed to all persons, acknowledging their rights to autonomous choices, while simultaneously protecting the incompetent. The second principle is to secure the welfare of others and provide protection from harm. The third - that of equity or justice, requires fair treatment and equal access. Both commissions emphasized the goal of common good rather than a utilitarian one in which people may be used for the benefit of society.

These principles find equal application when considering the ethics of preimplantation genetic diagnosis.

The primary purposes of preimplantation genetic diagnosis is to:

1. Select a genetically healthy embryo, one of the required gender (to expressly avoid X-linked disease), recognizing pari passu that the diagnostic focus at the present time is on one gene, the karyotype and perhaps on one gene product.

2. Avoidance of abortion.

Given the fundamental role of <u>in vitro</u> fertilization (IVF) in furthering the development of preimplantation genetic diagnosis, experience from that arena should be carefully analyzed and considered.

Preimplantation Genetics, Edited by Y. Verlinsky and A. Kuliev, Plenum Press, New York, 1991

SOME INSIGHTS FROM EXPERIENCE WITH <u>IN VITRO</u> FERTILIZATION

Walters, in a 1987 review of fifteen committees reporting from many countries[3] noted that all found IVF ethically acceptable.

* These committees all agreed that IVF was acceptable for married heterosexual couples.

* Eleven of fifteen committees agreed about the appropriateness of providing IVF for couples living together.

* Only five of fifteen committees held that <u>single</u> women should be allowed to have IVF.

* All committees specifically aimed to discourage use of IVF for eugenic or social reasons.

* Eleven of fifteen committees accepted the potential usefulness of embryo freezing.

Selected issues that defied consensus and remained unsettled included the following:

* The time limit for storing frozen embryos is yet to be defined.

* The rights or authority an individual or a couple have to decide about embryo disposal require further clarification. In a Tennessee case, following the divorce of Junior L. Davis and Mary Sue Davis, the woman sought and obtained custody of seven embryos that she and her ex-husband had conceived through IVF. The woman planned to use some of these embryos to have the children she originally planned. The ex-husband opposed implantation of any of these embryos. A lower court ruled in favor of providing the woman with custody of these embryos, equating such embryos to living children, and in so doing suggesting that life begins at conception. The Appeals Court in Tennessee, however, overturned the lower court ruling and directed that there should be joint custody of these embryos. This woman subsequently remarried. Her request that these embryos be donated to other couples has also been opposed by her ex-husband. In yet another case, a couple who had placed embryos in frozen storage, were killed in an airplane crash. No provisions had been made for disposition of these embryos in such an eventuality.

* While there has been acceptance of payment for sperm donors, the propriety of paying ovum donors curiously remains unresolved. It would seem that their risks of donation are quantifiable and one would, therefore, expect that in contradistinction to sperm donation, a significant premium could be charged for ovum donation.

* The question remains whether couples could choose their own gamete donors.

* Some uncertainty remains about whether or not donors should be allowed anonymity, even in the face of subsequent grave genetic disease (e.g. Huntington's disease) affecting the donor. The Benda commission in West Germany adopted the unique view that IVF offspring must be given access to the name and address of the donor when they attain the age of 17.[4] This position is similar to a West German policy of allowing adopted children, when they reach adulthood, to learn the identities of their biological parents.

THE MORAL STATUS OF THE EMBRYO

The issue concerning the moral status of the embryo arose only when IVF developed, and has been discussed thoroughly elsewhere.[5,6] Only then did it become possible to consider an embryo created _in vitro_, that might _not_ be implanted. The key controversy is of course "When does life begin?". The position of anti-abortion groups is well known:

* Life begins at conception.

* Hence, embryos have rights.

* Human embryos are alive and, therefore, should not be deprived of a chance for life.

* Since embryos are alive, we have a duty to protect them, for if we fail we will commit the sin of homicide.

* Human embryos should not, therefore, be used for any purpose, even to help infertile couples, because:

    a. we cannot obtain informed consent from an embryo

    b. without consent we obviously cannot experiment or treat

    c. we should not tamper with life

    d. scientists who do this work are unscrupulous and may have eugenic aims

* The anti-abortion group is the only moral view: all other views are amoral.

* Anti-abortion groups opine that their moral principles outweigh any possible benefits and they would happily aim to deprive others of any beneficial effects to satisfy these principles.

The opposing and majority view is considerably different. Embryos are _not_ persons under the law or the Constitution. _Potential_ persons cannot, and should not, be endowed with moral or legal qualities that do not exist. Human cells that exist "ex corporis" or _in vitro_, do not require recognition and respect any more than living cells that pass through our intestinal tracts, or exist as _in vitro_ sheets of fibroblasts grown to cover our burned bodies or blood cells (such as lymphocytes) grown _in vitro_, manipulated for gene administration, and transfused into our bodies. All these cells share human genes and cellular chemistry. Only embryos have the _potential_ to become human beings. Philosophical ethics would otherwise demand we respect and dignify cells with a potential for humanhood rather than endow putative humanness on a collection of cells. Indeed, even the potential for humanhood of such cells becomes possible _only after implantation_ in the uterus. If respect and dignity is to be accorded to the human embryo, then logic would dictate that such qualities obtain only after implantation. As it is, even those who believe in life at conception, do not hold burial and memorial services after first trimester miscarriage or fetal death. Why not?

There is another moral imperative often ignored by those who would ban the use of human embryos for any purpose. Great personal suffering of both individuals with genetic defects as well as their families, is well recognized (examples: the Lesch-Nyhan syndrome; adenosine deaminase deficiency). Should every effort _not_ be made to prevent or ameliorate such ills? Why should those with pain, suffering and certain early death be deprived of such opportunities? Is that moral? Is that ethical? Is that right?

There is an irrational argument that since an embryo has genetic potential - and is, therefore, a potential human being, it must be protected _as_ a human being. Aristotle in the _Meta-physics_ stated "As becoming is between being and not being, so _that_ which is becoming is always between that which is and that which is not". He went on "That which exists potentially and not actually is the indeterminate" (quoted by Dunstan[6]).

RELIGIOUS CONCERNS AND THE NEW REPRODUCTIVE TECHNIQUES

Baruch Brody of Baylor University has summarized the six major concerns religious communities have concerning the new reproductive techniques.[7] These new reproductive techniques, he states:

1. Disrupt conjugal intimacy and procreative potential that is morally required.

2. Introduce third parties into the process of reproduction, which is morally illicit.

3. Confuse lineage, children possibly being unaware of their biological parents. Again morally illicit.

4. May result in embryos being discarded. Such an action would be regarded as the equivalent of abortion, and hence, morally unacceptable.

5. Dehumanize the reproductive process and hence, be regarded as amoral.

6. May involve commercialization and exploitation that deem them amoral.

SELECTED MEDICAL ASPECTS

Indications For Preimplantation Genetic Diagnosis

The specific indications for preimplantation genetic diagnosis effectively overlap the indications for prenatal diagnosis in general through chorion villus sampling or amniocentesis, with the important exception of neural tube and other leaking defects.[8]

Contraindications to Preimplantation Genetic Diagnosis

In addition to the technological limitations of preimplantation genetic diagnosis, specific recognizable contraindications should be considered prior to the initiation of any plan. A few examples, which may be categorized as either obstetric or genetic, illustrate the need for prior consideration:

1. Obstetric Contraindications

* Advanced maternal age. Elsewhere in this volume is a discussion concerning the relation of advanced maternal age to the increasing spontaneous abortion rate as well as to the losses following IVF. Age limitations essentially reflect the problem of expense and the issue of scarce resources in a health insurance system. Further experience will be necessary to determine success rates, which are likely to fall at or below the rates for IVF programs.

* Considerable reservation will be exercised for habitual aborters.

* A seriously adverse obstetric history, including recurrent premature ruptured membranes, early and severe preeclamptic toxemia, may prejudice the decision to proceed with this high technology direction. Anatomical uterine abnormalities, either acquired or congenital, would be a clear and specific contraindication.

* Persistent or serious pregnancy-induced hypertension may be a serious contraindication.

* Insulin-dependent diabetes which is difficult to control is not likely to be favored.

* Maternal epilepsy requiring anticonvulsant therapy is likely to be regarded as a contraindication.

An extensive list of maternal illnesses or disorders exist in which preimplantation diagnosis is likely to be contraindicated. Examples include morbid obesity, chronic pelvic infection, systemic lupus erythematosus, serious heart, kidney or liver disease, etc.

2. Genetic Contraindications

Disorders which may threaten either the fetus or the mother as a consequence of maternal genetic disease may be regarded as a contraindication to preimplantation diagnosis. A few selected examples follow:

* Maternal cystic fibrosis or sickle cell disease are likely to become problematic during pregnancy with potential serious consequences for both mother and fetus.

* Maternal myotonic muscular dystrophy when associated with an affected fetus is highly likely to result in serious problems for mother, fetus or child during pregnancy, labor or delivery.[9]

* Maternal connective tissue disorders resulting in tissue friability or weakness of fibrous tissues are associated with significant increases in rupture of vascular or other structures and/or bleeding. The Marfan's syndrome, and some of the Ehlers-Danlos syndrome subtypes are two important examples.

PREREQUISITES FOR PREIMPLANTATION GENETIC DIAGNOSIS

A team approach is essential in developing an effective preimplantation genetic diagnosis facility. Obstetricians with expertise in IVF and embryo transfer, Clinical Geneticists, Clinical Cytogeneticists and Molecular Geneticists, and Ultrasonographers, as a group constitute the key elements of the team. Legal and ethical review would be recommended. With multiple specialty representation, a formal system can be established for patient education with special regard to genetic risks and modes of inheritance, procedures related to IVF and embryo transfer, discussions of success rates and costs as well as careful delineation of limitations concerning laboratory studies that might be anticipated. Clinical and laboratory risks and limitations will require formal informed consent documents and an appreciation by patients of their inherent background risk of having a child with congenital defects or mental retardation.

Additional prerequisites point to the need for excluding maternal diseases that could otherwise damage the embryo/fetus e.g. maternal phenylketonuria, alcoholism, AIDS, epilepsy, diabetes mellitus. Cautionary discussions would be required to emphasize the possible significance of factors which could complicate pregnancy e.g. specific viruses, alcohol, narcotics or other drugs of abuse. The maternal state of mental health could be important and should not be ignored. Opportunities to protect against neural tube defects by use of preconception and periconceptional folic acid supplementation[10] could be emphasized. Development of a register of births following preimplantation genetic diagnosis should be initiated in order to secure follow up data focused especially on normality and the frequency of defects. If donated ova or embryos are used, genetic information from donors would be required.

TECHNICAL CONSIDERATIONS

Attention to procedural detail regarding planned laboratory studies is essential. Patients will need to understand the potential sequence of IVF, harvesting embryos for direct testing or for freezing, followed by later thawing and implantation following normal results of DNA or chromosome study. Limitations concerning, for example, the potential possibility of missed mosaicism or inaccuracies related to DNA contamination during the use of the polymerase chain reaction technique will need discussion. Moreover, because of the high risk of chromosome abnormalities in human embryos, after DNA tests or micro-enzyme assays, there would seem to be a need for chromosome analysis before embryo transfer. Laboratory goals would be to establish policies, procedures and quality control systems, develop standard protocols, as well as normal ranges and established controls. The development of licensing authority and oversight will be required. At least in the early years of experience with preimplantation genetic diagnosis, confirmatory chorion villus sampling or amniocentesis could provide additional reassurance.

GENERAL CONSIDERATIONS

In the USA, institutional review board involvement will be required for the further development of preimplantation genetic diagnosis. The usual discussion concerning the allocation of scarce resources and the potentially prohibitive costs of combined IVF and preimplantation diagnosis may limit the application of these new possibilities. Ownership and dispositional authority should be ironed out formally in order to obviate later legal and ethical tensions.

Precision in prenatal diagnosis has become the rule and the expectation. Polar body diagnosis, as opposed to direct blastomere biopsy, may deviate from this principle in relying upon inferential diagnosis. Careful study, follow-up and oversight are the necessary ingredients to establish a promising new dimension that heralds man's first venture into total avoidance of specific genetic defects.

REFERENCES

1. National Commission for the Protection of Human Subjects of Biomedical and Behavioral Research, The Belmont Report: Ethical principles and guidelines for the protection of human subjects of research. U.S. Government Printing Office. DHEW Publication No. 78-0012-78-0014, (78).
2. Presidents Commission for the Study of Ethical Problems in Medicine and Biomedical and Behavioral Research. U.S. Government Printing Office, Washington, DC, (83).
3. L. Walters, Ethics and New Reproductive Technologies: An International Review of Committee Statements <u>in</u>: "Hastings Center Report", 17(3), 1987.

4. H.M. Sass, Moral Dilemmas in Perinatal Medicine and the Quest for Large Scale Embryo Research: A Discussion of Recent Guidelines in the Federal Republic of Germany, J. Med. & Phil. 12:279-290 (1987).
5. G.R. Dunstan, The Human Embryo in the Western Moral Tradition, in: "The Status of the Human Embryo. Perspectives from Moral Tradition", G.R. Dunstan & M.J. Seller, eds., King Edward's Hospital Fund for London, London (1988).
6. G.R. Dunstan, The Moral Status of the Human Embryo, in: "Philosophical Ethics in Reproductive Medicine", D.R. Bromham, M.E. Dalton and J.C. Jackson, eds., Manchester University Press, Manchester (1988).
7. B. Brody, Current Religious Perspectives on the New Reproductive Techniques, in: "Beyond Baby M", D.M. Bartels, R. Priester, D.E. Vawter, A.L. Caplan, eds., Humana Press, New Jersey (1990).
8. A. Milunsky, Edit. GENETIC DISORDERS AND THE FETUS: Diagnosis, Prevention and Treatment. Second Edition. Plenum Press, New York, (1986).
9. A. Milunsky, J.C. Skare, J.M. Milunsky, T.A. Maher, J.A. Amos, Prenatal Diagnosis of Myotonic Muscular Dystrophy With Linked DNA Probes. Am. J. Obstet. Gynecol. (in press).
10. A. Milunsky, H. Jick, S.S. Jick, C.L. Bruell, D.S. MacLaughlin, K.J. Rothman, W. Willett, Multivitamin/Folic Acid Supplementation in Early Pregnancy Reduces the Prevalence of Neural Tube Defects. JAMA 262:2847 (1989).

LEGAL AND ETHICAL ISSUES IN PREIMPLANTATION GENETIC DIAGNOSIS

John Robertson

The University of Texas at Austin
School of Law
727 E. 26th Street
Austin, Texas 78705

INTRODUCTION

The ability to obtain genetic information about human gametes and preembryos offers an important option to couples at risk for offspring with genetic disease. Rather than aborting an already established pregnancy after prenatal diagnosis, a couple at risk could undergo IVF and transfer only those embryos[1] free of the genetic disorder of concern.

Although preimplantation diagnostic techniques are still in the experimental stage, they are likely to generate enormous controversy. Many people will claim that genetic screening is unethical because it will lead to the destruction or discard of embryos. Others will oppose it as a dangerous eugenic practice that opens the door to germline gene therapy and Brave New World engineering of human beings. Opponents may urge strict regulation or even prohibition, thus preventing embryo screening techniques from becoming available to all who could benefit.

This paper will first examine whether genetic screening of gametes and embryos is ethical. It then discusses current legal restrictions in the United States and constitutional limits on state regulation of preimplantation genetic diagnosis. It concludes that separate regulation of this technology is not needed at the present time.

ETHICAL OBJECTIONS TO GENETIC SELECTION OF EMBRYOS

Although some persons (and some ethicists) will claim that screening and discard of embryos on genetic grounds is ethically suspect, none of the claims of ethical unacceptability withstand scrutiny. Preimplantation diagnosis is close enough to generally accepted practices of

[1] The term "embryo" is used here to refer to all stages from fertilization to implantation, rather than technically more accurate terms such as zygote, blastocyst or preembryo.

Preimplantation Genetics, Edited by Y. Verlinsky and A. Kuliev, Plenum Press, New York, 1991

genetic screening and in vitro fertilization (IVF) to be deemed ethically acceptable in its own right.

The End: Avoiding Offspring With Genetic Defects

The purpose of preimplantation screening of embryos is to avoid offspring with serious genetic defects. Some persons might argue that such an explicitly eugenic purpose is itself unethical, and for that reason alone oppose preimplantation genetic diagnosis.

But this claim does not stand up to analysis. Surely people do not have an obligation to have genetically affected offspring. It is within their procreative liberty to avoid such births if they choose. Indeed, people exercise control over the genes of offspring in a variety of acceptable ways, such as avoiding marriage or conception and terminating pregnancies that test positive for serious genetic defects. Given that private eugenic decisions, e.g., decisions to avoid genetic disease in offspring, is in many forms widely accepted as ethical, genetic selection of embryos is not unethical simply because it serves a eugenic purpose. Publicly imposed eugenics, however, would raise serious ethical problems.

A contrary judgment would call into questions many decisions that are clearly a part of marital and procreative liberty. It would also make genetic selection of gametes unethical. Yet the public reactions to news of gamete selection appear to be positive. For example, right to life groups that oppose IVF and embryo selection seem to have no problem with gamete selection, because nothing is done to a fertilized egg or embryo[2] (Time, 1989). If gamete selection, which is done to avoid genetic disease in offspring, is ethically acceptable, genetic selection of embryos cannot reasonably be opposed because of its eugenic character.

The Means Used: Manipulation and Discard of Embryos

Given the acceptability of genetic selection *per se*, ethical objections may still arise from the means used to pursue that goal. Genetic selection of embryos requires that women undergo IVF, though analysis of coitally created blastocysts that have been removed by uterine lavage is theoretically possible.[3] The resulting embryos are then biopsied and genetically diagnosed. Embryos without the deleterious genes are placed in the uterus, in the hope of producing offspring. Genetically affected embryos are not

---

[2] James Bopp, Jr., general counsel to the National Right to Life Committee, says that he sees nothing wrong with gametic selection, since it does not involve "the taking of human life".

[3] See paper by B. Brambati in this volume for an account of uterine lavage as a method of obtaining blastocysts for preimplantation genetic diagnosis.

transferred. Is this procedure an unethical way to pursue the ethically acceptable goal of genetic selection of offspring?

Egg Retrieval and IVF

The need to resort to egg retrieval and IVF does not make this procedure unethical. Women do have to undergo hyperstimulation and surgical removal of eggs for preimplantation diagnosis of embryos (and eggs) to occur. But the risks of IVF are not so disproportionate to the benefits, nor to the burdens of alternative selection procedures that preimplantation screening of embryos is unethical because of the burdens imposed on women.

The choice between preimplantation and postimplantation prenatal screening does present women with a different array of burdens and benefits, as is evident if we compare a 100 women who go through preimplantation screening with a 100 who undergo prenatal analysis. The first group will undergo ovarian stimulation and egg retrieval to prevent placement in the uterus of the 25% of produced embryos that on average will have the genetic defect in question. The second group will undergo chorion villus sampling or amniocentesis, with 25% of them testing positive and most likely terminating the pregnancy (Michael et al, 1990).

But it is not obvious which alternative will be preferable for the women at risk. While some might prefer coital conception and later prenatal diagnosis, with the 25% risk that pregnancy termination will occur, many others will prefer to undergo IVF rather than risk termination of a wanted pregnancy. The choice of option that best serves the welfare of women at risk should be made by the women in this situation. In terms of women's welfare, either option is ethically acceptable.

Embryo Manipulation

Some persons will also argue that preimplantation diagnosis is unethical because it involves research with or manipulation of embryos. But embryo research is not itself unethical, anymore than is research with infants or other human subjects. If the research occurs within existing ethical guidelines for embryo research, the purpose of improving genetic diagnosis would not make it unethical (Robertson, 1986a).

Nor is the manipulation of embryos involved in genetic diagnosis--blastomere biopsy and polymerase chain reaction of DNA--itself unethical. Removing a blastomere does not ordinarily harm the embryo. Even if some apparently normal embryos failed to implant after biopsy, it would not be certain that the failure was due to the biopsy, rather than to other factors that prevent implantation. In any event, the purpose of the biopsy is not to destroy healthy embryos, but to enable them to be placed in the uterus. Without embryo biopsy, the embryos may never have been created at all.[4] Taking a risk that screening itself will prevent some healthy embryos from implanting does not itself show such disrespect for human life that the screening process itself becomes unethical.

Discard of Embryos

A likely ethical objection to genetic selection of embryos is that it leads to the discard or nontransfer of embryos. A major problem with this objection, which is a distinct minority view, is that embryos are too rudimentary in development to have rights or interests (Grobstein, 1988). Because they are not sufficiently developed to be harmed, it is not unethical to deselect some embryos for transfer.

While controversy exists over the moral status of preembryos, a widespread consensus has emerged that preembryos are not themselves persons or moral subjects with rights. Although they have the potential to become persons, and therefore deserve greater respect than other human tissue, they are not already persons and are not owed the same respect as persons (American Fertility Society, 1986). They may therefore be created, frozen, used in research and even discarded.

On this view of embryos, which has been accepted by legal authorities and ethical advisory bodies in the United States, Great Britain, France, Canada, Spain, Australia and many other nations, there is no obligation to have all preembryos placed in a uterus (Robertson, 1986b; 1987). Failure to transfer embryos violates no moral duty owed to embryos. Thus, infertile couples seeking to become pregnant from IVF may have extra embryos not transferred or discarded,[5] as may occur when there are too many embryos to be safely placed in the uterus, or after periods of storage.

If discard or nontransfer of preembryos is generally acceptable regardless of their genetic characteristics, then selection for transfer on genetic grounds should also be acceptable. There is no greater harm to embryos or symbolic concerns for human life if nontransfer occurs because of deleterious genes, rather than because they are not needed to start a pregnancy. Indeed, genetic selection of embryos is preferable to later abortion because it is less burdensome to the woman and occurs at such an early developmental stage.

Some persons who object to abortion on genetic grounds

---

[4] That is, the couple would not have been willing to create these embryos, or transfer any of them, unless genetic diagnosis were available to identify the healthy ones.

[5] The terms discard or nontransfer refer to the decision not to place an embryo in the uterus. The embryo may then be kept in storage or allowed to divide until it no longer is alive. The term "discard" does not mean that the embryo is literally thrown away while still alive.

might accept genetic selection of embryos because of the earlier developmental stage at which it occurs. If they object to genetic abortion because of the advanced developmental stage of the fetus, they might distinguish genetic selection of embryos because the preimplantation embryo is too undeveloped to suffer harm. Thus, some women who have moral objections to abortion, even of offspring with genetic handicaps, might find preimplantation selection of embryos for transfer to be morally acceptable.

Of course, persons who believe that fertilized eggs and preembryos are already persons or moral subjects in their own right will object to discard or nontransfer of embryos. Persons holding this view will argue that there is a duty to place all embryos in a receptive uterus, even if the embryo has a serious genetic defect. Yet, as noted, this argument cannot be sustained in light of the biologic facts of preembryos. While transfer of all embryos may serve symbolic goals or otherwise demonstrate commitment to protection of all forms of human life, it is not a position required by a moral duty to embryos. Embryos are too rudimentary in development to have interests, and thus cannot be the subject of moral duties.

In any event, persons who take this strict view of embryo status are a minority. Ethical commentary on embryo status overwhelmingly rejects this position (Robertson, 1986b, 1987). It is noteworthy that persons holding such views also object to abortion on genetic grounds, and usually oppose IVF itself because of the risk that some embryos might be discarded. Unless one holds such beliefs, the claim that embryo discard on genetic grounds is unethical is unpersuasive.

Slippery Slopes

Ethical objections to genetic selection of embryos also arise from the fear that it will open the door to selection on inappropriate grounds, that it will lead to germline gene therapy, or that it will eventually lead to genetic engineering of offspring.

These objections are all of the slippery slope variety. They accept that current uses of the technique are not themselves objectionable, but argue that they nevertheless are unethical and should be banned because they will inevitably lead to other uses which are objectionable. Like all slippery slope claims, however, this argument is not adequate to show that an otherwise ethical practice becomes unethical merely because of the risk that it opens the door to more objectionable practices.

Selection on Inappropriate Genetic Grounds

Persons who would accept embryo discard for serious genetic defects may still oppose the practice because they fear that it will lead to deselection on inappropriate grounds, such as gender, minor genetic traits, susceptibility disorders and the like. Screening for trivial conditions for which abortion could not be morally justified may become acceptable at this earlier stage.

There are two responses to this claim. One is to question whether selection for these additional purposes is itself unethical. If genetic selection is not _per se_ unethical, and is occurring at a stage at which embryos are themselves too rudimentary to be harmed, is it clearly unethical to deselect embryos on grounds that would not justify abortion? For example, if a couple has produced five embryos, and two of the embryos were determined to have the gene for osteoarthritis, for neurofibromatosis, or a contingency disorder of variable expressivity, it is not clear that it would be unethical to place the three embryos without those genes in the uterus, and not transfer the others (New York Times, 1990). One may also question whether gender selection at this stage would also be unethical. In any event, a stronger case against screening for less serious genetic conditions is required to persuade that such practices are unethical.[6]

The second response would arise if screening and discard on the basis of less serious genetic conditions were clearly seen to be unethical. In that case, the question is whether unethical uses of preimplantation screening could be prevented without stopping the ethical uses. That is, does screening embryos on ethically acceptable grounds necessarily mean that screening on unacceptable grounds will also occur?

Yet if the feared uses are so clearly unethical, then physicians would probably be very hesitant to screen for them. No one argues that all amniocentesis and abortion should be stopped just because some persons may use them for sex selection. We are content to rely on the reluctance of most physicians to abort for sex selection purposes after amniocentesis. It is reasonable to expect that similar constraints would apply to using preimplantation genetic diagnosis for unethical purposes.

Germline Gene Therapy

Another slippery slope objection to genetic selection of embryos is that it will make germline gene therapy possible. Because germline gene therapy depends on genetic diagnosis of embryos, that diagnosis should be banned in order to prevent germline therapy from occurring.

Again, the same two responses are relevant. First, it is far from clear that germline gene therapy is ethically unacceptable. At present germline therapy is not a practical possibility. While ethical objections have been raised and current National Institutes of Health guidelines would not permit it, there may be good reasons for engaging

[6] Opposition to gender selection by any means must be distinguished from opposition to particular means of gender selection such as abortion.

in germline therapy, e.g., if the disease is so horrible, there is no reason why it should not be prevented in all later offspring (Fletcher, 1983; Robertson, 1985). When demand for germline therapy arises, a closer analysis of the case for and against germline therapy may show that it is ethically acceptable for certain conditions. Thus, fears about germline gene therapy should not stop preimplantation screening of embryos when it has not been clearly established that all germline gene therapy will be ethically unacceptable.

The second response questions whether the clearly unethical practice would follow from doing the ethical practice. It questions whether doing A necessarily requires doing B. The answer would appear to be no, as amniocentesis and abortion for some conditions but not sex selection show. Genetic diagnosis for selection of embryos and genetic diagnosis for germline therapy are sufficiently different that the former could occur without also inevitably producing the latter. If germline gene therapy is clearly seen as unacceptable, then it could be banned without also banning genetic diagnosis for all other purposes. At this stage of development, it is overkill to stop couples at risk for offspring with genetic disease from using preimplantation genetic diagnosis in order to prevent the development of germline gene therapy.

Genetic Engineering of Offspring

A final slippery slope objection is that genetic diagnosis of embryos should be banned in order to prevent a Brave New World scenario of genetic engineering of other offspring traits. Since genetic alteration of embryos to produce desirable multifactorial traits is an idea very far from reality, it suffices to present a brief response to this slippery slope claim.

First, it is unclear that genetic engineering of other traits would necessarily be bad. We need to know more about how and why that might occur, the risks and benefits involved, and how those practices would differ from common rearing practices now undertaken to improve the welfare or opportunities of offspring, which are accepted as part of parental discretion in rearing offspring.

Second, opponents of genetic diagnosis would have to show that there is no way to engage in screening of embryos without also sliding down the slippery slope to multifactorial genetic engineering. Speculation that such might occur, when the feared use is far in the future, does not justify banning a technique that is itself ethically acceptable.

Summary of Ethical Issues

Selection of preembryos on genetic grounds is ethically acceptable. The goal of avoiding the birth of offspring with severe genetic handicaps is part of procreative liberty and parental discretion. Pursuing this goal prior to implantation enables at risk couples to avoid the birth of a handicapped child without having to undergo prenatal diagnosis and abortion.

The means employed to achieve this goal--preimplantation genetic diagnosis and nontransfer of affected embryos--is also ethically acceptable. Because embryos are so rudimentary in development, they are not generally viewed as having interests or rights. Thus, they have no right to be placed in a uterus, and may be discarded if they carry the gene for serious disease. Indeed, it is preferable to discard at the early embryonic stage rather than abort fetuses that are more fully developed.

Finally, the fear of slippery slopes is not sufficient to render unethical a practice that is otherwise ethically acceptable. Deselection on grounds of serious genetic disease does make possible deselection on less serious grounds. It also opens the door to germline gene therapy and is no doubt a necessary step if positive genetic engineering of other offspring traits is to occur. Yet there is no showing that any of these possibilities is so clearly unacceptable nor so likely to occur, that the potential for such outcomes justifies denying at risk couples the benefits that genetic selection of embryos makes possible.

While these are the main ethical objections that will be raised in public controversy over preimplantation genetic diagnosis, other ethical issues in the use of these techniques remain. These issues arise with any new medical technology. They concern the ethics of offering a procedure whose safety and efficacy has not been clearly established. The danger is that patients will be led into thinking that the procedure offers a realistic possibility of real net benefit, when it is still highly experimental. Close attention to the ethics of human experimentation and fully informed consent are necessary to avoid these problems. However, if these concerns are overcome, then it cannot reasonably be asserted that preimplantation genetic diagnosis is unethical.

LEGAL ISSUES

Having seen that preimplantation genetic diagnosis and selection of embryos is ethically acceptable, we now examine legal restrictions on this technique at the present time in the United States. Comparable legal issues may arise in other jurisdictions. Three legal questions are addressed: (1) is preimplantation diagnosis of embryos now legally permissible; (2) could such procedures constitutionally be banned in the United States; and (3) is legal regulation needed.

Current Legal Restrictions on Preimplantation Genetic Diagnosis

Aside from restrictions that apply to human subjects research and to the practice of medicine generally, there are no specific restrictions on preconception and preimplantation genetic diagnosis. No jurisdiction has seen fit to pass laws specifically aimed at these practices, though the federal government does not now fund research with human embryos for any purpose.[7]

However, a few states have laws that could restrict genetic diagnosis and selection practices with embryos. The laws in question limit embryo research and prevent discard of embryos.

Fetal and Embryo Research Laws

In the middle 1970's 25 states enacted laws to restrict fetal research. Although these laws were inspired by concerns about research on fetuses, fifteen of the states used such terms as "embryo," "conceptus," "product of conception," "unborn child," or other language that might appear to apply to research activities with preembryos (Office of Technology Assessment, 1988). In some cases those laws might limit preimplantation genetic diagnosis research on embryos.

A closer look at these statutes, however, leads to the conclusion that they are most likely to apply to preimplantation embryo research in only four states. First, they were written with the purpose of restricting fetal not embryo research. Second, terms that could be read to include preimplantation embryos are either specifically defined as or used in a context where implanted embryos, fetuses or aborted fetuses are contemplated, not preimplantation embryos.

In four states (Minnesota, Massachusetts, North Dakota, and Rhode Island), one could argue that the operative definitions of the fetal research laws would include prohibitions on embryo research. However, even here the ban is not as broad as it might seem. Massachusetts (North Dakota and Rhode Island copy its statute) does not prohibit "diagnostic or remedial procedures the purpose of which is ... to preserve the life or health of the (embryo) involved".[8] Thus, embryo research designed to enable that embryo to be transferred to a uterus and brought to term would not be prohibited, if absent the research the embryo would not be transferred. Arguably, then, genetic diagnosis research that will enable those embryos to be transferred would be permitted in Massachusetts, even if research on

---

[7] See 45 C.F.R. 46 (Subpart B-Additional Protections Pertaining to Research, Development, and Related Activities Involving Fetuses, Pregnant Women, and Human In Vitro Fertilization). This subpart requires the approval of the Ethics Advisory Board for projects involving IVF or embryos, but no Ethics Advisory Board exists at the present time, so no such projects can be funded.

[8] Mass. Stat. Ann. c112 #12J. See also Rhode Island 11-54-1E; North Dakota 14-02.3.

embryos that will be discarded is prohibited.

Even that limit might not apply in Minnesota, which exempts from its ban on "any type of ... research or experimentation ..." on "a living human conceptus"[9] research "which verifiable scientific evidence has shown to be harmless to the conceptus." Since early embryos do not have a nervous system and are clearly not sentient, research could not harm them, and thus might be permitted. Of course, persons who view the embryo as having interests and rights would argue the opposite. Given Minnesota's current ban on embryo discard, the dispute on this point may be academic.[10]

Two other states--New Mexico and Pennsylvania--have laws that restrict fetal research with such broad language that some kinds of embryo research, including preimplantation genetic diagnosis, might not be permitted. For example, Pennsylvania, which defines "unborn child" to include fertilized eggs, bans "any type of nontherapeutic experimentation upon any unborn child".[11] New Mexico, which has a comparably broad definition of fetus, specifically excludes IVF research that does not "insure that each living fertilized ovum, zygote or embryo is implanted in a human female recipient" (New Mexico Stat. Ann., 1978).

Both Pennsylvania and New Mexico, therefore, would appear to prohibit nontherapeutic embryo research. Thus, preimplantation diagnostic research on embryos that will not result in the transfer of the embryos may be illegal in those states, though research that does lead to transfer may be acceptable.

A reasonable conclusion is that preimplantation genetic diagnosis research with embryos, whether followed by transfer or discard, is legally permissible in all but four to six states at the present time. In those states, however, such research might be permissible if designed to enable transfer to the uterus to occur. While research on embryos to be discarded is more clearly prohibited in those states, there may be some leeway for research on spare embryos created in the IVF process that will not be transferred.[12] Such research might be viewed as "harmless" to the embryo and be permitted in Minnesota. It may even be acceptable in Pennsylvania and New Mexico if the embryos

---

[9] Minn. Stat. Ann, #145.421-422.

[10] If embryos may not be discarded, research on embryos that will be discarded is unlikely. Also, a Minnesota court might find that harm to embryos would occur from research.

[11] 18 Pa. C.S.A. 3203, 3216.

[12] However, there are serious constitutional problems with such bans, both on grounds of vagueness and interference with reproductive rights.

were not created specifically for research purposes.[13]

It should be noted that these restrictions apply only to research or experimentation with embryos. They would not apply to preimplantation genetic diagnosis of embryos that is done primarily for the benefit of the couple at risk, with no research intended. In any event, researchers in any of the fifteen states that might conceivably ban embryo research should seek further legal advice before embarking on genetic research that uses human embryos.

Embryo Destruction and Discard Laws

The other possible barrier to genetic selection of embryos arises from laws in four states that appear to require that embryos be transferred or not discarded.

At common law, and by court or statutory law, embryos are not persons or legal subjects, and are not subject to homicide or other laws (Robertson, 1990). Legally, a new person does not exist until there is a live separation from the mother. Thus embryos have no legal status as such, and have no legal right to be placed in a uterus in the absence of state legislation specifically requiring transfer.[14]

Three states have statutes that appear to ban or restrict the discard of embryos. Minnesota includes within the proscriptions of its homicide laws unborn children, defined as "the unborn offspring of a human being conceived but not yet born" (Minn. Stat. Ann., 1987). Illinois also makes it a crime to kill an unborn child, defined as "any individual of the human species from fertilization until birth" (Ill. Ann. Stat. 1989). Louisiana requires that embryos that a couple wishes to discard to be donated so that they may implant and come to term (La. Rev. Stat. Ann., 1990). Its IVF statute states that "an in vitro fertilized human ovum is a biological human being which is not the property of the physician which acts as an agent of fertilization, or the facility which employs him or the donors of the sperm or ovum" (La. Rev. Stat. Ann., 1990). The preamble to Missouri's abortion law may also restrict discard of embryos.[15]

In those states, a person would still be free to screen

---

[13] If the embryos were otherwise going to be discarded (which is not prohibited in either state), then one could argue that using them for research first would not harm them.

[14] Thus, a Tennessee appellate court reversed a trial judge's finding that frozen embryos were children and therefore had to be implanted (Davis v. Davis, 1990).

[15] Mo. Stat. Ann. states that "the life of each human being begins at conception," and "unborn children have protectible interests in life, health, and well-being." Missouri courts could rule that this provision prevents discard of embryos (Mo. Stat. Ann., 1989).

the genes of preimplantation embryos as long as the diagnostic procedure did not itself destroy or harm the embryo. If biopsied embryos survived or were still capable of implanting, and no discard occurred, then the diagnostic procedure itself would not be illegal. Genetic diagnosis followed by donation, or long storage, or possibly even by passively allowing embryos to die would not itself be illegal under those laws.[16]

In sum, aside from a few states that ban some forms of embryo research, or which require that embryos be transferred or donated, there are no existing legal barriers to preimplantation diagnosis and selection of embryos.

Constitutional Limits: May the State Prohibit Genetic Screening of Embryos

While only a few states have enacted laws that could restrict preimplantation genetic diagnosis of embryos, more laws may be enacted as controversy grows. An important question, therefore, is whether there are constitutional limits on state regulation of genetic screening of embryos. The requirement that statutes be precise and that they not interfere with reproductive freedom are the main constitutional barriers that are relevant.

Vagueness

Laws must be sufficiently precise that they give a reasonable person fair notice of what is permitted. Precise laws also prevent arbitrary enforcement. At least two state bans on fetal and embryo research have been struck down on grounds of vagueness (Margaret S. v. Edwards, 1986; Lifchez v. Hartigan, 1990).

The most recent decision was handed down in April, 1990 by a federal district court in Illinois confronted with a challenge to an Illinois law that banned "experiment (s) upon a fetus produced by the fertilization of a human ovum by a human sperm unless such experimentation is therapeutic to the fetus thereby produced" (Ill. Rev. Stat., 1989). Although the court could have found that "fetus" was vague, in that it is unclear whether that term includes fertilized eggs and embryos, the court found vagueness in the terms "experimentation" and "therapeutic." A reasonable physician would have no way of knowing whether prenatal diagnostic procedures, extensions of basic IVF, and experimental procedures for the benefit of a pregnant woman were banned or permitted (Lifchez v. Hartigan, 1990). Accordingly, the law was void for vagueness.

The decisions in Margaret S. v. Edwards and Lifchez v.

[16] One could argue that these prohibitions apply only to active killing of embryos, as might occur with actual discard. Merely not transferring or allowing to continue to divide until the embryo no longer survives in the laboratory could be distinguished as permissible.

Hartigan suggest that there are major constitutional problems with existing embryo research laws. None of them appear to give a clear idea of which procedures are permitted and which are not. Indeed, the court in Lifchez even mentioned preimplantation genetic diagnosis of embryos as a procedure that might inappropriately be banned under such a vague statute (Lifchez v. Hartigan, 1990). The reasoning of these cases suggests that it would be extremely difficult to enact a law that defined "experimentation," "research," "therapeutic," and "nontherapeutic" precisely enough to meet constitutional requirements of precision.[17] Researchers fearful that state fetal or embryo research laws restrict their work should consider a constitutional challenge.

Procreative Liberty

Laws that prohibit embryo biopsy or embryo discard impinge on the procreative liberty of couples at risk for genetic disease. They bar access to information material to the couple's decision to reproduce. If the information is available, they may require that embryo transfer occur and thus prevent a couple from avoiding reproduction.

Although not mentioned specifically in the Constitution, some degree of procreative liberty seems to be constitutionally protected. A person's right, prior to conception, not to procreate is well-established, and continues until viability if pregnancy has occurred.[17] A right to reproduce no doubt would be recognized for married couples seeking to reproduce coitally, and arguably would exist for noncoital reproduction with their own gametes (Robertson, 1983). A cogent argument for extending the right of noncoital reproduction to gamete and embryo donors, and even surrogates can also be articulated (Robertson, 1983).

State restrictions on the use of preimplantation diagnostic techniques directly interfere with procreative liberty by limiting a couple's access to information essential to make procreative decisions. Such laws limit information that is material to a couple's decision whether to proceed with or to avoid reproduction.[18] Embryo research laws that prevent preimplantation genetic research are thus constitutionally suspect, even if they are precisely drawn. Maintaining symbolic respect for early embryos or preventing

---

[17] A state could solve the precision problems by specifying the exact procedures that could not be done on embryos or fetuses, rather than use the broad terms of existing statutes. Such a tactic would then raise directly questions of procreative liberty.

[18] Here the information is necessary for them to decide whether to go ahead with IVF, and if they do, to have the embryos placed in the uterus or not transferred at all. Whether or not they end up with offspring may depend upon access to the genetic information in question.

eugenic decisions are not sufficiently compelling interests to justify such restriction. Indeed, the court in Lifchez also found the Illinois ban on fetal research to be invalid because it intruded upon constitutionally protected choices about procreation (Lifchez v. Hartigan, 1990).

State restrictions on nontransfer or discard of embryos do prevent a couple from avoiding genetic reproduction, but such laws are probably valid under Roe v. Wade (U.S., 1973). Roe gives a right to terminate an unwanted pregnancy. It does not give a right to dispose of or kill aborted fetuses or embryos outside of the body. It, therefore, does not protect decisions to discard embryos (though a woman would not be obligated to accept embryos into her body) (Robertson, 1990).

A right to discard embryos would exist only if the right not to reproduce included the right to avoid genetic offspring *tout court*--the right not to have genetic offspring reared by others, even though no contact or rearing demands are made on the biologic parents.[19] Since this interest involves no bodily or physical burdens, it is largely psychosocial. It is unlikely that a Supreme Court disinclined to find new unwritten rights would find this interest to have fundamental right status (Robertson, 1990). If this prediction is accurate, then laws that prohibit embryo discard in order to protect all stages of human life would be constitutional even if Roe v. Wade remains good law. A contrary conclusion, on the other hand, would mean that embryo discard laws are also invalid.

Is Regulation Needed?

Research in preimplantation genetic diagnosis is now occurring at many centers. Rapid progress toward making this technique clinically available is likely to occur, with many genetic conditions discoverable at the gametic or embryonic stage. An important question is whether restriction or regulation of preimplantation genetic diagnosis is needed.

The analysis provided above suggests that specific regulation of preimplantation diagnosis of embryos beyond existing regulations for human experimentation and the practice of medicine is not needed at the present time. As we have seen, it is ethical to permit couples to make reproductive decisions on genetic grounds. It is also

---

[19] The situation here would arise if the couple that did not wish to use the embryos to procreate were required to have them donated to an infertile couple who would then have all rearing rights and duties in the offspring.

If a state constitutionally banned embryo discard, then it might also be able to ban embryo research that would lead to decisions to discard. Thus the validity of its embryo research laws might depend in part on its position on embryo discard generally. (Mass. Stat. Ann. 499-501.)

ethical to permit embryo discard to occur, including discard for genetic reasons. Given existing practices of genetic screening, prenatal diagnosis and embryo discard, preimplantation screening of embryos should also be permitted. No morally significant difference exists between existing techniques and preimplantation screening.

Some persons, however, might argue that genetic selection of embryos should be limited to certain well-defined cases of severe genetic disease, and should not be permitted for gender selection and other questionable preferences. The ethical arguments pro and con sex selection are too complex to rehearse here. However, objections to abortion for sex selection would not necessarily apply, because the selection is occurring at the early embryonic level, and thus does not injure a more developed fetus.[20] Similarly, objections to abortion for genetic defects with variable expressivity such as cystic fibrosis, or for defects that are manageable, such as PKU, also do not apply when the selection occurs at the gametic or preembryonic stage.

Moreover, there is no showing that genetic selection for less serious conditions occurs once genetic selection for more serious conditions occurs. While it is possible that there will be a greater willingness to screen for these other conditions because abortion is not required, the question of whether those uses should be restricted should await the development of such practices. At that time the need for and the constitutionality of such restrictions can be considered.[21]

Finally, there is no need at the present time to restrict preimplantation genetic diagnosis to prevent germline gene therapy or other genetic engineering of offspring. These extensions will not be feasible in the near future. Given existing federal oversight of gene therapy, there is no need for separate state efforts to restrict germline gene therapy at the present time.

CONCLUSION

Preimplantation genetic diagnosis of gametes and embryos may soon be technically feasible. These techniques offer special advantages to couples at risk for offspring with serious genetic disease, enabling these couples to have healthy children without undergoing prenatal diagnosis and possible termination of pregnancy.

Preimplantation diagnosis is a further extension of

---

[20] It is less clear that sex selection of offspring should be foreclosed to parents, if the means of selection are not themselves inherently objectionable, e.g., preconception or preimplantation means.

[21] Any restriction of preimplantation genetic diagnosis would have to avoid interfering with procreative rights to be constitutional.

existing techniques of hyperstimulation, in vitro fertilization and prenatal genetic diagnosis. It combines IVF and genetic diagnosis so that the genetic status of the embryo can be determined before implantation and pregnancy occurs. Although it is a necessary step to even more radical procedures, preimplantation genetic diagnosis is not itself such a radical departure from existing practices as to be ethically unacceptable.

The legal system now places few barriers in the way of use of this technique, though more may occur as controversy increases. Constitutional rights of procreative liberty may prevent an absolute ban on preimplantation genetic diagnosis, even if the state may prohibit embryo discard. However, there do not appear to be such special dangers or risks from these practices that an independent system of regulation is necessary at this time. It is reasonable to expect that persons who carefully analyze the competing considerations will share this view.

REFERENCES

American Fertility Society, 1986, Ethical Considerations of the New Reproductive Technologies, 46 Fertility and Sterility, 1S, 29S-30S.

Davis v. Davis, 1990, C/A No. 180, Court of Appeals of Tennessee, Eastern Section, September 13.

Fletcher, Moral Problems and Ethical Issues in Prospective Human Gene Therapy, 1983, 69 Virginia Law Revue, 515.

Grobstein, C., 1988, "Science and the Unborn", 21-58.

Illinois Statutes Annitated, 1989, ch. 38, para. 9-1.2(3) (b), "Smith-Hurd Supp."

Louisiana Revised Statutes Annitated, 1990, #9:126, "West Supp."

Lifchez v. Hartigan, 1990, 735 F. Supp. 1361, 1365-68, N.D. Ill.

Margaret S. v. Edwards, 1986, 794 F2d 994 5th Cir., Margaret S. III.

Michael M., Buckle, S., 1990, Screening for genetic disorders: therapeutic abortion and IVF, Journal of Medical Ethics, 16:43-47.

Minnesota Statutes Annitated, 1987, #609.266 (a), "West Supp."

Missouri Statutes Annitated, 1989, #1.205.1(1), (2), "West Supp."

New Mexico Statutes Annitated, 1978, 24-9A-1.

New York Times, September 5, 1990.

New York Times, September 8, 1990.

Office of Technology Assessment, 1988, Infertility: Medical and Social Choices, 324.

Robertson, J., 1983, Procreative Liberty and the Control of Conception, Pregnancy and Childbirth, 69 Virginia Law Revue, 405-7.

Robertson, J., 1985, Genetic Alteration of Embryos: The Ethical Issues, in genetics and the Law III, A. Milunsky & G. Annas eds., 115.

Robertson, J., 1986(a), Embryo Research, 24 Ontario Law Revue, 15, "West Supp.", 30-33.

Robertson, J., 1986(b), Embryos, Families and Procreative Liberty: The Legal Structure of the New Reproduction, 59 Southern California Law Revue, 939, 971-975.
Robertson, J., 1987, Ethical and Legal Issues in Cryopreservation of Human Embryos, 47 Fertility and Sterility, 371, 380.
Robertson, J., 1990, In the Beginning: The Legal Status of Early Embryos, 76 Virginia Law Revue, 437, 450-451.
Time, 1989, Nov. 27, at 56.
U.S., 1973, 410:113.

PART VII

TECHNICAL ASPECTS OF PREIMPLANTATION GENETICS

TECHNIQUES FOR MICROMANIPULATION AND BIOPSY OF HUMAN GAMETES AND PREEMBRYOS

Yury Verlinsky, Jeanine Cieslak, and Sergei Evsikov

Reproductive Genetics Institute
Illinois Masonic Medical Center
836 W. Wellington
Chicago, IL 60657

INTRODUCTION

The aim of these procedures is to provide a technical guide for scientists willing to learn micromanipulation techniques which can be applied to research on human gametes and embryos.

TOOL MAKING

The first step in microtool construction is the preparation of microneedles from glass coagulation tubes. This can be done with the use of the Flaming-Brown mechanical pipette puller (Model P-87, Sutter Instrument Co.). The program and heating filament should be adjusted to give the desired length of the tapered end of the microneedle.

The holding pipette is prepared using the DeFronbrune type microforge (Model MF-1, TPI Instruments). Under the control of the microforge stereomicroscope equipped with eyepiece reticule, the tip of the needle is broken by placing the section of the needle which approximates 200mcm in diameter on top of the glass ball which surrounds the heating filament. The temperature of the filament is increased until the needle is fused to the glass ball. The power is switched off causing an abrupt decrease in temperature and shrinkage of the heating filament, leading to the breakage of the tip of the needle fused to the glass ball. Then, the pipette tip is flame-polished by being placed approximately 0.5mm away from the glass ball while the temperature of the filament is increased to achieve melting of the glass. Finally, the attenuated portion of the micropipette is bent by being placed above the glass ball. The filament is heated until the glass begins to soften so that the pipette tip bends by its own weight. This is done to all microtools so that they are parallel to one another and to the microscope stage during micromanipulation.

To prepare beveled micropipettes, the KT Brown type micropipette beveler (Model BV-10, Sutter Instrument Co.) is used. Before the microneedle is attached to the beveler, the plane of beveling is indicated by marking the shoulder of the microneedle with a small dot. This dot should be facing up when the microneedle is attached to the beveler. The time required for beveling depends upon the pressure applied to the needle, the diameter desired, and the type of glass. This should be determined experimentally. Immediately after beveling, the diameter and the quality of beveling are checked under the microscope fitted with an eyepiece reticule, and the needle is thoroughly rinsed with deionized water. The micropipette is bent on the microforge, the dot indicating the plane of beveling facing the operator. For pronuclei removal, the beveled micropipette is siliconized (with Sigmacote) to prevent the oocyte plasma membrane from adhering to the glass.

The microtools are stored in a petri dish (150 x 15mm). Clay or foam rubber can be used to hold the pipettes in place.

MICROMANIPULATION SET UP

All equipment used for micromanipulation and handling of human embryos is placed in a laminar flow hood. An inverted microscope such as the Nikon Diaphot allows any type of micromanipulation to be performed. Micromanipulators are mounted to the microscope stage which reduces vibration caused by laminar flow hood. The Narishige MO202 micromanipulator provides good control of the beveled microneedle while the less expensive Narishige MN151 can be used to control the holding pipette. Hydraulic control for the holding pipette can be performed using the Narishige IM-6 microsyringe; the Stoelting #51218 microsyringe can be used for the beveled microneedle. The microsyringes are connected to the micropipettes using polyethylene or teflon tubing (.040 I.D. X .070 O.D.). The whole hydraulic system is filled with light paraffin oil, with special attention paid to eliminate any air bubbles. A stage warmer should be attached to the microscope stage to maintain the needed temperature during micromanipulations.

Gametes and embryos are transferred under the control of a dissecting microscope. Special measures are taken to maintain the pH and the temperature of the culture media when outside the incubator. An immersion heater circulator regulates temperature, while the gas flow meters control the passage of tri-gas mixture (5%CO_2, 5%O_2, 90%N_2) to the working surface.

Micromanipulations are performed in a Petri dish chamber. Several drops of media are placed around the center of the 60 X 15mm petri dish, and covered with filtered paraffin oil. Two media, Ham's F-10 supplemented with 10% serum or INRA Menezo B2, can be used for micromanipulations. As with PZD (Malter, this volume) the culture media used for Polar Body Removal is supplemented with 0.1 molar sucrose. This causes shrinkage of the oocyte, which increases the perivitelline space and helps to avoid oocyte damage.

Before any micromanipulation an empty petri dish is placed on the microscope stage to prevent leakage of oil from the pipette onto the objectives and the instruments are aligned in such a way that they are parallel to the microscope stage. Hyaluronidase (200 IU/ml of media) is prepared in order to remove cumulus cells from the oocytes.

Materials required for micromanipulation and biopsy are listed in table 1.

POLAR BODY REMOVAL

The polar body of an oocyte is removed for undertaking preconception genetic diagnosis (Verlinsky, this volume).

To perform Polar Body Removal, a holding pipette and a beveled micropipette approximately 12-15 mcm in diameter are used. The beveled pipette is filled with silicone oil before it is attached to the micromanipulator holder. The micropipettes are lowered into the petri dish chamber, excess oil is removed from the outer surface of the instruments and a small amount of media is aspirated into the beveled pipette. Once the oocyte has been gently aspirated into place by the holding pipette, the oocyte is turned using the beveled pipette until the Polar Body is positioned at 12 o'clock. It is important <u>not</u> to aspirate any corona cells which may still be attached to the zona pellucida. The beveled micropipette is passed through the zona, aspirating the polar body into the pipette. If the polar body is still attached to the ooplasm, further incubation is required before it is completely detached. The beveled pipette is lifted out of the dish and handed to an assistant (gloves should be worn by an assistant to avoid

Table 1. Required Materials

150 mm Coagulation tubes	Fisher #02-667
60x15 Petri dishes	Fisher #08-757-100B
150x15 Petri dishes	Fisher #8-757-14
Paraffin oil	Fisher #0121-4
Silicone oil	Sigma #DMPS-12M
Fluorinert (FC-70)	Sigma #F-9880
Sucrose	Sigma #S-9378
Cytochalasine B	Sigma #C-6762
INRA Menezo B2 media	IMV International ZA146
HAM'S F-10 media	Gibco #430-1200
Hyaluronidase	Sigma #H-3506
HPLC H20 (Mallinckrodt)	Scient. Prod. 6795-1NY
Sterile pasteur pipets (601-050)	Invitro Scient. Prod.
Sigmacote	Sigma #SL-2

contamination with normal DNA). The polar body is expelled by lowering the tip of the beveled micropipette into the micro-centrifuge tube containing HPLC grade acrodisc filtered water. After expelling, the pipette tip is broken on the bottom of the micro-centrifuge tube.

The oocyte is released from the holding pipette, washed in media to remove sucrose and transferred back to the original dish.

EMBRYO BIOPSY

Embryo biopsy can be performed as soon as the embryo reaches the 3-cell stage, but it is preferable to biopsy at a later stage (up to compaction) when a loss of one blastomere will cause relatively less decrease in cell number (Hogan et al, 1986; Handyside et al, 1990).

The microtools used for embryo biopsy can be the same as for polar body removal, although it is preferable to use a beveled pipette with a larger diameter (approximately 18-25mcm). As during the aspiration the membrane of the blastomere breaks in the pipette, it is preferable to choose a blastomere with a clearly visible nucleus so that its aspiration into the pipette can be controlled. Once aspirated, the blastomere nucleus is expelled into the microcentrifuge tube as in polar body removal.

The procedure itself is easier than polar body removal, but as the sperm is attached to zona pellucida, it is very critical to avoid its aspiration into the beveled pipette.

BLASTOCYST BIOPSY

When an embryo develops to an expanded blastocyst, a few trophectoderm cells can be taken for genetic analysis (Summers et al, 1988).

The procedure is performed in two steps: (1) Using the same set up as for partial zona dissection (PZD) (Malter et al, this volume), the zona is dissected above the trophectoderm cells opposite the inner cell mass. After PZD, the blastocyst is placed in culture until it begins to hatch through the opening made in the zona. (2) Using a microneedle, herniated trophectoderm cells are pressed to the bottom of the petri dish. While moving the needle back and forth, assurance is made that the "neck" of the herniated blastocyst is sealed. The herniated piece is cut away by a razor blade attached to a metal arm held by the micromanipulator. This procedure can also be performed under a dissecting microscope by hand using a scalpel blade (N15). The biopsied cells are transferred into a microcentrifuge tube for genetic analysis using a Pasteur pipette.

SPERM MICROINJECTION

Microinjection of sperm into the perivitelline space of an oocyte is performed to enhance fertilization in cases where male factor infertility is present. The diameter of the beveled needle should be 6-10 mcm. In most cases, unless the sperm grade of movement is extremely poor, the needle should be filled with fluorinert, instead of silicone oil.

A small number of sperm are expelled into the drop of media containing the oocyte(s). One oocyte is picked up by a holding pipette. The sperm is aspirated into the beveled pipette using a quick twist of the microsyringe. It is easier to catch the sperm along the edge of the bubble, where the most motile sperm may usually be found. Once the sperm is in the micropipette, it is essential to equilibrate the pressure so that there is no flow of medium from (or in) the needle. The sperm should be taken into the micropipette tail first to allow it to swim out by itself. By the time the sperm is near the tip of the needle, the zona pellucida should be punctured so that the sperm swims out of the needle, and into the perivitelline space. The oocyte is checked for the presence of pronuclei after a minimum of six hours.

PRONUCLEUS REMOVAL

For research purposes, the polyploid zygote can be transformed to a diploid state by microsurgical removal of the extra pronuclei. The beveled needle used for enucleation should be 18-20 mcm in diameter, flame-polished and siliconized.

10-15 minutes before the beginning of micromanipulation the zygote is placed in a medium containing 1-2 micrograms per milliliter of cytochalasine B, enucleation being performed in the same medium. The main principle is that the plasma membrane of the zygote is not punctured: the pronucleus is removed as a karyoplast. The method, proposed by McGrath and Solter (1983) still remains the only reliable approach of zygote enucleation. However, no method exists to distinguish male and female pronuclei, presenting the risk that the female pronucleus may be removed leaving both male pronuclei in the zygote. As such an androgenetic embryo may be at risk for congenital disorder including transformation into a hydatidiform mole, restored diploidy does not make it acceptable for transfer.

REFERENCES

Handyside, A.H., Kontogianni, E.H., Hardy, K., and Winston, R.M.L., 1990, Pregnancies from biopsied human preimplantation embryos sexed by Y-specific DNA amplification, Nature, 344:768-770.

Hogan, B., Constantini, F., and Lacy, E., 1986, "Manipulating the mouse embryo. A Laboratory Manual," Cold Spring Harbor Laboratory, New York.

McGrath, J., and Solter, D., 1983, Nuclear transplantation in the mouse embryo by microsurgery and cell function, Science, 220:1300-1319.

Monk, M. (ed.) 1987, "Mammalian development. A practical approach." IRL Press, Oxford.

Summers, P.M., Campbell, J.M., and Miller, M.W., 1988, Normal in-vivo development of marmoset monkey embryos after trophectoderm biopsy, Human Reproduction, 3:389-393.

TECHNIQUES FOR MICROSURGICAL FERTILIZATION

Henry E. Malter, Beth E. Talansky and Jacques Cohen

The Gamete and Embryo Research Laboratory
The Center for Reproductive Medicine and Infertility
The New York Hospital - Cornell University Medical College
New York, New York

MICROSURGICAL FERTILIZATION TECHNIQUES

During the past two years, pregnancies have been reported following the use of two microsurgical fertilization techniques: Partial Zona Dissection (PZD) and Subzonal Insertion (SI or MIST for microinsemination sperm transfer) (Cohen et al, 1988; Ng et al, 1988). To date, 25 pregnancies have been established with 12 healthy babies born from the application of PZD in 7 clinics worldwide. The pregnancy and birth rates from SI are not definite, although at least four pregnancies and one birth have been reported (Ng et al, 1990; Fishel et al, 1990).

Both techniques seem to provide a method for improving fertilization in cases of male factor infertility. Hopefully, as more clinics begin reporting on success (or failure) with microsurgical fertilization, an understanding will emerge concerning the most appropriate mode of application and the true efficacy of these techniques. The following is a technical description of PZD and SI as they are being practiced in our program.

Equipment and Supplies

1. Micromanipulation system: This must be a high precision system with joystick type controls and should be mounted to the body (not the stage) of the microscope. In our experience, the best system for embryological micromanipulation is the Narishige. At least two micromanipulators are required. A microsyringe system may also be required, although a mouthpiece system is the easiest for controlling the holding pipette.

Preimplantation Genetics, Edited by Y. Verlinsky and A. Kuliev, Plenum Press, New York, 1991

2. <u>Microtool Fabricating Equipment</u>: The fabrication of microtools requires a Microforge and a Pipette Puller. Once again, Narishige makes an excellent Microforge and a good basic Pipette Puller that is suitable for pulling PZD needles. A more sophisticated puller may be appropriate for advanced tools such as SZI needles. A Microgrinder is needed for SZI needles. Capillary tubing of the correct size and type will also be needed. SZI needles are made from a very thin walled tubing.

3. <u>Microscope</u>: Most workers use an inverted scope, although a fixed-stage upright scope is also satisfactory. The scope must be able to accommodate the micromanipulator mounting. Objectives and condenser will probably have to be of the ultra-long or long working distance type. Micromanipulation is generally done at the 100-200x magnification range. Nomarski or Hoffman optics is highly recommended but not required. The scope should be fitted with some type of temperature control (I.E. a warm stage or incubation chamber).

4. <u>Supplies</u>: The gametes must be placed in some type of holder during micromanipulation. Glass depression slides work quite well and are easily washed, sterilized, and reused. Some workers prefer disposable tissue culture dishes (or dish tops). PZD requires the use of a standard IVF media (I.E. Earles, etc.) with the addition of Sucrose - and equilibrated oil. Cumulus cell removal is done using Hyaluronidase and mechanical dissection with needles or fine bore pipettes.

PARTIAL ZONA DISSECTION (PZD)

<u>Microtools</u>

Partial zona dissection uses very basic microtools. The only consideration for the holding pipette is its size. If the holding pipette is too large, it will be in the way of the needle and insertion through the zona will be difficult. A tip diameter of approximately 100 microns works well.

The needle is a simple sharp probe that can be used directly from the needle puller. The needle should not be too thin as it may not create a sufficient opening in the zona during dissection. In general, any sharp pointed needle with a gradual taper will be suitable.

<u>Preparation</u>

Mature oocytes with a first polar body are selected for PZD. Often only a proportion of oocytes is micromanipulated if it is felt that the final sperm preparation may be suitable for a standard insemination.

Preparation for PZD commences just before normal insemination would take place. Oocytes for micromanipulation are placed in a solution of 0.1%

Hyaluronidase in culture media and incubated for a few minutes. At this time, mechanical dissection of the cumulus is done either with hypodermic needles or with a fine bore pipette. The majority of PZD related damage occurs at this step. It is often impossible to remove all the cumulus. Dissection should proceed with caution and last for a brief time.

The oocytes are washed through several (5) drops of culture media and placed in a drop of 0.1M Sucrose solution in culture media. Up to three eggs are then placed in a microdrop on the well of a sterile glass depression slide covered with a layer of warm equilibrated oil.

Prior to or during the egg preparation, the micromanipulator is set up with the tools properly aligned for use.

Micromanipulation

The slide is placed on the warm stage and the oocytes are located and brought into focus. The tools are lowered carefully into the drop and micromanipulation commences.

Suction is pulled through the holding pipette causing an egg to be aspirated and held against its face. The egg is positioned and re-positioned until a view of the perivitelline space (PVS) is obtained and some PVS is visible along the top edge of the oocyte. The needle is positioned and "focused" at the "12 to 1 o'clock" position and carefully inserted through the zona pellucida into the PVS. Needle insertion continues, possibly with some adjustment of the needle "focus", until the needle passes across the top of the oocyte and out through the zona pellucida on the opposite side. At this point, the egg is released from the holding pipette. The egg is held on the end of the needle while the holding pipette is positioned along the top edge of the egg. The holding pipette is placed so that it is in contact with the portion of the zona pellucida that is pierced by the needle. The egg is pushed back along the needle slightly and the zona is rubbed with the holding pipette in a left-to-right motion. During this rubbing, the focus position of the two tools must be constantly adjusted to ensure good contact between the holding pipette and the area of the zona that is being rubbed. Eventually, the holding pipette forces the needle through the zona pellucida creating a slit-like opening. When the egg "drops off" of the needle, the procedure is finished.

Problems occur when the egg is surrounded by a dense layer of corona cells. These cells may adhere to the tools and clog the opening on the holding pipette. With experience, it becomes easier to deal with this problem. After micromanipulation, the oocytes are removed from the slide and washed in several drops of culture medium to remove the sucrose. The eggs should return to their normal size very quickly. Eggs are then placed in final drops for insemination (Figure 1).

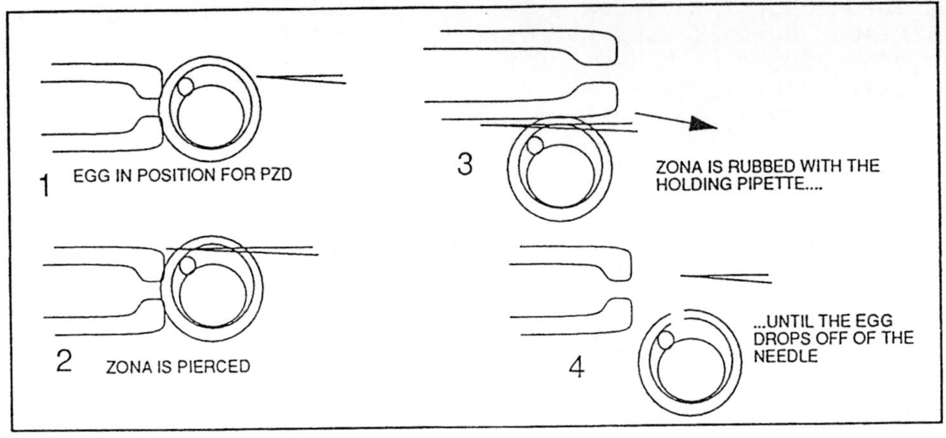

Figure 1. (explained in the text)

Insemination

Incubation in sucrose media may have an effect on oocyte activation (Malter et al, 1989). If eggs are fertilized immediately after sucrose exposure the rate of polyspermy may be higher than if insemination is delayed slightly. A period of 30 minutes between sucrose exposure and insemination may be indicated. All insemination is done in small drops under oil.

SUBZONAL INSERTION (SI)

Microtools

Subzonal insertion involves the use of a holding pipette and a bevelled microneedle through which sperm are aspirated and introduced into the perivitelline space of the oocyte. Since the holding pipette must provide leverage for the oocyte during sperm insertion, it is optimal to have a holding pipette with a tip diameter which is slightly wider than that used for PZD (150 μ instead of 100 μ).

The microneedle which is used to handle the spermatozoa is made by a multistep process. The needle is pulled, broken on the microforge, bevelled on a grinder (for about 10 sec) and washed with hydrofluoric acid and sterile water. The size of the aperture is an important factor in manufacturing the subzonal needle. For example, the initial break in the pulled needle should be made at approximately 12-15 μ. However, the bevelling process will increase the size of the opening to a diameter of about 20 μ. Variations in the size of the microneedle may be made according to the quality of the sperm sample to be manipulated. For instance, when sperm are of poor "quality" and motility and are few in number it is much safer and more efficient to use a larger microneedle. Since spermatozoa are susceptible to physical damage, a large aperture will be helpful in avoiding direct contact with the microneedle.

Preparation

As with PZD, mature oocytes exhibiting a first polar body are suitable for manipulation by SZI. Depending on the number of oocytes retrieved and the particular case, a proportion of eggs may be set aside as non-manipulated controls.

Subzonal insertion is performed according to the same general schedule (re: oocyte retrieval) as PZD. All preliminary steps to micromanipulation, including hyaluronidase treatment, removal of excess cumulus cells, and preparation of sucrose-containing microdrops (in depression slides) are all performed as described above for the PZD procedure. A maximum of two oocytes are placed in a microdrop on a single slide for SZI.

Micromanipulation

The most tedious step involved in SZI is loading the microneedle with sperm. The method by which this is accomplished varies with the quality of the final sperm preparation. It is efficient to front load the SZI needles by applying suction (see description of aspiration system-below) near the periphery of a 15-30 μl droplet which is placed in a depression slide under oil. Alternatively, in cases when sufficient motile sperm fractions cannot be secured by standard preparatory techniques it may be necessary to aspirate sperm directly from the (sedimentation) dish (this step will depend on the method of sperm preparation). If the microneedle becomes obstructed during this procedure, it is helpful to move the microneedle through the oil and back into the drop while applying positive pressure. It is sometimes helpful to preload several microneedles in this manner, and carefully place them in a humidified incubator until use.

In contrast to PZD in which the holding pipette is controlled by direct mouth aspiration, SZI is more easily performed with the holding pipette attached to a syringe system. The sperm aspiration microneedle, however, is directly controlled by a mouthpiece. Before securing the oocyte, the holding pipette is partially filled with medium in order to maintain better flow and avoid turbulent movements which may detriment the oocyte. After drawing the oocyte against the holding pipette, the microneedle is brought into close view. Positive pressure is applied to the mouthpiece in order to clear the microneedle of debris and excess medium before entering the PVS. The microneedle is brought near the "12-1 o'clock" position of the oocyte and is inserted through the zona pellucida. After entering the PVS, it is necessary to refocus on the tip of the microneedle in order to clearly visualize the sperm as they are released. It is sometimes necessary to retract the microneedle to clear it of excess medium and bring sperm to the tip. Gentle positive pressure is maintained until the desired number of spermatozoa are released. Slight pressure is continued as the microneedle is very carefully withdrawn through the initial point of entry, in order to avoid re-aspiration of the sperm. Following micromanipulation, oocytes are immediately removed from sucrose and transferred through several drops of culture medium (Figure 2).

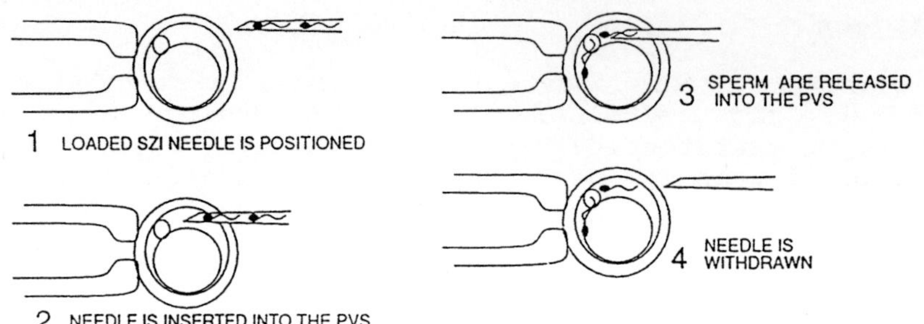

Figure 2. (explained in the text)

REFERENCES

Cohen et al (1988) Implantation of embryos after partial opening of the oocyte zona pellucida to facilitate sperm passage. Lancet 2:162.

Fishel et al (1990) Twin birth after subzonal insemination. Lancet i:722-723.

Malter et al (1989) Monospermy and polyspermy after partial zona dissection of reinseminated human oocytes. Gam Res 23:377-386.

Ng et al (1988) Pregnancy after transfer of multiple sperm under the zona. Lancet 2:790.

Ng et al (199) Subzonal transfer of multiple sperm (MIST) into early human embryos. Molec Reprod Devel. 26:253-260.

METHOD FOR OBTAINING HUMAN SPERM CHROMOSOMES

Renee Martin, Evelyn Ko, and Leona Barclay

Division of Medical Genetics, Department of Pediatrics, University of Calgary and Genetics Clinic, Alberta Children's Hospital 1820 Richmond Rd., Calgary, Alberta, Canada T2T 5C7 Canada

A technique for obtaining preparations of human sperm chromosomes was introduced by E. Rudak et al (1978) which is based on the use of hamster oocytes to reactivate human sperm. Detailed descriptions of the methods used in our laboratory have been published elsewhere (Martin, 1983; Martin, 1988). The following procedures currently used in our laboratory are presented below.

HAMSTER CARE AND SUPEROVULATION

Female golden hamsters (Mesocricetus auratus) are obtained from Charles River Breeding Laboratories (Laval, Quebec) at 6 to 8 weeks of age. They are kept, 4 to a cage, at 23°C, with a light cycle of 14 h of light (8 a.m. to 10 p.m.) and 10 h of darkness. The hamsters are given a week to adjust to these conditions and are 2 to 6 months of age when used for experiments.

Two mornings prior to the experiment day, hamsters are superovulated with an intraperitoneal injection of 25 to 30 IU of pregnant mare's serum gonadotropin (Sigma, G-4877), followed by an intraperitoneal injection of 25 IU of human chorionic gonadotropin (A.P.L. Ayerst Laboratories, New York, N.Y.) at either 6 p.m. (for 10 a.m. experiments) or 10 p.m. (for 2 p.m. experiments) on the evening prior to the experiment day. The cumulus cell mass containing eggs is in the oviduct 15 to 17 h after HCG injection; each hamster yields approximately 40 to 60 eggs.

MEDIA

Biggers-Whitten-Whittingham medium (BWW) (Biggers, Whitten, and Whittingham, 1971) with some modifications is used for sperm washing, egg preparation, and for sperm capacitation in fresh experiments. A 10X stock salt solution

Preimplantation Genetics, Edited by Y. Verlinsky and A. Kuliev, Plenum Press, New York, 1991

is prepared by dissolving the following components in water to a total volume of 1 litre:

NaCl	55.40 gm
KCl	3.56 gm
$CaCl_2$ pellets	1.89 gm
$KH_2\ PO_4$	1.62 gm
$MgSO_4\ 7H_2O$	2.94 gm

This can be stored at 4°C for an indefinite period of time.

A 1X BWW stock solution is prepared by combining the following components and adding distilled water to a total volume of 1 litre: (1X stock is kept at 4°C for a maximum of two weeks)

10X stock salt solution	100 ml
D-glucose	1.0 gm
sodium pyruvate stock solution (0.28 gm sodium pyruvate in 10 ml water)	1.0 ml
Antibiotic stock (penicillin G, 10^5 IU/ml and streptomycin sulfate, 50 mg/ml)	1.0 ml
Phenol red (0.5%)	0.5 ml
Hepes, free acid (2M) (Sigma)	9.5 ml
Hepes (2M in 3M NaOH)	10.5 ml

The BWW working solution is prepared daily by adding 0.2016 gm $NaHCO_3$, 0.37 ml DL-lactic acid (DL-V. 60% syrup, Sigma), and 0.5 gm human serum albumin (Fraction V, Sigma) to 100 ml of the 1X BWW stock solution. This mixture is filtered through a cellulose acetate membrane (pore size 0.22 micrometers) to sterilize it, and the pH is checked at room temperature, and adjusted if necessary to 7.4 to 7.5 with a small volume ($\leq$ 0.1 ml) of acid or base Hepes.

The medium used for egg culture is Ham's F10 (Flow Laboratories), prepared daily by supplementing with 15% fetal bovine serum (Flow Laboratories, heat inactivated at 56°C for 1 h), 100 IU penicillin G/ml, and 50 micrograms streptomycin sulfate/ml. The pH of the solution is adjusted to 7.2 with approximately 2 drops of 1N HCL.

PREPARATION OF SPERM

The human semen sample is collected in a sterile container and is processed as soon as liquefaction has occurred (10 to 20 min at 37°C). Sample volume, concentration of sperm, total sperm count, motility, and forward progression are determined: a wet mount (one drop of liquefied semen on a microscope slide, covered with a cover slip), is used to visually estimate the percentage of sperm moving (motility), and the efficiency at which the moving sperm travel on a scale of 0 to 10 (forward progression). To determine the concentration of sperm, 0.02 ml of liquefied semen sample is added to 0.38 ml of sperm count diluent (5 g $NaHCO_3$, 1 ml 35% formamide, in 100 mls 0.9% NaCl), and a count is taken of the number of sperm in 5 of the 25 squares in the grid of a hemacytometer (Bright-Line, American Optical, Buffalo, N.Y., 14215). The number of sperm counted in this manner is equal to the number of millions of sperm per milliliter of the original semen sample. If the sample is large, a portion of it may be frozen, at this point, for use in future experiments.

Depending on the type of experiment (fresh, frozen, or TEST-yolk buffer) to be carried out, the sample is further processed in one of three ways:

(1) for fresh experiments: the liquefied semen sample is diluted to 10 ml with BWW in a 15 ml conical centrifuge tube and is centrifuged at 600 x g for 6 min. The supernatant is decanted and the pellet is resuspended in 10 ml BWW. This is centrifuged two more times as described above, in order to remove seminal fluids and debris which inhibit capacitation, and, therefore, egg penetration. The final pellet is resuspended to a volume of 1 ml and another sperm count is taken, as described above. Depending on the sperm parameters and previous fertilization experiences with the same donor, the sperm concentration is adjusted to between 1×10^6 to 40×10^6 sperm/ml (average is about 10×10^6) with BWW. Using an Eppendorf Pipettor, four 50-microliter drops are placed in each of 4 tissue culture dishes (Falcon, 60 X 15 mm), and are covered with paraffin oil (Fisher, 0-121). The sperm drops are incubated at 37°C, 5% CO_2, and 95% humidity for 4-6 hours to effect capacitation before co-incubation with eggs.

(2) for frozen experiments: the liquefied semen sample is diluted slowly with an equal volume of Ackerman's cryoprotectant (Behrman and Ackerman, 1969). The mixture is drawn into 0.5-ml plastic freezing straws (United Breeders, Guelph, Ontario) which are then plugged with polyvinylpyrrolidone (PVP) powder. Five straws are bundled in a plastic tube and are placed in a wire basket. Freezing is accomplished by lowering the basket at a controlled rate into a Minnesota Valley engineering tank (MVE Cryogenics, model TA-60), filled to a depth of 2 inches with liquid nitrogen. Frozen specimens are stored on canes submerged in liquid nitrogen at a temperature of -196°C in a large Dewar flask. Specimens are thawed by removing the tubes from liquid nitrogen, separating the straws, and leaving them to thaw at 37°C for 5 to 10 minutes. Thawed sperm may then be washed and centrifuged as described above for fresh sperm, (allowing the sperm in drops to capacitate at 37°C, 5% CO_2, and 95% humidity for 0 to 6 hours before co-incubation with the eggs), or may be mixed with an equal volume of TEST-yolk buffer, and be processed as a TEST-yolk buffered sample (see below). The sperm appear to capacitate somewhat in the frozen cryopreservative and often require less capacitation time at 37°C than fresh samples.

(3) for TEST-yolk buffer (TYB) experiments: the liquefied semen sample (or thawed frozen sample) is mixed with an equal volume of TYB (prepared as per Brandriff et al., 1985), and is sealed in a small screw-capped tube, which in turn is sealed in a jar of room temperature water that is then plunged into a bucket of ice and refrigerated for approximately 24 h, until experiment time the following day. At that time, simultaneous to the eggs being collected (see below), the sperm are washed with BWW and centrifuged as described for fresh experiments, above. Sperm drops are put

into tissue culture dishes and covered with paraffin oil, as described above, and then are put into the incubator at 37°C, 5% CO_2, and 95% humidity for a few minutes, until the eggs are ready for co-incubation. The sperm appear to capacitate while refrigerated in the TYB, and tend to become less (or non-) viable if they are left to capacitate for too long at 37°C.

PREPARATION OF EGGS

Approximately one hour before dissection time, four staining watchglasses (Canlab W2517-3), are filled with 3 ml BWW, two watchglasses are filled with 1.5 ml BWW, two watchglasses are filled with 3 ml of F10, and two tubes of hyaluronidase (Type 1-S, Sigma, 1.5 ml of a 2 mg/ml solution in BWW) are prepared, as is one tube of trypsin (Type XII, Sigma, 3 ml of a 1 mg/ml solution in BWW). Colcemid solution (Gibco, 10 micrograms/ml) is added to F10 to a concentration of 4 micrograms/ml (for 10 a.m. experiments) or 8 micrograms/ml (for 2 p.m. experiments). Using an Eppendorf Pipettor, 50 microliter drops of F10 are made in tissue culture dishes (Falcon, 60 X 15 mm) and are covered with paraffin oil. All are warmed to 37°C until experiment time.

At experiment time, approximately 16 h after HCG injection, two groups of 4 hamsters are stunned with ether and are killed by cervical dislocation. Oviducts from each group are removed to one of the prepared (3 ml) watchglasses of BWW. In order to keep the time spent from dissection to sperm and egg co-incubation between 30 and 45 minutes, two persons routinely dissect four hamsters each.

The following procedures are carried out with the aid of a dissecting microscope equipped with a red light filter. Low incandescent room lights are used when handling eggs, since short wave lengths from fluorescent bulbs disturb the completion of normal meiosis and pronuclear formation (Hirao and Yanagimachi, 1978). The oviducts are transferred to another of the prepared (3ml) watchglasses of BWW to wash off excess blood, and are then transferred, one at a time, to the prepared (1.5 ml) watchglass of BWW, where the oviduct is punctured and the cumulus cell mass is removed. The prepared tube of hyaluronidase solution is added (final concentration is 1 mg/ml), and the hyaluronidase disperses the cumulus cells in a few minutes, freeing the eggs. Using a micropipette (made by drawing out a 5 3/4 inch pipette) attached to rubber tubing and a mouthpiece, the eggs are immediately given three BWW washes in a prewarmed (37°C) 9-well plate, in order to remove the cumulus cells. The eggs are then transferred to another well in the same plate, containing the prepared trypsin solution which dissolves the zona pellucida. The eggs are given three further BWW washes after the first polar body has detached from the egg (indicating that the zona has been completely dissolved by the trypsin). If eggs are removed from the trypsin before the polar body is free, there are occasionally cases where the zona is not clearly visible, and yet is able to hinder fertilization. Eggs are subsequently put into the prepared sperm drops and are co-incubated at 37°C, 5% CO_2, 95% humidity for 20 min to 3 h, depending on the sperm sample.

FERTILIZATION CHECK

In order to determine the most appropriate time to transfer the eggs from sperm drops to F10, fertilization checks are carried out. The timing of these checks depends on the sperm parameters: a vigorously-swimming, 50+% motile sample often has eggs checked for fertilization after 20 minutes of co-incubation, whereas a sluggish or largely non-motile sample may not require a fertilization check until after 1 h or more of co-incubation.

Slides used in the fertilization check (Fisher, 12-550-12) are prepared with four small blobs of vaseline (precooled to 4°C) positioned so that the four corners of a 22 mm coverslip (Corning) will rest on them. Groups of 5 to 10 eggs are removed from the sperm drops, washed in one of the prepared watchglasses of F10, and then transferred in a drop of medium (approximately 5-8 mm in diameter) to the center of the prepared slide. The cover slip is placed on the vaseline blobs and the eggs are compressed by gently pressing on the cover slip until the eggs are flat enough to be observed clearly with a phase contrast microscope at 400 x magnification.

Swollen sperm heads, indicating fertilization, appear as a clear round area with a tail, which is much larger than sperm which have not penetrated the egg. Attached sperm appear to have the half of the sperm head which is distal to the tail "dissolved" in the egg cytoplasm. Sperm which appear whole (often swimming) are merely in close proximity to the egg, and are not attached.

EGG CULTURE

When it has been determined that there are swollen sperm heads (indicating fertilization) in 50% of the eggs sampled, (<u>OR</u> when there are 2 or 3 attached sperm per egg in a sperm sample that is TYB-treated and very vigorous), OR after a maximum of 3 hours of co-incubation, the eggs are washed in the prepared watchglass containing F10 and are transferred to F10 drops, distributing the eggs evenly throughout the drops. The eggs are incubated at 37°C, 5% CO_2, and 95% humidity, for 7 h (10 a.m. experiment) or overnight (15-17 h) (2 p.m. experiment), at which time 50 microliters of 4 microgram/ml colcemid in F10 (10 a.m. experiment) or 8 microgram/ml colcemid in F10 (2 p.m. experiment) are added to each F10 drop. The eggs are cultured overnight (15 h) in the resultant 2 microliter/ml colcemid solution (10 a.m. experiment) or 5-7 h in the resultant 4 microliter/ml colcemid solution (2 p.m. experiment).

SLIDE PREPARATION AND FIXATION

Since fixation seems to be the portion of the experiment which is the key to the success or failure of the entire technique, we have prepared an <u>extremely</u> detailed protocol, described below.

Relative humidity is important in the area where fixation takes place. A relative humidity of 35 to 45% is maintained. It is also essential that standard fluorescent lighting is available above the area where fixation will take place so that eggs can be observed by the reflection of the light and solution spreading on the slide can be observed. The following steps are essential to obtain reproducible results.

- Prepare a number of slides by cleaning with isopropyl alcohol and wiping dry with a lint-free tissue.

- Using 2 watch glasses place 1-2 drops of fetal bovine serum on the bottom of each. Add 1% Na Citrate (about 2.5 ml). Leave for 15-20 minutes at room temperature to coat the bottom of the glass to prevent eggs from sticking.

- Just before the start of fixation tip out Na Citrate solution from both watch glasses and add fresh room temperature Na Citrate to each. Two watch glasses are used to prevent the transfer of oil to the slide which can interfere with the proper fixation of the eggs.

- Prepare fixation solution: 3 mls 95-100% ethanol (no ketones) and 1 ml glacial acetic acid, place in small tube.

- Transfer a number of eggs from F10/colcemid drops to watch glass #1. Return culture dish with remaining eggs to incubator. (We usually start with 10 eggs/slide. If there is exceptional fertilization we will fix fewer eggs per slide so that there are not multi-spreads to photograph on the fluorescence microscope. Chromosome banding fades rapidly and the chromosomes deteriorate under the ultraviolet light of the fluorescence microscope. If there are few karyotypes we increase the number of eggs per slide.)

- After approximately 3-5 minutes (we visually check eggs to make sure they have swollen 1-1/3 to 1-1/2 times their original size) these eggs are transferred to Na Citrate dish #2. Make sure pipette is free of oil inside and out. At this time transfer more eggs from incubator to Na Citrate dish #1 and allow to sit until eggs in dish #2 have been fixed.

- With pipette in dish #2, allow capillary action to fill the pipette, then slowly draw up eggs in as small a volume as possible following eggs with a small volume of sodium citrate (1-2 μl). This final volume will act as a cushion to eggs.

- Expel the small volume last sucked into pipette (1 mm drop), then slowly expel the eggs onto this drop. The size of the final drop with eggs is approximately 1.5-3 mm in diameter.

- Using an automatic pipettor set at 20 μl, one drop of fix is dropped directly on citrate drop from a height of about 0.5 cm.

- A very gentle moist breath is used to stop eggs from rolling around on the slide. This will add moisture to the slide and stabilize the eggs.

- As the drop starts to spread out on the slide the eggs will become quite visible, standing out in relief, visible under the fluorescent light. It is at this point that a diamond marker can be used to circle the eggs on the underside of the slide.

- Draw the circle and immediately add a second drop of fixative from a height of 0.5-1 cm above the area where the eggs were circled (do not allow the slide to dry out). The drop will spread out along the slide toward the edges and begin beading. With a tissue wipe fix off in a square of about 1" around the eggs. Keep wiping beading away as it accumulates in a 1" square (1-2 times). A rainbow formation visible under the standard ceiling fluorescent lighting will begin moving in toward the center of the slide where the eggs are. Allow the rainbow to come within approx. 1/2" on either side of where the eggs are (do not allow the slide to dry out).

- At this point, add a 3rd drop of fix onto the eggs from 1 cm above the slide and repeat wiping procedure as described above.

- When rainbow effect appears, a 4th drop is added and the same procedure is carried out.

- With this final 4th drop when the beading has stopped and the rainbow effect has appeared, a long gentle dry breath is used to dry the last fix from the slide. The slide is held about 6" from pursed lips and the breath is <u>very</u> gentle and slow. The rainbow is drawn in toward the eggs in a smooth fashion, carrying on to blow gently for about 2 seconds even after the rainbow disappears.

- Using a phase contrast microscope, check within the etched circle for chromosomes. They should be very dark, compact, evenly spaced chromosomes. If chromosomes are scattered you may be: (1) dropping fix from too high; (2) blowing too hard in either instance, or, (3) your eggs may have burst as they were expelled onto slide.

- If chromosomes are too compact and tangled: (1) let eggs swell more in hypotonic solution, or, (2) don't wipe slide so close to the eggs, i.e. let the drop of fixative spread further out on slide to 1-2". Occasionally, chromosomes do not develop properly and very little can be done to untangle them.

SCANNING

After fixation, slides are scanned using a Zeiss phase contrast microscope to locate chromosome spreads, which are then circled using a microscope-mounted diamond etcher. This circling reduces the time that the chromosomes are exposed to ultraviolet light (which tends to fade and then melt the chromosomes) when photographing with the fluorescent microscope (see below).

STAINING AND FLUORESCENT PHOTOGRAPHY

Spreads are allowed to age for exactly 2 weeks. Then a 0.5% solution of quinacrine dihydrochloride is prepared in a Coplin jar (0.25 gms quinacrine in 50 mls distilled water). The pH is adjusted to 4.4-4.5 with 0.1M HCl. In addition, 3 Coplin jars containing distilled water pH 4.4-4.5 are prepared. Slides are stained for 25 minutes in the quinacrine dihydrochloride, rinsed in Coplin jar #1, placed in coplin jar #2 for 5 minutes and then in Coplin jar #3 for 5 minutes. Slides are removed and air dried, placed in slide boxes and stored in freezer at 0 to -4°C until fluorescent photos are taken.

Slides are removed from freezer and placed in refrigerator at +4°C. Then one at a time slides are removed, 2-3 drops of pH 4.4-4.5 distilled water are placed on slide, coverslip is added and excess water is removed by gently blotting between 2 filter paper layers. Using the back of a bent dissecting needle, melted paraffin wax is used to seal the edges of the coverslip to the slide. Circled karyotypes are found under fluorescence and photographed on Kodak technical Pan black & white film with a Zeiss fluorescent microscope with a D.C. powered HBO W2 mercury lamp. The barrier filter is set at 47 and the excitation filter is set at BG3.

REFERENCES

Behrman, D.J. and Ackerman, D.R., 1969, Observation of human sperm, Am. J. Obstet. Gynecol., 103:654-664.

Biggers, J.D., Whitten, W.K., and Whittingham, D.G., 1971, The culture of mouse embryos in vitro, in: "Methods in Mammalian Embryology," J.C.Daniel, ed., W.H.Freeman and Company, San Francisco.

Brandriff, B., Gordon, L., and Watchmaker, G., 1985, Human sperm chromosomes obtained from hamster eggs after sperm capacitation in TEST-yolk buffer, Gam. Res., 11:253-259.

Hirao, Y. and Yanagimachi, R., 1978, Detrimental effect of visible light on meiosis of mammalian eggs in vitro, J.Exp.Zool., 206:365-369.

Martin, R.H., 1983, A detailed method for obtaining preparations of human sperm chromosomes, Cytogenet.Cell Genet., 35:253-256.

Martin, R.H., 1988, Cytogenetic analysis of sperm from a male heterozygous for a 13;14 Robertsonian Translocation, Hum. Genet., 80:357-361.

Rudak, E., Jacobs, P.A., Yanagimachi, R., 1978, Direct analysis of the chromosome constitution of human spermatozoa, Nature, 274:911-913.

RELIABLE TECHNIQUE FOR CHROMOSOMAL PREPARATIONS FROM MAMMALIAN OOCYTES, PREIMPLANTATION EMBRYOS AND ISOLATED BLASTOMERES

A.P. Dyban

Department of Embryology, Institute for Experimental Medicine, Academy of Medical Sciences, Leningrad, USSR

INTRODUCTION

The method for chromosome preparation from preimplantation mouse embryos devised by Tarkowski (Tarkowski, 1966) is widely used in experimental and clinical cytogenetics. The main advantage of this method is simplicity, but considerable skill is needed to obtain high quality preparation. Moreover, it is difficult to use this method when chromosomal preparations are made from zona free embryos or isolated blastomeres obtained by biopsy of preimplantation embryos. In this paper, the method suitable for the above mentioned materials obtained from different animal species (mice, rats, rabbits, monkeys, etc.) as well as humans is presented. This method was successfully used for karyotyping isolated blastomeres from the four-cell stage mouse embryos (Severova, Dyban, 1984) and the eight-cell stage human embryos (Handyside, 1989).

The main principal of this technique (Dyban, 1983) is fixation of embryos (oocytes, isolated blastomeres) after hypotonic treatment in a large volume of cold fixative (as in standard method of chromosome preparations from somatic cells), followed by spreading the cells by "softening solutions", which are the mixtures of acetic acid and ethanol. This approach improves the quality of preparation as it preserves the chromosome structure and leads to a sufficiently high spreading of metaphase plates.

EQUIPMENT AND REAGENTS

The results are mainly dependent on the appropriate equipment, especially on stereomicroscope (dissecting microscope). It should have the optical system which can be adjusted for better view of oocytes or embryos during fixation and spreading on slides. It is preferable to have a stereomicroscope with Zoom system, while a frosted mirror and clean glass stage are obligatory. The position of the

Preimplantation Genetics, Edited by Y. Verlinsky and A. Kuliev, Plenum Press, New York, 1991

mirror must be changeable, as embryos are better seen when the angle of the mirror is adjusted to place them on the boundary between the dark and the light parts of the visual field. For working with tiny materials such as isolated blastomeres, the best optic with a high resolution is needed. These criteria are not satisfied by some types of dissecting microscopes. The best stereomicroscope for this purpose is SMZ-2B from Nikon.

Microscope with phase contrast is required for examining the unstained chromosome preparation, i.e. for checking spreading and the quality of fixation. Any microscope with good optics and three phase objectives (6,3; 16; 40 X) can be used. This microscope should be located close to the working place where the chromosomal preparations are made.

Because the technique requires using a very cold fixative, it is necessary to have a refrigerator with two cameras (freezer sets for -20°C and cooler for 0-+4°C). It is advisable to locate the refrigerator in the same room where the chromosomal preparations are made.

In some cases the embryos must be cultured in media with colcemide before being used for chromosomal analysis, so the CO_2 incubator is required.

The working place for making micropipettes is needed, requiring a gas burner and a supply of glass tubes. Some of the micropipettes for handling the embryos can be made from commercially available disposable Pasteur pipettes. For working with isolated blastomeres it is better to make micropipettes from commercially available disposable microcapilares (100-200 μl). Watch glasses, small glass bottles and Copplin jars are also required. As the results depend on the quality of glass slides, commercially available preclined slides should be carefully selected (some slides may have small scratches which can interfere considerably with spreading if metaphases are located close to or on the scratched field). Slides are used after storage (1-2 hours or longer) in mixture of absolute alcohol:diethyl ether (1:1). Immediately prior to use, the slides are taken out with forceps and wiped dry (not with a textile but with coarse paper because some textiles carry traces of grease).

The following reagents are needed: methyl alcohol, glacial acetic acid, tri-sodium citrate, potassium chloride, hyaluronidase, colcemide, Giemsa stain, salts for making standard ovum culture media (M2, M16, etc.) or ready made media, buffer tablets or ready made buffer (pH 6,8). All reagents should be of highest purity. A supply of pure water (deionized or distilled) is also essential.

CHROMOSOME PREPARATION PROCEDURE

Collection of Material

Immature oocytes at the first metaphase stage are collected by puncturing ovarian follicles in any culture medium which holds pH on air (i.e. M2 medium) and cultured in any appropriate medium (M16, F10, etc.). The oocytes must be put in culture medium before ovulation takes place. The mature oocytes at the second metaphase stage are obtained from oviduct after ovulation. For removing cumulus cells the oocytes must be incubated for ten minutes or longer in a medium with hyaluronidase (300 i.u. per ml).

Embryos at the pronuclear stage are obtained in mice after induced or natural ovulation. This stage may be found in most strains of mice 28-32 hours after injecting hCG or 16-20 hours after mating. It is advisable that one-cell embryos are put into culture medium with colcemide (0.035 g/ml) 10-12 hours before pronuclear mitosis, i.e. before chromosomal preparations are made.

Cleaving embryos are collected from oviduct or uterus by flushing by any culture medium. Before making chromosomal preparation it is advisable that the embryos are cultured in a medium with the same concentration of colcemide for 2-3 hours. Isolated blastomeres are obtained (after zona pellucida is removed) by gently pipetting of the embryo in media which do not contain calcium and magnium. Microsurgery of embryo can also be used because it does not interfere with mitosis and has no influence on the quality of chromosomal preparation of isolated blastomeres (Handyside, 1989). Isolated blastomeres must be cultured in media with colcemide with the time of culture and concentration of colcemide appropriately adjusted for every stage of development when the blastomere is removed. In case this is not achieved either the mitosis in blastomere will not be found or a very contracted metaphase chromosomes unsuitable for cytogenetic analysis will be obtained.

The description of our technique for chromosomal preparation in comparison with Tarkowski's method is presented in Fig. 1.

Hypotonic Treatment

Two solutions are prepared in deionized water (0.56% KCl and 1.93% tri-sodium citrate) and kept in a cooler for not longer than 10-15 days. One part of 1.93% tri-sodium citrates and three parts of 0.56% KCl are mixed immediately before use.

The embryos are transferred to the above hypotonic solution in precooled watch glass and left at room temperature. The oocytes at the metaphase 1 or 2 are treated by hypotonic solution at room temperature for 20-30 min. (or longer if in cooler), while isolated blastomeres for 3-5-7 min., morulae - for 10-15-20 min., blastocysts for 25-30 min.

Fixation

The fixative (mixture of methanol:glacial acetic acid, 3:1) is prepared on the day of use and cooled in a freezer for at least 1-2 hours (it may be kept in closed bottles in a freezer (-20°C) for up to five-seven days).

The embryo must be added to a few ml of fixative and placed in a precooled watch glass. It is important to

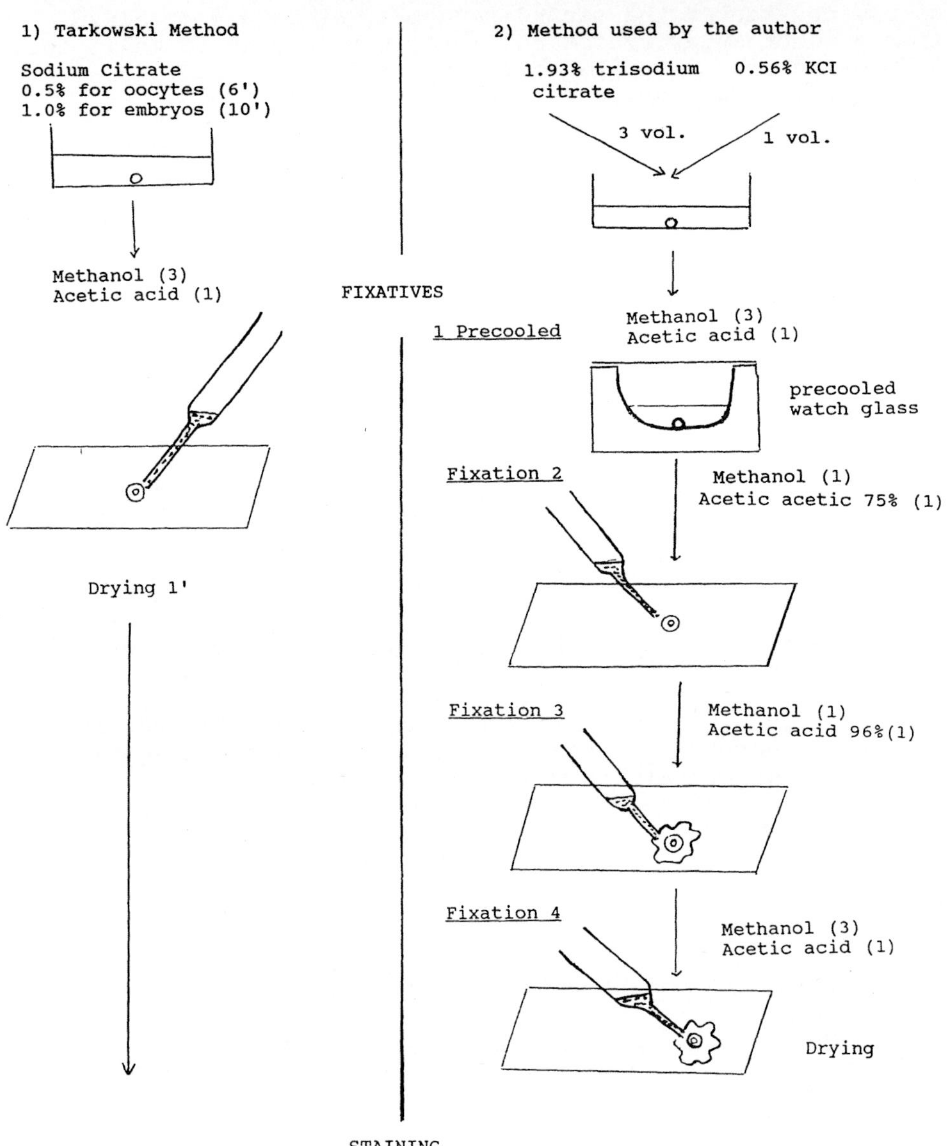

Figure 1. The scheme of two main approaches for chromosomal preparation (description in the text).

transfer the embryos to fixative with a minimum volume of hypotonic solution. This is achieved by placing the embryo in the tip of micropipette and expelling it under the control of the dissecting microscope. In the cold fixative, the embryos sink to the bottom of the watch glass and do not burst. However, the fixative should not be cold to the extent of freezing the tip of the micropipette. The procedure is the same for oocytes and particularly critical for isolated blastomeres as they can be easily lost in fixative. To avoid this, blastomeres should be carefully observed during their transfer to the fixative from the micropipette: isolated blastomeres are of brown color at the beginning (like the color of intact embryos) which facilitates the observation. The best time to remove the blastomeres from the fixative for spreading on slides is 3-5 min. after the cells lose color and become transparent.

The duration of fixation is not critical for the quality of chromosomal preparation. Oocytes, morulae, blastocysts are usually fixed for 10-15 min. until they lose dark brown color and become transparent. Therefore, they should be observed continuously under the dissecting microscope, also to avoid the problem of finding them for spreading. If the fixative is cold enough the embryos will not move during the observation. Otherwise, the watch glasses should be covered by coverslips and stored in the refrigerator for 20-40 min. (particularly appropriate for blastocysts).

Spreading

Three solutions are prepared before usage:

(1) one part of methanol and one part of 75% acetic acid.
(2) one part of methanol and one part of glacial acetic acid.
(3) three parts of methanol and one part of glacial acetic acid.

These solutions are kept in small bottles and placed very close to the dissecting microscope in the order of use. Every bottle should be supplied with a pipette with a thin tip and a rubber bulb. The diameter of the pipette is critical because the drop used for spreading must be small.

The fixed materials (oocytes, blastomeres or embryos) are removed from fixative with mouth controlled micropipette and placed in a small drop of fixative onto a grease free slide under the control of dissecting microscope. Before the fixative evaporates, a small drop of solution 1 (methanol:75% acetic acid, 1:1) is added and the outline of the embryo is carefully observed to monitor the degree of spreading under dissecting microscope. As the cells start spreading the drop of solution 1 flattens and floats from the embryo location to the periphery. Sometimes, when the spreading is finished the "wave" of fluid, floating from periphery to the center, can be observed, which is usually the result of using a big drop of solution 1. This is a critical moment for adding solution 2 as the "wave" can cause the breakage of the metaphase plates and the loss of chromosomes. On the other hand, if solution 2 is added prematurely the spreading of the metaphase will not be adequate. Following a drop of solution 2 the final spreading of metaphase may be fastened by gentle blowing which speeds up quick drying. Following this step it is useful to ring the embryo with a glass marker, which is better than a diamond pencil as it avoids producing glass chips on the slides. After the embryo is marked, a drop of solution 3 (methanol:glacial acetic acid, 3:1) is added to remove the traces of acetic acid from the slides. Slides must be dried quickly by the flow of compressed air. This step is performed after removing the slides from the stand of dissecting microscope.

In some cases (especially when isolated blastomeres are fixed) good spreading can be obtained without solution 1, i.e. with solutions 2 and 3 or just with solution 3, the procedure being the same: the blastomeres are transferred from fixative to the slides, followed by the addition of the drop of solution 2 and gentle blowing to facilitate spreading. This approach is particularly useful when hypotonic treatment of isolated cells lasts too long.

The slides can be stained by any conventional methods, including chromosome banding or hybridization in situ.

REFERENCES

Dyban, A.P., 1983, An improved method for chromosome preparations from preimplantation mammalian embryos, oocytes or isolated blastomeres, Stain Technology, 58:2:69-72.

Handyside, A.H., Pattinson, J.K., Penketh, R.J.A., Delhanty, J.D., Winston, R.M.L., Tuddenham, E.D.G. (1989) Biopsy of human preimplantation embryos and sexing by DNA amplification, Lancet i:347-349.

Severova, E.L. and Dyban, A.P., 1984, Selection of mice embryos in living condition according to sex and peculiarities of karyotype, Ontogenesis, (Sov.J.Devel.Biol.), 15:6:585-591.

Tarkowski, A.K., 1966, An air-drying method of chromosome preparations from mouse eggs, Cytogenetics, 5:394-400.

DNA AMPLIFICATION OF Y-SPECIFIC REPEAT SEQUENCES FOR SEXING PREIMPLANTATION HUMAN EMBRYOS

Alan H. Handyside and Eleni H. Kontogianni

Institute of Obstetrics and Gynaecology
Royal Postgraduate Medical School
Hammersmith Hospital
Du Cane Road
London W12 0NN, UK

INTRODUCTION

DNA amplification using the polymerase chain reaction (PCR) of a Y chromosome-specific repeat sequence in single cells biopsied from cleavage stage embryos allows the sex of the embryos to be accurately identified (Handyside et al., 1989). For couples at risk of transmitting X-linked recessive disease only affecting hemizygous males, this allows the selection and transfer of normal or carrier female embryos and several pregnancies have been established using this approach (Handyside et al., 1990; Handyside, this volume).

The main advantage of amplifying repeated sequences from single cells is that enough product is generated to be easily visualised on ethidium bromide stained gels without reamplification (with or without nested primers) which is generally required for detection of unique sequences (eg. Coutelle et al, 1989). In practice, it also appears to be no more or less susceptible to sporadic contamination but low levels of 'carry-over' contamination with the amplified product can generally be distinguished from the levels of product generated from single cells. Nevertheless, carry-over contamination is in our experience the most frequently encountered problem and all of the recommended procedures for separating sample preparation from product analysis should be rigorously applied (Kwok and Higuchi, 1989). As with any DNA amplification, dilutions of DNA down to single cell levels (2pg) and medium controls with no added DNA should always be run concurrently as controls. In addition, we would recommend considering the use of reamplification with nested primers, if time allows, which also guards against this form of contamination because the first set of primers are unable to anneal to the final shorter amplified fragment.

Initially, we used a sequence repeated 800-5000 times on the long arm of the Y chromosome (Cooke, 1976) using the primers described by Kogan et al (1987) for sexing fetuses from chorion villus samples. In rare cases, however, this portion of the chromosome is absent in males or conversely can be translocated to females (Bobrow et al, 1971; Cooke and Noel, 1979). Therefore, we routinely test lymphocytes from both partners to confirm the specificity of this fragment. To avoid this problem, we have recently been amplifying another Y-specific alphoid repeat which is located near the centromere (Witt and Erickson, 1989). Although this sequence is only repeated about 100

Preimplantation Genetics, Edited by Y. Verlinsky and
A. Kuliev, Plenum Press, New York, 1991

times, it appears to be equally efficient and it may be possible to co-amplify this fragment with an X chromosome specific alphoid repeat (Willard and Waye, 1987; Kontogianni and Handyside, this volume).

REAGENTS

Oligonucleotides

Careful preparation of oligos under sterile conditions prevents initial contamination. As far as possible, the whole procedure should be carried out under sterile conditions ie. in a laminar flow cabinet and with sterilised tubes or disposable plasticware to prevent contamination of the oligos. All tubes should be siliconised to prevent absorption and positive displacement pipettes and siliconised tips used instead of conventional Gilson or Eppendorf pipettors. The water used throughout should be both autoclaved and millipore filtered. Restriction enzymes have been used to decontaminate oligos (Handyside et al, 1989) but this approach is not always satisfactory. Oligos should be eluted from the column as soon as possible and within a few days of synthesis. If immediate elution is not possible they should be stored at 4^0C. For storage, oligos eluted from the column in ammonia can be stored at 4^0C indefinitely since the alkaline pH prevents hydrolysis.

Elution from column:

1. Insert a disposable 2ml syringe into one end of the column.
2. Take up 1ml of ammonium hydroxide into another 2ml syringe and insert into the other end of the column. 1ml aliquots of millipore-filtered ammonium hydroxide (4:1 ammonia:dH_2O) are stored at -20^0C. Allow to reach room temperature before use.
3. Slowly draw the ammonia solution through the column taking care to avoid bubbles.
4. Depress the plunger on one syringe and then on the other to force the solution through the column. This should be repeated 4-5 times.
5. Leave at room temperature and repeat this procedure after 45 min and 1h 30 min. Then take up all the fluid into one syringe and transfer to a sterile (optionally siliconised) screw top 1.5 ml Eppendorf tube.
6. Incubate for at least 16h or overnight at 55^0C. The oligos can be stored at this stage indefinitely at 4^0C. The concentration cannot be measured at this stage as ammonia absorbs UV light.

Ethanol precipitation:

1. Transfer the oligos in ammonia to suitable siliconised glass centrifuge tubes and add 3 volumes ethanol and 0.1 volumes of 2M sodium acetate, both millipore-filtered and pre-cooled on ice, to precipitate the oligos. At this stage, there should be a milky precipitate.

2. Cover or cap the tubes and leave at -70^{0}C for 1h (or overnight).
3. Spin down precipitate in pre-cooled centrifuge (rotor temperature about 4^{0}C) at 14,000 rpm for 1h, or alternatively, aliquot into 1.5 ml Eppendorf tubes and spin on a benchtop minifuge for 10 min.
4. Pour off the supernatant, which should be clear, into a sterile container and retain at 4^{0}C until the yeild has been measured. Invert the tube and gently tap out most of the remaining drops of supernatant without disturbing the pellet which should be visible.
5. Resuspend the pellet in 1ml of dH_2O by carefully vortexing the recovered tube.
6. Repeat 1-5.

Dilution:

1. Measure the optical density (OD) of the oligo suitably diluted (10 µl of the oligo made up in 1 ml dH_20) using quartz cuvettes and a spectrophotometer with a UV light source. Each oligo should have an OD of between 0.150 to 0.300 depending on yield.
4. Calculate the concentration as follows:

   OD x 0.033 x dilution factor (100 for above dilutions) = mg (in 1 ml)

   The yield for 20mers is generally 0.3 to 0.9 mg. If the yield is greater than 1.2 mg from a 0.2 µmole synthesis, reprecipitate the oligo.
5. Make approximately 20-30 100 µl aliqots in small Eppendorf tubes at a concentration of 0.5 or 0.25 mg/ml, depending on the yield, and store at -20^{0}C. If the aliquots are protected from repeated transient thawing, it should be possible to store the oligos for long periods.

PCR buffer

The standard PCR buffer recommended for use with the Taq DNA polymerase by Perkin Elmer Cetus can be used successfully. However, we now use a modification of a buffer designed for multiplex PCR (Chamberlain et al, 1990) which appears to optimise yeild of the specific fragment. In both cases, the $MgCl_2$ is made up as a separate 10x stock to allow for optimisation with different primers. The use of salmon sperm DNA to reduce 'primer-dimer' formation is not very effective and so this is now ommitted.

Standard 10x PCR buffer

Stock A: 100 mM Tris-HCl pH 8.3, 500 mM KCl, 0.1% (w/v) gelatin

Stock B: 15 mM $MgCl_2$

Each stock is made up in autoclaved distilled water, autoclaved, millipore filtered, aliqoted and stored at -20^{0}C. Equal volumes of stocks A and B are added to the PCR mix to give a final dilution of 1 in 10.

Multiplex 10x PCR buffer

Stock A: 166 mM $(NH_4)_2SO_4$, 670 mM Tris-HCl pH 8.8, 1.7 mg/ml bovine serum albumin.

Stock B: 67 mM $MgCl_2$

Stock C: ß-mercaptoethanol

Each stock is made up in autoclaved distilled water. Stocks A and B are autoclaved. Sufficient stock C is added to stock A to give a final concentration of 100 mM and this combined stock and stock B are millipore filtered, aliquoted and stored at -20^{0}C. Equal volumes of the combined stocks A and C and stock B are added to the PCR mix to give a final dilution of 1 in 10.

Deoxynucleotides (dNTPs)

We use 100 mM aqueous stocks of the four dNTPs (Pharmacia) and prepare an intermediate stock of a mixture diluted 1 in 10, ie. 10 mM each dNTP. This is aliquoted and stored at -20^{0}C.

Taq DNA polymerase

We use cloned Taq DNA polymerase (AmpliTaq; Perkin Elmer Cetus).

PREPARATION OF CELL SAMPLES

Single cells biopsied from cleavage stage embryos are first examined for nuclei using phase or interference contrast optics and if there is any doubt vitally labelled with Hoechst 33342 (Sigma). An aliquot of a 1mg/ml stock is thawed, vortexed and diluted 1:200 (5 μg/ml) in medium. Cells are incubated for 5-10 min, washed in medium and examined by fluorescence microscopy for the presence of a nucleus using UV excitation filters.

Cells with single nuclei are washed twice in PCR buffer and transferred to 0.5 ml Eppendorf tubes loaded with 10 μl distilled water using a dissecting microscope which allows observation of the transfer of the cell into the tube. It is important to select tubes with smooth surfaces before attempting to transfer cells. (Some batches of tubes have very rough surfaces which make it impossible to see the cell being transferred). If the cells are transferred in a minimal volume of medium washing may not be necessary; we have not detected any inhibition of PCR. Cells are, therefore, transferred using drawn out capillary tubing (Leitz micromanipulator tubing; heat sterilised) with the pipette held at the top of the tube for maximum visibility. Tubes loaded with water are kept at 4^{0}C until immediately before use to minimise condensation on the sides of the tubes and help ensure the cell is placed at the bottom of the tube. The caps on the tubes are then closed and left at room temperature for 10 min to allow the cells to lyse before freezing and thawing twice to -20^0C.

PREPARATION OF MASTER MIX

A master mix of 10x PCR buffer, 10x $MgCl_2$, dNTPs and primers sufficient for all samples is prepared initially and kept at 4^0C while the samples are denatured (see below). The final volume of each reaction mix is 100 µl so the master mix is made up to the equivalent of 90 µl per sample with distilled water to take account of the sample volume. Two µl of mixed dNTPs (10mM each), 3 µl of each primer (0.25 mg/ml - 20mers) and 2.5 units Taq polymerase are used per sample. The Taq polymerase is added immediately before use and the mix vortexed gently. All manipulations involved in setting up the PCR are carried out in a Class II laminar flow cabinet which has not been exposed to amplified product; the tubes are kept shut as far as possible; gloves are changed frequently; and positive diplacement pipettes used.

POLYMERASE CHAIN REACTION (PCR)

Each sample in 10 µl of distilled water is overlaid with 100 µl of millipore filtered light liquid parafin and heat denatured. For DNA, 6 min at 99^0C is sufficient but for cells we incubate at 95^0C for 35 min to inactivate DNAses and proteases. After addition of Taq polymerase to the master mix, 90 µl is added to each sample through the oil without disturbing the sample at the bottom of the tube. The tubes are then tranferred to the thermal cycler programmed as follows:

1.	94^0C	4 min	initial denaturation
2.	94^0C	30 sec	denaturation
3.	65 or 55^0C^*	1 min 30 sec	annealing/extension
4.	72^0C	10 min	final extension
5.	4^0C	indefinite	storage prior to analysis

Steps 2 and 3 repeated for 40 cycles. The timing would have to be adjusted if using a different thermal cycler.

* 65^0C for Y repeat on long arm; 55^0C for Y alphoid repeat.

POLYACRYLAMIDE GEL ELECTROPHORESIS

Amplification products are analysed on ethidium bromide stained 12% polyacrylamide minigels according to standard methods as described 'Molecular cloning: a laboratory manual' Maniatis, Fritsch and Sambrook, Cold Spring Harbor Press and examined and photographed using a UV transilluminator.

EQUIPMENT

Gilson 'Microman' positive displacement pipettes and tips.
DNA Thermal Cycler, Perkin Elmer Cetus.
Biorad Mini Protean II minigel apparatus.

REFERENCES

Bobrow, M., Pearson , P.L., Pike, M.C. and El-Alfi, 0.S. (1971) Length variation in the quinacrine-binding segment of human Y chromosomes of different sizes. Cytogenetics 10:190-198.

Chamberlain, J.S., Gibbs, R.A., Ranier, J.E. and Caskey, C.T. (1990) Multiplex PCR for the diagnosis of Duchenne muscular dystrophy. In 'PCR protocols: a guide to methods and applications' (eds. Innis, M.,Gelfand, D.H., Sninsky, J.J. and White, T.J.) pp.272-281 Academic Press, San Diego.

Cooke, H.J. (1976) Repeated sequences specific to human males. Nature 262:182-186

Cooke, H.J. and Noel, B. (1979) Confirmation of Y/autosome translocations using recombinant DNA. Hum. Genet. 67:222-224.

Handyside, A.H., Kontogianni, E.H., Hardy, K. and Winston, R.M.L. (1990) Pregnancies from biopsied human preimplantation embryos sexed by Y-specific DNA amplification. Nature 344, 768-770.

Handyside, A.H., Pattinson, J.K., Penketh, R.J.A., Delhanty, J.D., Winston, R.M.L. and Tuddenham, E.D.G. (1989) Biopsy of human preimplantation embryos and sexing by DNA amplification. Lancet: 347-349.

Kogan, S.C., Doherty, M. and Gitschier, J. (1987) An improved method for prenatal diagnosis of genetic disease by analysis of amplified DNA sequences. N. Eng. J. Med. 317:985-990.

Kwok, S. and Higuchi, R. (1989) Avoiding false positives with PCR. Nature 339:237-238.

Willard, H.F. and Waye, J.S. (1987) Hierarchical order in chromosome-specific human alpha satellite DNA. Trends Genet. 3:192-198.

Witt , M. and Erickson, R.P. (1989) A rapid method for the detection of Y chromosomal DNA from dried blood specimens by the polymerase chain reaction. Hum. Genet. 82:271-274.

DNA ANALYSIS OF GAMETES AND EMBRYOS: ANALYSIS OF DELTA-F508 CYSTIC FIBROSIS MUTATION IN SINGLE CELLS

Svetlana Rechitsky*, Gloria Enriquez and Charles M. Strom

Reproductive Genetics Institute
Illinois Masonic Medical Center
836 W. Wellington
Chicago, IL 60657

INTRODUCTION

The delta-F508 mutation is the most common mutation causing cystic fibrosis (CF) (Kerems et al, 1990). A method to consistently identify the presence of the delta-F508 mutation in single cells has been developed using decontamination of buffers and reagents by restriction digestion with MseI followed by heat inactivation before Polymerase Chain Reaction (PCR) (Strom et al, 1990; Verlinsky et al, 1990). After PCR, the amplified products are allocated to three tubes. PCR products from a homozygous normal individual are added to one tube, and products from a CF patient homozygous for delta-F508 are added to the second tube. These tubes are heated to 95°C for 10 min. and allowed to reanneal at 68°C for 10 min. The procedure permits heteroduplexes to form between delta-F508 and normal sequences (Rommens et al, 1990). The third tube is untreated. The DNA from the three tubes is then analyzed by polyacrylamide gel electrophoresis (equipment and reagents required are presented in annexes 1 and 2).

The polymerase chain reaction can efficiently amplify DNA from minute amounts of DNA. The difficulties in reliably analyzing DNA obtained from single cells derive from contamination of reagents with extraneous DNA rather than the ability of the PCR to amplify the sequences.

We have developed protocols to test reagents for contamination and for their ability to amplify DNA prior to their use in preimplantation or preconception genetic analysis. If contamination occurs, we use decontamination procedures to eliminate low level contamination. These decontamination procedures will only eliminate double stranded DNA contamination so they are not always successful in eliminating contamination. Under these circumstances, it will be necessary to discard this reagent and make new ones.

* In previous publications Svetlana Milayeva

Preimplantation Genetics, Edited by Y. Verlinsky and A. Kuliev, Plenum Press, New York, 1991

Avoiding contamination: General considerations

We have found that having a separate laboratory for Preimplantation genetics has been immensely helpful in preventing contamination. The laboratory has its own laminar flow hood, refrigerator, microcentrifuge, and PCR machine. The following rules are strictly adhered to:

(1) Tubes containing PCR products are never opened in the preimplantation genetics lab!!!!

(2) No men are allowed in the preimplantation genetics laboratory for gender determination.

(3) Gloves are worn at all times when performing preimplantation genetic diagnosis until after the PCR is completed.

(4) Positive displacement pipettes must be used for all preimplantation genetic procedures including buffer preparation.

(5) No traffic is allowed in the preimplantation genetics laboratory when analysis is being performed.

(6) The door to the preimplantation genetics laboratory remains closed at all times.

(7) Only new bags of eppendorf tubes are used. These are opened with gloves in the preimplantation genetics laboratory and are never removed from the laboratory.

(8) Reagents are to be stored as small volume aliquots to minimize the number of times the tubes are opened.

(9) Only HPLC grade water is used that has been autoclaved followed by millipore filtering.

(10) Sterile paraffin oil is used after millipore filtering.

(11) All solutions with the exception of Taq polymerase and dNTP's are millipore filtered.

Sources of contamination

IVF - Embryology reagents

- Paraffin oil used for microscopy
- Parental serum used for gamete and embryo culture

PCR Reagents

- Oligonucleotides
- Mineral oil
- Water
- dNTP's
- Taq polymerase
- Salt solutions

Reagents

Cloned Taq DNA Polymerase	5 units/mcl (AmpliTaq;Perkin-Elmer Cetus)
10x Reaction Buffer	100 mM Tris-HCl (pH8.3) 500 mM KCl 15 mM $MgCl_2$ 0.01% (w/v) gelatin

Stock is made up in HPLC water, autoclaved, millipore filtered, Aliquoted and stored at -20°c.

10xdNTPs	2.0mM of each of four dNTPs (Perkin-Elmer Cetus). Stock is aliquoted and stored at -20°c, thawed only for immediate use.

Oligonucleotides

Oligonucleotides were synthesized on an Applied Biosystem 381-A DNA Synthesizer and eluded from the column in ammonium hydroxide solution (3-5 times), incubated overnight at 55°c, vacuum dried in speed vac concentrator.

Pellet was resuspended in 1ml of HPLC H_2O and optical density was measured using quartz currettes and a spectrophotometer with UV light source.
Oligos were diluted to 10mcM.

Molarities calculated as:

363(#of Gs) + 323 (#of Cs) + 347 (#of As) + 322 (#of Ts) =gram grams/mole

primer C16B 5', GTT,TTC,CTG,GAT,TAT,GCC,TGG,CAC,3'

primer C16D 5', GTT,GGC,ATG,CTT,TGA,TGA,CGC,TTC,3'
(based on Kerem et al, 1989)

Restriction enzyme Mse-I 10u per 100mcl of 10mcM oligo in 1x buffer (manufacture supplies 10x buffer) have been incubated for 2.5 hrs at 37° to decontaminate oligos - then incubated at 95° for 10 minutes to inactivate the enzyme.

Reaction Mix (for 100mcl reaction):

10x Reaction Buffer	10.0 mcl
10x dNTPs	5.0 mcl
Taq Polymerase	1 unit (0.2 mcl)
primer C16B	1.0 mcM
primer C16D	1.0 mcM
H_2O	to 80 mcl

The final volume of each reaction mix is 100 mcl, therefore cocktail is made up to 80 mcl per sample because we already have 20 mcl of HPLC water in each tube.

Decontamination Procedures

For decontamination of reactions for delta F-508 cystic fibrosis PCR, the final reaction buffer was constituted. Ten units of Mse-I restriction enzyme are added for each 100 mcl of solution immediately before amplification. The buffer is then aliquotted into separate sterile tubes (100 mcl for genomic DNA and 80 ml for single cells) and incubated at 37°C for 2 hrs in the thermo-cycler. The restriction enzyme is then inactivated at 95°C for 15 minutes. Mse-I was chosen because it cleaves inside the region flanked by the C16B and C16D primers (Kerem, 1989).

Preparation of Cell Samples

(1) For every DNA amplification it is necessary to include positive (DNA that is known to successfully amplify for CF) and negative (no DNA or cell) controls to check for PCR efficiency and contamination.

(2) First polar body, blastomere or single cell is transferred into sterile 0.5ml Eppendorf tube with 20 mcl of millipore filtered HPLC water and left for 5-10 minutes at room temperature to allow the cells to lyse before freezing.

(3) The tubes containing the cells, positive and negative controls are then frozen for at least 30 minutes at -80°C. It should be possible to store them indefinitely.

(4) After thawing the tubes are spun for 10 seconds in Microfuge and 40 mcl of millipore filtered mineral oil is added to each tube.

(5) Each tube is heated for 15 minutes at 95°C in Thermo-cycler before polymerase chain reaction buffer is added.

(6) 80 mcl of reaction cocktail is added into each PCR tube.

(7) The tubes are spun for 10 seconds in a Microfuge to collect all reaction components under the oil overlay.

(8) Tubes are then put in Thermo-cycler (PCR)
Start PCR

Thermocycling conditions are as follows:

1 - initial	denaturation	95°C	- 7 minutes
2 -	denaturation	94°C	- 1 minute
3 -	annealing	62°C	- 45 seconds
4 -	extension	72°C	- 2 minutes
5 - final	extension	72°C	- 10 minutes
6 - storage		4°C	- indefinite

Step 2,3 and 4 repeated for 40 cycles.

Analysis of PCR Products By Polyacrylamide Gel Electrophoresis

Bromphenol blue dye:	50% sucrose 0.1m EDTA (pH 7.0) 0.05% Bromphenol blue dye
Ethidium bromide:	5mg/mcl in distilled H_2O Use 0.5 mcL/ml
Electrophoresis buffer:	10xTBE (pH8.3) 1M tris HCl 1M Boric Acid 20mM EDTA
40% Bisacrylamide:	acrylamide 38% N,N'-methylene bisacrylamide 2%
Acrylamide gel:	(6ml per 1 gel)

	8%	12%
40% Bis Acrylamide	1.2	1.8
10xTBE	0.6	0.6
10% Ammonium persulfate	0.075	0.075
TEMED	0.005	0.005
H_2O	4.2	3.6

(1) 15 mcl of PCR product is put in 0.5 ml sterile Eppendorf tube and 8 mcl of material derived by amplifying DNA of known noncarrier of the F508 mutation (N/N - first set) or homozygous carrier of the mutation (CF/CF - second set) is added.

(2) The cocktail is denatured for 10 minutes at 95°C and reannealed for 10 minutes at 68°C.

(3) 4 mcl of Bromphenol blue dye is added.

(4) 8% Acrylamide mini gel is loaded and run in 1xTBE at 90v for 1 hr.

(5) Staining with Ethidium bromide 5mcl/1ml.

(6) Pictures are taken under ultraviolet light.

(7) Results interpretation:

Single cell DNA that is homozygous for the normal allele will show two bands when PCR products of delta-F508 DNA are added but only one when homozygous normal

DNA is added. Conversely, single cell DNA homozygous for the delta-F508 mutation will show two bands when normal DNA is added and only one when delta-F508 DNA is added. Heterozygous DNA will show two bands in untreated DNA and when either normal or delta-F508 DNA is added.

REFERENCES

Kerem B-S, Rommens J, Buchanon JA, Markiewicz D, Cox TK, Chakravarti A, Buchwald M, Tsui, L-C 1989, Identification of the cystic fibrosis gene: genetic analysis. Science, 245:1073-1080.

Rommens J, Kerem B-S, Greer W, Chang P, Tsui L-C, Ray P, 1990, Rapid nonradioactive detection of the major cystic fibrosis mutation. Am J Hum Genet., 46:2:395-396.

Strom CM, Verlinsky Y, Milayeva-Rechitsky S, Evsikov S, Cieslak J, Lifchez A, Valle J, Moise J, Ginsberg N, Applebaum M 1990, Preconception genetic diagnosis for cystic fibrosis by polar body removal and DNA analysis. Lancet, 336:306-308.

Verlinsky, Y., Ginsberg, N., Lifchez, A., Valle, J., Moise, J., Strom, C.M. (1990): Analysis of the First Polar Body: Preconception Genetic Diagnosis. Human Reproduction 5(7):826-829.

Annex 1. Required Materials and Reagents

Ampli Taq DNA Polymerase, 250 units	Perkin Elmer Cetus N801-0060
GeneAmp dNTPs (10mMdATP, 10mMdCTP, 10mMdTTP, 10mMdGTP)	Perkin Elmer Cetus N808-0007
DNA size Marker D8397	Perkin Elmer Cetus N930-2106
DNA marker:123 base pair ladder	BRL
Mineral Oil	Perkin Elmer Cetus N0186-2302
Restricted enzyme Mse-I	New England Biolabs #525
HPLC water Chromatography grade	(Mallinckrodt Inc.) Scient. Prod. 6795-1NY
Ethidium bromide tablets	Bio-Rad 161-0430
Bromphenol blue	Bio-Rad 161-0404
Ammonium persulfate	Bio-Rad 161-0700
TEMED	Bio-Rad 161-0800
Acrylamide	BRL 5512 UA
N,N^1-metilenebisacrylamide	BRL 5516 UA
Sucrose	BRL 5503 UA
Potassium chloride	Fisher P-217
Tris HCl	Fisher Biotech BP152-500
Boric acid	Fisher Biotech BP168-500
Gelatin	Sigma N G2500
Magnesium chloride	J.T. Baker Chemical Co. 2444-1
EDTA	J.T. Baker Chemical Co. 8993-1
Ammonium Hydroxide	Fisher A669-500

Annex 2. Required Equipment

Labware, glassware including beakers, graduated cylinders, Erlenmeyer flask	
Laminar flow (biological) biosafety cabinet	Edge gard Hood Eg-3252 Sanford, Main
Cetus Thermal Cycler	Perkin Elmer Cetus
Microfuge™ 12	Beckman
Freezer Revco (-80°c)	
Refrigerator (2-8°c)	

Annex 2 (continued)

Freezer (-20°c)	
UV Transilluminator C-62	Ultra-violet Products, Inc.
Camera FCR-10	Fotodyne, Inc.
Polaroid film Type 667	
Gel apparatus mini-protean II Dual slab cell	Bio-Rad 165-2940
Electrophoresis power supply	Bio-Rad model 100/200
PH meter	Fisher, model 955
Positive displacement pipettors	Rainin Instrument Co. micromen M-25, M-50, M-250
Pipettors, adjustable to deliver to 1-1000me	Rainin Instrument Co. Gilson P-20, P-200, P-1000
Serological Pipets disposable (5ml)	Fisher brand #13-678-3114
Serological Pipets disposable (10ml)	Fisher brand #13-676-10J
Polycarbonate Microcentrifuge tube opener	USA Scientific Plastics USA 050
Microcentrifuge Tube, Flat Cap (1.5ml)	USA Scientific Plastics USA 505-FT
Microcentrifuge Tube, Flat Cap (0.5ml)	USA Scientific Plastics USA 502
50 Place (5x10) Polypropylene Storage Box	USA Scientific Plastics PB50-V
Microcentrifuge Tube Racks	Nalgene #5973-0315
Transfers Pipets	USA Scientific Plastics USA 304
DNA Synthesizer	Applied Biosystem 381-A DNA Synthesizer
Speed Vac Concentrator	Savant
Refrigerated Condensation Trap	Savant
Cameo IIS 0.22mm filter unit	MSI West Borough, MA 01581
Vortex - Genie 2	Scientific Industries

<u>Supplies</u>

Gloves Disposable

Lab Wipes

Parafilm
Waterproof Marking Pen

UV Goggles

Syringes

SUMMARY

Joe Leigh Simpson

University of Tennessee, Memphis
Department of Obstetrics and Gynecology
853 Jefferson Avenue
Memphis, TN 38163

In his introduction, Robert Edwards (Edwards, this volume) stated the challenges with which we should wrestle developing preimplantation genetic diagnosis. The first general concern dealt with extrapolation, concerning the analysis of a polar body biopsy lending to replacing the complementary oocyte based on the assumption that the latter must be normal. A second general concern dealt with contamination, in particular the possibility that DNA of sperm attached to the zona pellucida would prove diagnostically confusing. A third concern was the stage most applicable for preimplantation genetics: Edwards believed that at present this was the 8 cell stage. However, a fourth concern - necessity for replicate assays - would better be met by trophectoderm biopsy.

The papers presented systematically covered all stages of preimplantation genetics, providing the data relevant to overcome some of the above concerns.

ABNORMALITIES IN GAMETES

The frequency of structural chromosomal abnormalities in sperm of normal men is 9.4%; the frequency of numerical abnormalities is 1.4% (Martin, this volume). Sperm disomy (n + 1) for all chromosomes has been recovered, but it was a surprise to find an excess of not only chromosomes Nos. 21 and the sex chromosome but also chromosome number 1, trisomy for which has never been observed even among abortuses. The sex ratio in sperm is very nearly one.

The frequency of chromosomal abnormalities in oocytes is 13.6% (Plachot, this volume), being complementary to those data on sperm. Especially relevant was finding a maternal age effect, based on stratifying subjects before and after age 35 years. Fortunately, data do not show a deleterious effect of ovulation induction.

Preimplantation Genetics, Edited by Y. Verlinsky and
A. Kuliev, Plenum Press, New York, 1991

ABNORMALITIES IN EMBRYOS

Cytogenetic data on gametes are consistent with our knowledge of the chromosomal status of embryos - human and mouse: murine haploids and nullisomies are lost prior to implantation, perhaps at the two to four cell stage; monosomies and trisomies are more likely to be lost after implantation, the former immediately and the latter later but usually before parturition (Dyban, this volume).

Murine studies are consistent with the high frequency of human pregnancy losses observed around the time of the ß-hCG detection, that is at implantation. In humans, early losses have been known for many years to be the result in part of morphologically abnormal embryos. Such abnormalities might be chromosomal in nature, consistent with murine studies. Of relevance to devising a priori loss rates is a further belief that most morphologically abnormal preimplantation embryos (at least those polyploid and monosomic, although perhaps not nullisomic) implant and thus temporarily secrete ß-hCG. If true, overall loss rates in human development would be far lower than the rate if preimplantation embryos fail to implant (Simpson, this volume). In turn, the rate would be consistent with the human loss rate from ß-hCG detection onward being no greater than 30%.

At least half of the losses occurring after ß-hCG detection do so during the fourth gestational week (second embryonic week). The basis for this conclusion is a study showing loss rates 28-30 days after menses to be only 16%, compared to 32% ascertained at 21-25 days (see Simpson, this volume). After 8 weeks the loss rate is only 3%, and after 16 weeks 1%. If loss rate calculations are based solely on clinical criteria, losses will appear to be occurring later in pregnancy because deceased fetuses are usually retained in utero for weeks (missed abortion).

DIAGNOSTIC ACCURACY

Cytogenetics

The high background frequency of chromosomal abnormalities is relevant not only to our hopes of verifying safety of preimplantation biopsy, but in assuring that we are replacing chromosomally normal embryos. The overall frequency of chromosomal abnormalities in 8 cell IVF embryos was calculated to be 29% (Plachot, this volume). Even in embryos with two pronuclei the rate is about 20%. A special caveat for preimplantation cytogenetic diagnosis is finding a relatively high frequency of diploid mosaicism. This in turn is consistent with the well known CVS discrepancies.

The ability to karyotype a single cell in the majority of mouse embryos is arguable, although this was considered possible by Dyban (this volume). The approach to preimplantation cytogenetic diagnosis may need to involve utilizing in situ hybridization (Grifo, this volume) or studying DNA polymorphisms that will assure an informative result (i.e., detection of trisomy).

Enzyme Analysis

Consensus about the feasibility of enzyme analysis on single cells is that much work is needed as it is difficult to distinguish embryonic from maternal enzyme (e.g., HPRT) at the 4-8 cell stage (Braude, this volume). Enzyme diagnosis may therefore not likely be completely reliable.

DNA Accuracy

DNA accuracy appears to be more achievable, although formidable problems still remain. In optimal hands some very lovely work has been done, but the spectre of cell contamination is very wide (Strom, this volume). This very serious issue needs to be addressed cautiously by anyone with diagnostic aspirations. On the other hand, sperm contamination would probably not be a problem because its DNA is highly compact *in vivo* (Arnheim, this volume). Sperm DNA can indeed be extracted, but actually only with difficulty; thus, diagnostic confusion may not be likely when analyzing blastomeres. Whether DNA is still as compact after capacitation *in vitro* is, however, not yet clear.

RECOVERY OF EMBRYOS

All are well aware of the promise and limitation of IVF, and its low efficiency at present. The optimum is represented by the few groups who reported a 15% success rate *per embryo* (Van Steirteghem, this volume). (Data are usually presented not per embryo but rather per cycle.) On the other hand, patients undergoing IVF for preimplantation genetic diagnosis are likely to be younger than women who undergo IVF solely for infertility, and as well to have less uterine pathology.

Uterine lavage by superovulation has not yet been expanded beyond the initial early success achieved on natural cycles. The groups in Milan, Memphis and Los Angeles have been unable to obtain blastocysts consistently after superovulation induction. Reasons for this lack of success remain unknown. A new catheter, which will recover instilled fluid cc per cc was proposed (Brambati, this volume), for use in a program of preimplantation detection of ß-thalassemia, using natural cycles.

EMBRYO AND GAMETE BIOPSY

Both animal model systems as well as limited human data suggest preimplantation biopsy will be relatively safe.

Polar Body Biopsy

Data were presented on polar body biopsy and safety (Verlinsky, this volume). Embryonic development to the blastocyst stage was 64% after polar body biopsy. Polar bodies diagnosis has been performed in pregnancies at risk for cystic fibrosis (CF), α-1-antitrypsin deficiency, and hemophilia A. To address the issue of recombination, analysis of both the first as well as the second polar body was proposed if necessary.

Blastomere Aspiration - 8 Cell Embryo

Before attempting preimplantation diagnosis, it was first shown that there was no deleterious effect on subsequent development when not just one but two cells and sometimes three were removed from an eight cell embryo (Handyside, this volume). Glucose and pyruvate uptake were normal. Thirteen cycles in 7 women at risk for X-linked recessive disorders were reported, yielding five pregnancies. Mean number of oocytes per cycle (N = 11), mean fertilizations per cycle (N = 6), mean embryos suitable for biopsy (N = 5) and mean embryos suitable for transfer after diagnosis (N = 2) suggested that the approach is laborious but feasible. An intriguing finding was that pregnancies were achieved in 6 of 19 (31.6%) biopsied human embryos, a success rate actually higher than would be expected from IVF experience in infertile couples. Perhaps embryo biopsy even enhances IVF success! At any rate, biopsy at the 8 cell stage seems not to further depress the existing low IVF efficiency (Handyside, this volume).

Blastocyst Biopsy

In trophectoderm biopsy no human work on ongoing pregnancies has been reported. However, several potential biopsy techniques using a mouse model were compared by Carson (this volume). Optimal success was obtained with either spontaneous hatching or slitting followed by herniation. Aspiration was less safe, this technique thus being less successful than that in the 8 cell embryo. The same slitting technique was used in studying human IVF embryos (Dokras, 1990).[1] None were replaced, but subsequent development was considered similar to that theoretically expected.

The potential advantage of studying blastocysts was appreciated. Pregnancy rates after transfer are far higher (60%) than observed with 8 cell (IVF) embryos, and far more cells are available for analysis (10-30 at trophectoderm biopsy, 1-3 at the 8 cell stage). However, blastocysts can be obtained through IVF relatively inefficiently; thus, uterine lavage is generally considered necessary. Given difficulties posed by DNA contamination requiring replicate assays, diagnosis seems far more easily achievable with trophectoderm biopsy than through aspiration at the 8 cell stage.

GENE THERAPY

The potential attractiveness of somatic gene therapy in hematopoietic disorders was reviewed. New techniques to enhance efficacy have been proposed, but many previously touted improvements have not been confirmed (Bank, this volume). For germ cell therapy, the task is still very laborious and inefficient.

[1] Dokras, A., Sargent, I.L., Ross, C., Gardner, R.L., and Barlow, D.H. Trophectoderm Biopsy in Human Blastocysts, Human Reproduction 5(7):821-825.

With respect to the ethical arguments against germ line therapy, Elias has concluded that there were no new plausible objections over those already raised against for somatic cell therapy. It thus seems inappropriate to withhold germline therapy if somatic cell therapy is concurrently considered appropriate. The problem seems to be the public's erroneous conception of what germline therapy truly means. In the United Kingdom it was stated that it proved essential to bring the public into this debate. Prerequisites to be fulfilled before attempting germline gene therapy are thus worth restating: severe disease and prior somatic cell experimentations.

ETHICAL AND LEGAL CONSIDERATIONS

A general disquiet for the "water" of preimplantation genetics in which we are yet learning to swim was voiced by Milunsky (this volume). Informed consent and confirmatory chorionic villus sampling after preimplantation genetics seem to be important requirements. The bewildering series of laws that confront those in the U.S., although enviously not necessarily the problem in every country, was presented (Robertson, this volume). Although at least one U.S. House of Representatives Bill has proposed control over embryonic research and laboratories in the United States, investigators are now left to the devices of their 50 states.

With renewed awareness for the need to take into account the public's sensitivity, we did not lose track of the reason for initiating preimplantation genetics - to help patients at risk for serious genetic disorders for whom existing options are not satisfactory. Worth recalling in summary is that some of the fears now voiced regarding preimplantation genetics are reminiscent of these raised at initiation of amniocentesis and prenatal diagnosis (circa 1968). Prenatal diagnosis for the detection of Down syndrome was said by some to be inappropriate, the alarm raised that the process would inexorably lead to positive eugenics. Soon we would be offering prenatal diagnosis for hair color and the like! Of course, the subsequent 22 years verified that the scientific community is indeed fully cognizant of the public's concern, thoughtful in its laboratory explorations, and cautious in introducing new technology to patients. Both the scientific community and the public as a whole have used good sense, and surely will continue to do so as preimplantation genetics evolves.

INDEX